T0256376

Routledge Revivals

Arid Lands

In 1951 UNESCO launched an Arid Zone Programme with the object of promoting research into arid regions from every relevant scientific point of view. This book, originally published in 1966, represents the range of research undertaken and gives a general conspectus of arid zone geography. 17 authors from 8 countries contributed and the book deals comprehensively with all the main areas, with specific examples used to illustrate arguments. There are chapters on meteorology, geology, geomorphology, botany and zoology and almost 50% of the book is devoted to man's activities: irrigation and agriculture; industry; animal breeding and human survival in the desert.

Arid Lands

A Geographical Appraisal

Edited by E. S. Hills

Routledge
Taylor & Francis Group

First published in 1966 by UNESCO and Methuen & Co. Ltd

This edition first published in 2024 by Routledge
4 Park Square, Milton Park, Abingdon, Oxon, OX14 4RN
and by Routledge
605 Third Avenue, New York, NY 10158.

Routledge is an imprint of the Taylor & Francis Group, an informa business

© 1966 UNESCO

The right of E. S. Hills to be identified as the editor
of this work has been asserted by him in accordance with sections 77 and 78 of the
Copyright, Designs and Patents Act 1988.

ISBN 13: 978-1-032-73070-7 (hbk)
ISBN 13: 978-1-003-42663-9 (ebk)
ISBN 13: 978-1-032-73092-9 (pbk)
Book DOI 10.4324/9781003426639

Arid Lands

A GEOGRAPHICAL APPRAISAL

Edited by E. S. Hills

London : Methuen & Co Ltd   *Paris* : Unesco

Foreword

This book was conceived when the Major Project on arid lands under the Arid Zone Programme of Unesco was nearing its end. It was realised that there was no one book which was available for use as a reader for advanced geography classes or for the educated public, in which a general conspectus of arid zone geography is given, and the time was thought to be ripe to produce such a book.

The Arid Zone Programme of Unesco was started in 1951, its object being to promote and stimulate research in the many scientific disciplines which have a bearing on the problems of arid lands, with the ultimate aim of improving the living conditions of people living in desert or semi-desert regions. In 1957 the programme became a Major Project, which terminated in 1962.

It is true that there is as great a need for people whose lives are spent in humid climates to understand the problems of aridity, as for those who themselves live under arid conditions. Lack of appreciation is natural among those who normally need have no fear of shortages of water: indeed there is no single word in English meaning 'to die of thirst', whereas in Arabic all degrees of thirst have their verbal expression, as the following list shows:[1]

frequently used words	al-'atash	thirst
	al-Zama'	thirst
	al-Sada	thirst
	al-Ghulla	burning thirst
	al-Luhba	burning thirst
	al-Huyām	vehement thirst (or passionate love!)
	al-Uwām	burning thirst, giddiness
	al-Juwād	excessive thirst (this is the thirst which kills)

In the law, water is treated very differently in countries where it is scarce from those where it is plentiful. It is indeed only now, when growing populations in humid lands have created a relative scarcity of water as a commodity, that Europe and Great Britain, for example, are gaining the degree of sophistication in relation to water use that has been developed over centuries in dry lands.

Interest in the development of arid and semi-arid regions has been stimulated,

[1] I am indebted to Dr A. K. Kazi of the University of Melbourne for the list of Arabic words.

and knowledge concerning the proper ways of exploring and using arid land has increased and spread, during the course of the Arid Zone Programme, which still continues within the Natural Resources Programme of Unesco and has been widened in its regional scope. In many developing areas there is a sense of urgency in relation to the poor conditions of life of millions of people; but the whole world has a stake in the potential productivity and uses of arid and semi-arid land, in view of rapidly growing populations. To what extent these lands will be self-sufficient in foodstuffs or may help, through increased production, to supply the needs of other regions, will be a vastly more significant problem in the year 2000 than in 1964. Social and political as well as material and economic problems will be raised and must continue to be solved if the notion of increasing well-being for man is to be sustained.

The book has been planned with a view to presenting a broadly based conspectus, as little coloured by subjective concepts as may be possible, of the vast area of dry country in the world – over one-third of the land area of the globe. Full regional treatment is not possible within the ambit of one book, but illustrative examples have been taken from all major arid areas. Seventeen authors from eight countries and sixteen institutions have collaborated, and it is a great pleasure to record their willingness not only to prepare the chapters they have written, but also to permit the editing that was necessary in order to achieve a certain uniformity of literary style, and to avoid too much duplication in subject matter. Such overlap as remains in the contents of the various chapters was thought to be desirable so that each might be to some extent a self-contained unit and might be read fluently, without excessive cross-reference to other parts of the book. In Chapter VII, K. W. Butzer deals with recent climatic changes especially in the Mediterranean region, and C. R. Twidale with older climates and with the deserts of the geological past. Their contributions have been conjoined editorially.

The work of Dr Herbert Greene, a former member of the Arid Zone Advisory Committee of Unesco, who died suddenly in West Africa shortly after he wrote Chapter XII, will remain as a tribute to him in many lands, but especially in Africa and the Middle East where he spent long periods of his life. The editor in particular acknowledges Herbert Greene's friendship and his willing assistance in contributing to this book.

Permission to quote several passages in Chapters X and XI was freely given by the American Geographical Society and the Royal Geographical Society, and by the authors, F. Debenham and D. F. Thompson. Acknowledgement for the sources of illustrations is made in the appropriate captions. Moreover, it should be mentioned that the points of view, selection of material and opinions expressed thereon are those of the authors who contributed to this volume and not those of Unesco.

It will appear from perusal of this book that there is certainly no justification for over-optimism as to the rapid development of arid or even of semi-arid regions. The pitfalls are many, the capital outlay great, and the human problems imponderable but equally great. That the struggle to make use of any potentially productive area will be made is certain – the ideals and the economic forces that drive mankind will ensure this; and it is with a feeling that in making the strenuous efforts required we should not lose sight of the fact that the lives of the men and women involved must be made useful, satisfying and pleasant, that the editor concludes his task.

Melbourne, March 27, 1964   E. S. HILLS, *Editor*

Contents

List of Illustrations

TABLES

CHAPTER I

Arid Lands and Human Problems

The personal vision that one has of the deserts and semi-deserts and of the lives of the men and women who dwell in them is inevitably highly subjective. The Bushman, the Bedouin or the Australian aborigine who lives in and with the desert knows the ways of life in his own region but may have no understanding whatever either of other arid regions, or of ways of life in regions of affluence in the temperate zones of the world. The savant who has travelled and studied in many dry lands will certainly have gained much knowledge of them; but he too has his limitations, the chief of which is that, by being a savant, he has removed himself from the everyday affairs of the people. Moreover, in these days he is generally a specialist, and often a narrow specialist, in one or other of the many disciplines that are involved in maintaining the whole complex of the life of man in any region. Two fundamental considerations that affect the making of syntheses in regional geography must also be noted. The first is that, in the social sphere particularly, the act of observing and recording involves establishing contact between the observer and the observed. This itself means some change, however subtle, in both parties, and when as is the case today the number of observers may be large, the effect is multiplied and many new vistas are opened. What we are observing is also changing of its own accord, and this leads to the second consideration, that of time. Change is everywhere, even in those cultures that for centuries have changed but slowly, and any synthesis or conspectus the regional geographer may attempt will inevitably be dated. These are some of the difficulties in principle which the regionalist must face, apart from many others, physical and mental, that affect his work in the field.

When it is realised that over a third of the land area of the globe is to be classed as arid in some degree, it will be seen that the attempt made in this book to give even a broad oversight of the arid lands must indeed be subject to severe limitations as to the range, depth, and present accuracy of the information presented. Accordingly, descriptions of actual achievements, say as to the location and use of irrigated areas, the various types of primary production or industrial developments, are restricted to particular examples which illustrate general principles.

The possibility of establishing fundamental guiding principles for the development of arid land of various types and in various physical regimes has been demonstrated by achievements made with the scientific and technological knowledge and experience already available. Knowledge gained in one place about the principles of infiltration of water into soil may be applied, with due regard for the physical properties of local soils, in other arid lands; knowledge of plant physiology, soils and climate may encourage the introduction of useful plants from one country into another; and so likewise for very many other lines of potentially fruitful investigation and action.

Our ultimate interest and concern is, of course, humanity itself, and it may be that in venturing to generalise concerning the arid environments we offend those to whom the differences between human communities and their cultural achievements are paramount. It is indeed true that each region of the world is unique, both as to the actualities of the environment and the people who live in it. Anatomical, physiological and sociological differences between races and tribes are obvious, but little if anything has been reliably established about fundamental differences in intellect or emotions. There may well be such, but it is nevertheless demonstrable that physical similitudes between regions may be recognised, which afford a sound basis for analogous usages and practices in agriculture, stock raising, housing, and many other activities by which men support themselves and their families in arid lands – and equally, of course, in other climatic situations. Thus although absolute determination of human actions by environmental factors is ruled out by the facts of human and historical geography, the recognition of homoclimatic regions is of great practical value where guidance is sought as to the potential uses of any area of land. To put to such practical use the recognition of homoclimates, and, on a broader scale, the subdivision of the land areas into climatic regions, is a major step forward from the mere recognition and delineation of such regions – a step which has been taken as an exercise in the analysis and synthesis of regional geography by Köppen in his *Handbuch der Klimatologie*, and, using vegetation as an index of climate, by Herbertson in *The Major Natural Regions*. It had required a span of two thousand years or more during which the discovery of geographical and climatological data and their dissemination among educated people went on, to achieve such generalisations, which derive in essence from the ancient postulate and since proven fact, that the earth is a planet.

That there is some systematic variation in climate from north to south must, one realises, have been obvious to the early coastal or overland voyagers from the Mediterranean to the cold lands of the north, and to the Varangian Goths who took their boats by river, lake and portage from Sweden to Byzantium; but out of these, and the equally ancient east–west trade routes from the Orient to Europe, largely continental rather than marine, there arose little in the way of meteorological or climatic generalisation except the broad notion of the existence of Frigid, Temperate and Torrid Zones, until the great expansion of sail which depended so much on a knowledge of the global wind systems.

By the late nineteenth century the descriptive records of geographers and meteorologists had begun to be the basis for recognising various types of geographical regions, whether so-called 'natural regions' having several elements in common (including topography, climate, production and potentialities), or vegetation regions, or climatic regions. Climatic regions are of fundamental importance. They are a consequence of the variation of insolation with latitude, the rotation and revolution of the earth, the planetary circulation of the atmosphere, the

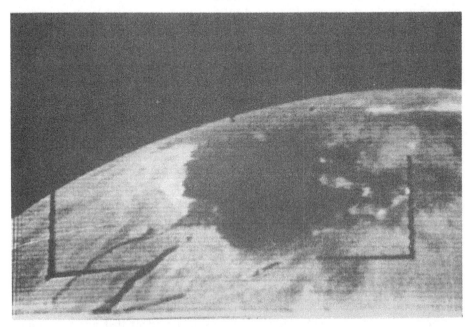

Fig 1-1. The Mediterranean Sea and surrounding lands, from Tiros V (Orbit 143R/O 142, 29 June, 1962; height 619 km). Note arid Africa, the Nile Valley and Delta, and the vegetated land of Europe. (*Photo Tiros V; by courtesy of the United States of America*)

distribution of land and sea, and the physiographic configuration of land masses, which factors influence temperature, rainfall and winds.

It is indeed remarkable that while technological advances have at this moment of time given us space travel and man has actually circled and viewed the earth as a planet, there remain in fact vast areas of the globe of which our knowledge, even as to descriptive aspects including topography, soils, vegetation and geology, is scant or even lacking, and where some peoples are still living in a virtual Neolithic culture. Apart from the frigid Arctic and Antarctic regions, the arid lands are least well known.

Our first space traveller might perhaps recognise these arid regions from afar by the lack of cloud cover and by the red and yellow colour as against the green of well-watered lands. He could obtain a visual impression of their extent, as is vividly suggested in the photograph (Fig 1-1) taken from the satellite Tiros V, clearly showing the cultivated land of Egypt set in the immensity of the Sahara.

But for precision we must rely on some form of cartographic representation, which in turn implies that we have available the data to permit the arid lands to be outlined and marked in by the draughtsman. The criteria that may be used have been determined by geographers, meteorologists, pedologists and plant ecologists. Boundary lines must be drawn on the map to separate the different climatic regions, but there is in nature a transition from one climatic type to another. Sharp changes are rare, although they do exist in places, particularly at the crests of mountain ranges and the margins of intermontane basins. It takes but a few hours of air travel to see displayed the zonation of landscape, from barren sandy desert wastes, treeless and torrid, to the sparse vegetation and irregularly distributed crop areas of the semi-arid and semi-humid areas, to the green and fertile humid lands. One of the first tasks that was undertaken when Unesco drew attention to the generally low level of living standards of dwellers in the arid lands, and resolved to embark on a programme which might ameliorate this, was to produce a world map showing climatic regions, from which the location and extent of land affected by aridity might be gauged [1]. The map is printed on the title page of this book.

Our main interest lies in the regions shown as extremely arid, arid, and semi-arid, but it must be admitted that a further category, that of semi-humid, is also at times subject in some degree to those influences of high temperature and low precipitation that characterise the arid lands, so that much that is written here about crop practices, water conservation and the like applies, in certain circumstances, to Mediterranean lands and to other parts which experience seasonal aridity. Indeed, as the world population grows, the shortage of water that is *par excellence* the feature of arid lands is more and more being experienced in naturally well-watered regions. Great Britain in 1959 and Java in 1960–1 came close to disaster through drought, and it is increasingly evident that water supplies, whether for agriculture or for conurbations and industry, place a limit on population growth, and that there will, in future years, be few areas of overabundance of water supply.

It is, however, in places which are normally deficient in rainfall that temporary deprivation causes severe drought and famine. Temporary excess leading to floods is, in arid regions, not so disastrous as in humid, densely populated regions and may, as with the annual flooding of the Nile, even be the basis for subsistence of the people, affording natural irrigation along the rivers. The floods of the Tigris and Euphrates watered the ancient irrigated fields of Mesopotamia; in Queensland (Australia) the flooding of the upper Diamantina River in the 'Channel Country' (Fig 1-2) provides pasture for cattle *en route* to distant markets, and the practice of 'water-spreading' is in regular use in Arabia and Africa. Mention of the ancient Chaldean civilisation, based on irrigation long since disrupted and overwhelmed by uncontrollable floods, by natural changes in the river regime, by the accumulation of salts in the irrigated soils and by conquests and the collapse of organised government, reminds us of the apparent uncertainty of outcome of man's interference with the natural scheme of things.

Our survey in this book, after reviewing the world's arid regions, climates and microclimates, the physical agents that mould the land surface, and their effects, the water on the earth's surface and beneath, and the plants and animals that live in a natural state in arid lands, has much to say about the people and their actions. The introduction and breeding of plants and animals, the methods of

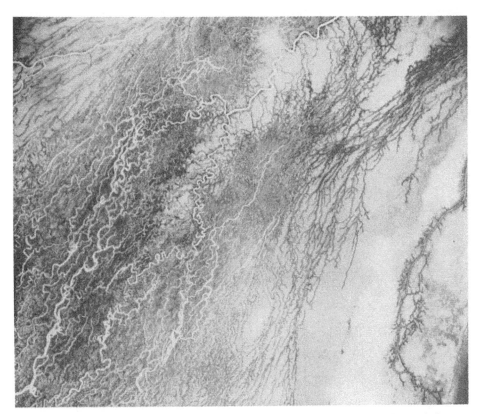

Fig 1-2. Flood waters in distributary channels of the Diamantina River, 'Channel Country', western Queensland. (*Photo by courtesy of the Division of National Mapping*)

husbandry used by farmers, the social structures and ways of life of sedentary and of nomadic peoples (Fig 1-3), and the lessons of history in arid lands in all the continents are discussed.

Despite all the researches of scientists and sociologists of recent years, we are still far from understanding or even fully appreciating the attitudes of people in relation to their mode of life and to their aspirations. Social and political systems, religious and philosophical beliefs, personal attitudes of man to woman, of woman to man, of man to man, all seem wellnigh infinitely variable.

If this book has an aim other than to instruct, it must be to provide a stimulus towards the recognition of the potentialities, whether great or small, of the arid zone as an environment in which human life can develop fully and freely. But if, in the inexorable march of 'development' and 'progress', the social and moral beliefs and standards that anchor the indigenous peoples in their environments are cast adrift by the impact of western civilisation, it may well be asked by future generations by what right we do this, unless we can provide immediately an acceptable substitute. There is little evidence that this can in fact be done, but the march of material advance goes on, and the many and little-understood

Fig I-3. Bedouin encampment near Kabul, Afghanistan. (*Photo J. L. Loder*)

wonders of ancient and primitive desert cultures are destroyed, or fixed in unlovely commercialised forms to make tourist attractions.

In material culture, the Australian aborigine lacks all that relates to permanence, other than verbal expression. Without writing, with no building more secure than a few strips of bark, and with no control of energy sources other than his own person and the use of fire, he has achieved a social culture so complex that few, if any, can claim fully to comprehend it, even for one tribe (Fig I-4).

In the varied arid environments of Africa, of India and Pakistan, of Peru and Mexico, or in other lands of ancient civilisations, the situation is different because the indigenous peoples have a highly developed material culture; but the problems of these arid regions as a home for increasing numbers of persons are correspondingly complex. Economic, political, religious and traditional factors are involved, with men of intense feelings, high intelligence and deep-rooted habits. Widely varying, from the nomadic Bedouin or the Berber, to the city dwellers of Damascus or Karachi or Isfahan, rich or poor, craftsmen or artists or poets, these people live and have their being in an arid clime: and the unmistakable *élan vital* of the Mediterranean and the Middle East intrigues us and adds to the interest of a study of these lands, great in tradition and in material and moral culture, in the arts, in mathematics, and in war.

Fig 1-4. The Australian aborigine in his primitive environment near Alice Springs (1901). Welcoming dance of the Arunta Tribe. (*Photo Baldwin Spencer; by courtesy of the National Museum, Melbourne*)

The deepest springs of life, nourishment, love, humour, the thrill of creative work or of competition, and all our human traits, are surely the same for the Bushman as for the Punjabi or the Parisian. In a different medium, the theme is the same. The nomadic weaver of rugs (Fig 1-5), the Australian aborigine painting on bark or on the walls of a cave (Fig 1-6), the Bushman decorating a gourd, have much in common as to basic notions about life, but little or nothing in common as to the ways and means of life. The responsibility is clearly great when educated, scientifically and technologically advanced groups modify the old and long-established ways of life of such indigenous peoples; but even in a different situation, where a nation has a more or less uniform ethnic composition, the problems of those who are called upon to live in rural areas remote from the culture of cities, and the problems of those who live in those cities, are many and different indeed, and it must certainly be true that these problems confront those whose task it is to organise and direct rural and urban growth in countries of diverse climate and regional development.

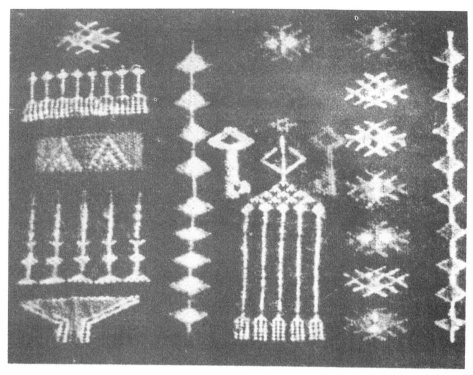

Fig 1-5. Mozabite rug, Sahara. Each motif has a symbolic meaning. (*Photo Jean Gabus*)

Fig 1-6. Australian aboriginal painting, from a cave in Arnhem Land (Northern Territory). Frieze of women with hands linked. (*Photo C. P. Mountford*)

That life in hot dry lands can be personally satisfying, the facts of history and geography demonstrate. That it can, equally, be hard in the extreme, especially for the under-privileged, is also a matter of observation. R. O. Whyte stresses this and sees in the sedentarisation of nomads, the development of urban centres and beneficial trends in land-use, some hope for betterment (Chapter xv). Nevertheless, there are those who would argue that unless the foul slums of the cities

Fig 1-7. A dairy 'farm' on the outskirts of Karachi (1957). Fodder is brought to the tethered cows. (*Photo N. C. W. Beadle*)

can also be eradicated, the life of even a poor nomad would be preferable (Fig 1-7). In this situation, it is desirable to have clear ideas as to the attitudes of the people who are the subject of social changes, and the extent to which their attitudes are superficial, imposed by circumstance, by decree or by precept, or may be more deep-rooted in physiological or basic mental equipment.

What then are the men and women with whom one has to deal? Enough has already been said to indicate that they are as many in type as the trees of the forest, nomad and city dweller, farmer or grazier, irrigator or dry farmer, miner; Muslim or Christian, Jew or Buddhist, of ancient lineage and local descent, or a newcomer from Europe or the Yemen to the Negev, from Chicago to the San Joachim Valley. Each an individual striving for himself and his family; each subject to the laws of the land; each either ignorant or educated, more or less according to his circumstances. These men and women cannot be regarded as inanimate unitary objects to be manipulated by statisticians, sociologists, scientists or administrators, the more so since in many practical matters our knowledge is still incomplete. If a Persian farmer using a qanat is told that good water can be got from a bore at 900 feet on his property, he may well wish to continue to use his qanat, feeling, with the conservatism of the farmer, that to use bore water might in some way interfere with his proven and tried qanat. The scientist must be very sure of his facts before he tries to persuade the farmer to install a bore; but if the facts are clear the persuasion must go on, to the benefit of the farmer. Nomadic herdsmen may be unwanted intruders in a newly developed irrigation area, and the government may, for various reasons, wish to sedentarise these nomads. What can they expect from sedentary life? Perhaps, from their present point of view, nothing of value. A life displeasing to them, contrary to their knowledge and experience, leading to changes in social habits with which they, as parents and neighbours, will be unable to cope. To enlighten the approach of the scientist and administrator is the task of the social scientist – a far harder task than that of the biologist or agricultural scientist, but one on which the future not only of the arid lands but of the whole world may eventually depend.

It is hoped that from this book the reader may at least glimpse something of the realities of life in arid lands: that he may in the mind experience the searing heat of summer, the desolation of drought or the welcome spate in a river whose bed for years may have been filled with nothing but sand and pebbles.

The population density of arid lands is in general low; hundreds of square miles may be devoid of human habitation. Yet, if present population growth-trends continue, the pressure to find some space in which to live, with a possible world population of 5–6,000,000,000 by 2000 A.D., will be such that, without doubt, areas of arid land vacant today will by then be settled. The economics of land-use, with water at a premium price (whether natural water or fresh water derived from the sea), may limit agriculture and farming, although there are many possibilities for increased production of some kind along conventional lines.

Already major discoveries of minerals and oil in arid regions have meant that hundreds of thousands of workers in all continents live in and endure aridity unenlivened by the green of irrigated land or of oases. The experience of many decades shows that, given adequate remuneration and housing, a full and rewarding life is possible for them, although, to be sure, persons brought up under the conditions of western civilisation rarely decide to live permanently in the true deserts. But the winning of minerals and oil is a transitory affair, for these are wasting assets and in time any one deposit must be used up. In the meantime, the wealth gained may lead to great benefits in the establishment of roads and railways, the building of dams and the discovery of ways of obtaining a profitable yield from the land, even if this be limited. Although new finds will extend the life of a mining community, in the long term mineral development will not give that permanent settlement of arid lands which is sought.

Just as it was technological developments that opened the gates to expansion and wealth in the cold-temperate regions, which languished while great civilisations flourished in and near the deserts of the Old World, so may technology in the coming years help in overcoming to an even greater extent than has yet been done the severities and disabilities of arid lands, as pressure to use them builds up with population growth.

Reference will be made in the appropriate chapters that follow to many of the scientific, technological and social problems that are being studied towards the better utilisation of arid lands. The rate of scientific discovery and technological application of scientific results is, however, now so great that it may confidently

Fig 1-8. Home built of Nile alluvium, Cairo. (*Photo E. S. Hills*)

be predicted that new fundamental knowledge and new techniques will be discovered in the coming years. The use of atomic power is a topic that is receiving much attention, but the economics of this great new source of energy are, in fact, not simple, and at the moment of writing it cannot be said that a clear picture has yet emerged of any practicable scheme of beneficiation of arid lands using atomic power. It is, perhaps, axiomatic that any development scheme which involves heavy capital expenditure must, in a broad view, be measured against the benefits that would accrue to mankind through similar expenditures in regions from which the returns might be comparable or better. And in many instances it is currently true that expenditure on semi-humid or even humid lands would yield a greater return. If such were true for any planned development, then other considerations, of a social or political kind, might in fact be those which would justify the planned programme for the arid area. Where the lives of people who already live in the area and must continue to do so are in question, then it is generally felt that social benefits must be given considerable weight, and any purely economic considerations may accordingly be subordinated. Such a view is, of course, not always easy to maintain, and it may well be that in the ultimate analysis the best use for arid land is in fact that for which it has been endowed by nature: that for some purposes its very emptiness may be an asset. To some extent this is seen already in the increasing numbers of people who visit desert and semi-desert regions for recreation, retreat and recuperation. Technological developments in which public safety or military security are strongly involved, as with atomic energy research stations or installations, or rocketry, have been located in deserts, and it is virtually certain that in the coming years further use will be made of these sparsely populated lands, for purposes as yet unforeseen. There is a challenge to architects to conceive and construct buildings well adapted to desert climates (Fig 1-8), and to medical science to advise us how to conduct our lives under such conditions. For while it may be true that it is possible, with adequate clothing, to withstand great cold more readily than great heat, it is also true that the frigid zones offer far less potential for productive growth than do the arid.

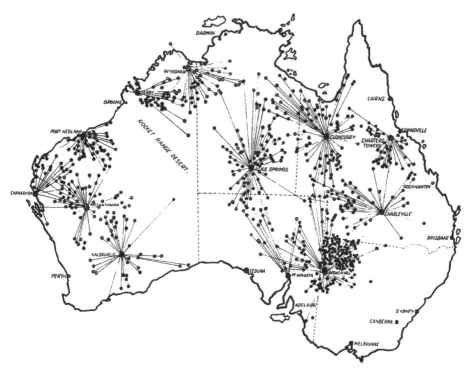

Fig 1-9. The distribution of Flying Doctor bases and contact points in the arid interior of Australia

A medico-social service of the utmost value in Australia, which might well afford a pattern for other countries, is the Flying Doctor service, the network for which is shown in Fig 1-9. This, and radio communication between widely separated homesteads, has brought succour and comfort to the lonely inhabitants of the 'Red Centre' of Australia.

E. S. HILLS

Reference

[1] MEIGS, PEVERIL, *Distribution of Arid Homoclimates*, Maps Nos. 392 and 393, United Nations Organisation, 1952.

ADDITIONAL REFERENCES

DICKSON, B. T. (ed.), *Guide Book to Research Data for Arid Zone Development*, Arid Zone Research IX, Unesco, Paris, 1957.

RESEARCH COUNCIL OF ISRAEL, *Desert Research*, Proc. International Symposium (Jerusalem, May 1952), Jerusalem, 1953.

STAMP, L. D. (ed.), *A History of Land Use in Arid Regions*, Arid Zone Research XVII, Unesco, Paris, 1961.

SUSLOV, S. P., *Physical Geography of Asiatic Russia*, transl. by N. D. Gershevsky, Chaps. XIII, XIV, San Francisco, 1961.

UNESCO, *The Problems of the Arid Zone*, Arid Zone Research XVIII, Proc. Paris Symposium, Paris, 1962.

WALTHER, JOHANNES, *Das Gesetz der Wüstenbildung*, Berlin, 1900.

WHITE, GILBERT, *Science and the Future of Arid Lands*, Unesco, Paris, 1960.

— (Ed.), *The Future of Arid Lands*, Amer. Assoc. Adv. Sci., Washington, 1956.

CHAPTER II

The World's Arid Areas

Arid areas essentially are lands of high risk and of uncertainty. Under their generally clear, bright skies the scarcity and variability of rainfall are dominant elements in the complex of physical factors that set the stage upon which man ventures his livelihood, but the precise kind of risk is immensely diversified, the range of uncertainty is great, and each sector of the zone provides its own special problems and opportunities for human occupation.

While the limits of arid lands are set by climatic conditions affecting the surplus and deficit of water for plant growth, the lands themselves differ greatly from place to place because of the variety of water balance, and of the terrain. To an already complex environment man has brought a variety of cultures, and has decided over the years to use various of its sectors in radically different ways: in some places, as in parts of the Argentine plains, he has managed to maintain the thin cover of soil and vegetation in a relatively stable state; in others, as in the Saharan oases, he has built up intensive uses by management of water and minerals; and in others, as on the flanks of the Baluchistan mountain ranges, he has violently disturbed and destroyed the natural cover. Characteristically, the equilibrium of water, soil, geological and vegetative processes in arid lands is a delicate one, so that a slight shift in one aspect – such as grazing of vegetation – may set in motion a disastrous chain of events. It also is pleasing but often misleading for development planners to think that another change in the balance – as by increasing the amount of water available to plants – may initiate a no less sharp improvement in economic production. The boundaries of human use change frequently and rapidly. Currently man is withdrawing in some sectors of the semi-arid margins, while advancing in others, and plunging deep into certain of the extremely arid sections.

Several themes thus run through the drama of man's occupation of arid lands. One is the unity of the area in terms of the general risks of rainfall deficiency and variability. A second is the diversity of the area in specific combinations of physical and cultural environment in which man deals with those risks. A third is the area's high sensitivity to radical readjustments triggered by a small change in the complex. The hazards and uncertainty are great, the possible solutions are many, the consequences of ill-advised adjustments are likely to be quick and drastic.

Man's decisions to avoid, use, develop or withdraw from arid lands may be thought of in terms of a slow, intricate process in which he assesses the resources with which he has to deal, applies the technology and organisation at his command to using them, winnows out the measures that seem inefficient on economic grounds, takes account of effects upon other areas, and progressively adjusts his institutions to advancing uses that seem desirable. But the results are not always happy: sometimes they are tragic and are written in abandoned projects or traces of forgotten farming, and oftentimes they continue in uneasy and uncertain adjustment, with failure always threatening.

The major dimensions of these human decisions are revealed in the distribution of climate classified by water balance (see Chapter III), and by five other distributions. A population density map shows where people now occupy arid lands. A classification of economic activity shows the major type of livelihoods (see Chapter X). Combinations of terrain and climatic characteristics divide the arid zone into a large number of sectors according to the kinds of water supply, slope, soil and vegetation that confront human occupation (see Chapters IV, V and VI). Fourth, the zone varies tremendously in terms of the degree to which the natural environment is being affected by different uses. Finally, the relation of national boundaries to aridity gives a gauge of political involvement and economic integration in the resulting resource management problems.

The arid zone as defined in Fig II-1 has the common characteristic that the climate everywhere is judged to be too dry to permit successful growth of crops in average years. Obviously, the outside line is not a sharp line: it is bound to be a transitional fringe because the rainfall decreases very gradually at the borders. Moreover, technological advances from time to time shrink the definition of what is unsuitable for crop growth. Three lines nevertheless may be taken as at present marking critical zones of rainfall; the outer margin of the semi-arid climate, the line between semi-arid and arid, and the limits of the extremely arid sectors. The first, as determined by the Thornthwaite moisture index of −20, roughly sets the outside limits beyond which the threat of inadequate rainfall is not generally the dominant aspect of agricultural risk. The second, as determined by a Thornthwaite moisture index of −40, divides lands where there is enough reliability of rainfall to warrant taking the chance of cropping in some years from the lands where the risk is too great and different strategies are required. The third, following the Emberger criterion of a record of twelve consecutive months without rainfall, indicates the extremely dry areas.

The cold deserts are excluded. These are taken by Meigs as areas where the mean temperature in the warmest month does not reach 10°C, and where, accordingly, agriculture is insignificant. Further subdivisions by (a) season of precipitation and (b) temperature of wettest and driest months, have been made by Meigs and are discussed in the next chapter.

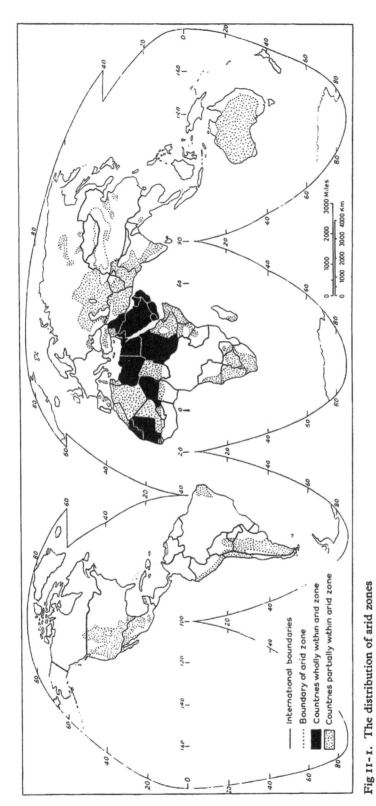

Fig II-1. The distribution of arid zones

A further note should be made on climatic classification. We take the climatic data as of the recent past, and disregard the controversial evidence as to shifts in aridity since the last glacial, making two major assumptions which further research, or history, may either bear out or repudiate. One is the assumption that there has been no pronounced change in the amount of precipitation in these areas since the Neolithic period. The second is that whatever short-term trends have been observed, as for example the rise in temperature on the borders of the Sahara since 1911, are reversible and will prove roughly recurrent (although not

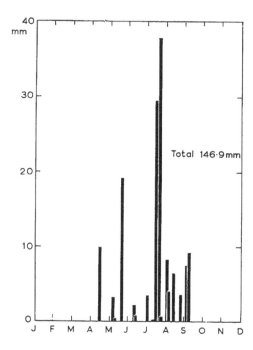

Fig 11-2. Daily rainfall during one year at Khartoum. The total rainfall for the year at this desert station fell during 17 days, and two-thirds of it fell in five of those days

cyclical) rather than secular. This is an important and possibly crucial assumption, because if there were in truth a secular trend towards increasing aridity much of what follows would require qualification.

The chief hazards which man faces in arid areas and which he tries to counter with many strategies are the risk of variation in precipitation and the uncertainty of its time of occurrence. It is not uncommon for more than 50 per cent of all precipitation in a year to fall in 10 to 15 per cent of the days with rains. A study of drought seasons in Puerto Rico, which has an arid belt on the southern leeward slope of the island, found two-thirds of annual precipitation occurred in ten of the storms [1]. In a sample year at Khartoum two-thirds of the moisture fell in five days (see Fig 11-2).

It has long been known that the reliability of occurrence of rainfall is a partial inverse function of total rainfall, but that there are large anomalies. Negative deviations from a linear relationship are greatest in the more arid regions [2], and there are other departures that vary with terrain and climatic conditions.

For example, in south-west Asia the variability is less in areas with maritime influences than in more continental areas, and it is greater in winter than in summer but has greater persistence in winter [3]. More than in temperate sub-humid regions or than in monsoon regions, the swing from crop success to utter failure is quick and frequent, with the greatest risk in the semi-arid areas.

The amount of annual drought still is unpredictable, so that while the hazard can be stated in probabilities or frequencies calculated from longer records, the next occurrence is completely uncertain, and the Bedouin who plants a field of barley in a dry valley, or the farmer who awaits a flood spate to carry moisture to a wadi on the south border of the Sahara, or a livestock rancher who moves his herds towards a grassy Patagonian steppe at the beginning of the wet season, has no way of telling whether the rains will come or not this year. He may know the risks over a decade with rough accuracy and yet be forced to make his decisions about land-use in a particular year in complete uncertainty.

Population[1]

While it is customary to think of the arid lands as being sparsely populated, the density in fact ranges from zero in some of the great uninhabited portions of the continents to several of the most densely settled concentrations. Indeed, sharp discontinuities are the rule. Four major densities are widespread (see Fig II-3).

In large segments of the South American coastal deserts and in the ergs of the Sahara as well as in portions of the Australian, south-west African and Arabian dry lands there are no permanent habitations and virtually no seasonal use. These generally are the extremely arid sectors.

At the other extreme, the arid zone is flecked with the green patches and strips of oases and mining settlements in which densities commonly run over 250 per square mile.

Greatest in area are the expanses in which population density is less than about 2 per square mile, the major steppes of central Asia, Arabia and North Africa, and the basin and range province of the western United States.

Densities of from 2 to 25 per square mile are widespread in the semi-arid margins of the zone. For example, the northern and southern borders of the Sahara, eastern Africa, Patagonia, the Great Plains of the United States and south-west Asia are predominantly in this density group.

Except for portions of Mexico, dry sectors of northern West Africa, Spain, the Fertile Crescent and much of western Iran, there are few arid lands with popula-tion densities of 25–250, the densities that are so common in adjoining humid and sub-humid regions. Commonly, the densely settled areas are scattered among very sparsely settled expanses, and the lightly settled areas of 2–25 per square mile are on the semi-arid borders. The zone has many sharp contrasts: the clean boundary between Egyptian city and Nubian desert, the line that an Iranian irrigation ditch cuts between rough grazing land and intensive market gardening,

[1] The distribution of population in arid lands is further treated in Chapter x.

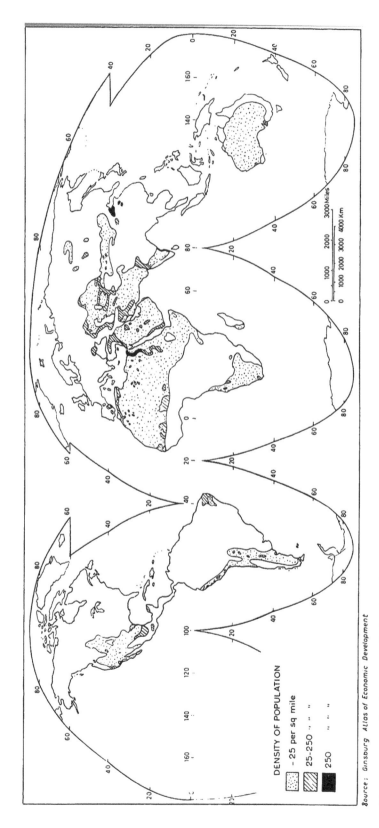

DENSITY OF POPULATION

– 25 per sq mile

25–250 " "

250 " " "

Source: Ginsburg *Atlas of Economic Development*

Fig 11-3. The density of population in arid zones. *(From Ginsburg: 'Atlas of Economic Development')*

the margin where grazing pasture is lost in Arabian erg or western Australian waterless area. The sectors in which the population density is 2–25 tend to be the especially risky ones.

Nations now are required by the circumstances of economic planning to make judicious assessments of the capacity of these arid lands to support population. Their answers must be cautious, and in the very act of seeking them may change the answer itself, but the question persists and is vital. Can we expect to witness a substantial permanent increase in population of the arid zone? Or is it now near or past its optimum? Is it unduly sanguine to expect, as is the case in economic development plans for Syria, Turkey and Rhodesia, to name only three, that the more effective management of arid lands can be an important vehicle, assuming wise fiscal and administrative hands at the helm, for major advances in production? In Syria, for example, the IBRD mission in 1955 recommended a development plan in which approximately one-tenth of the government expenditures for a five-year period would be for irrigation and drainage, chiefly in the Ghab of the Orontes Basin [4, pp. 41–51, 175]. New irrigation is expanding rapidly in Turkey but plays a less prominent role both there and in Southern Rhodesia, where it nevertheless attracts a good deal of public attention, especially when linked with large dams such as in the Sabi valley. These questions lead to the strategy man follows in gaining a living from arid lands.

Livelihood

Closely related to population densities but not directly coincident with them is the pattern of predominant livelihoods. The uninhabited areas except for a few oases within them may be ruled out as not now providing for human needs, and the remaining area may be divided into seven major types.

Most widespread is the nomadic herding of the Sahara, Arabia and south-west and central Asia. In effect, this has two sub-types, the complete nomad such as the Ouled Biri of Mauretania who with no permanent base follows the rains, and, much more common, the nomads who maintain control or attachments in irrigated and dry fodder-producing areas or who are based on commercial centres. This is wholly missing in the Americas and is one of the drastic distinctions between arid lands of the New and Old Worlds. Either form of nomadism requires the pastoralist to order his livelihood so that he can survive the extreme droughts if he is to retain his crops, and in some places he may do some dry farming when the rains permit.

Sedentary livestock ranching where pastures are grazed from an established farmstead occupies a large part of the arid lands in the Americas, South Africa and Australia. In some areas, such as in the Upper Colorado Plateau, ranchers also cultivate fodder in dry fields or in irrigated plots, but livestock is dominant and stability of production as well as of soil depends a good deal upon either having sufficient fodder reserves to carry over dry spells, or means of marketing and replenishing herds at those times.

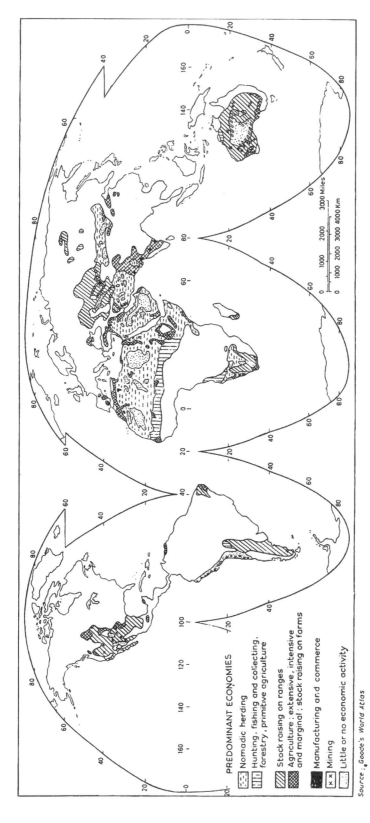

Source : *Goode's World Atlas*

Fig 11-4. Predominant economies in arid zones. (*From Goode's 'World Atlas'*)

A shifting type of cultivation in which farmers move crop and livestock from place to place at intervals of two to seven years is widely practised in the southern Sahel margins of the Sahara, in eastern Africa, and in dry segments of central Africa. Unlike ranching, the operating units are small and the farmsteads move frequently within a village's district.

Dry farming where land is cropped without artificial application of water accounts for the major wheat and barley production of the arid zone. In semi-arid areas of the USSR, south-west Asia, northern Africa, south-east Australia and the Plains of the United States, techniques of gaining harvests with moisture-conserving practices and fallowing have been especially adopted to mechanisation of large operating units. Yields are low, risks of crop loss from drought, hail and locusts are substantial, but the large units often are able to weather these hazards.

Irrigation farming is, of course, the great stable source of agricultural production in the arid zone. Sources of water and methods of applying it differ greatly within the zone, and range from simple diversion of flood flows from alluvial channels, as in the Ghash basin in the Sudan, to highly organised surface-water storage and distribution works as in the Sind and the Punjab or in Turkestan.

Mining is a principal source of income in some arid lands, and mining is the explanation for population concentrations in the Chilean copper and iron areas, in the Persian Gulf petroleum concessions, in Australian mining camps and elsewhere. Although these may make demands upon nearby areas for foodstuffs, they often are rather independent of agricultural use, and they may affect it by opening up transport facilities or new labour markets.

Finally, the city with its government, commercial and manufacturing functions is the most rapidly growing place of livelihood in the arid zone. Breakdowns in subsistence economies and the extension of trade and transport relations within countries are leading with other factors to booming city development in and on the margins of the desert. The most dramatic growth is taking place in dry parts of Arizona and California in the south-western United States, where amenities of living in a dry, warm climate are responsible in part for commercial and manufacturing development.

Of these seven types of livelihood only irrigation farming and urban occupations are not highly subject to vicissitudes of precipitation variability and of technological shifts. Even irrigation which depends upon floodwaters and diversion without storage may suffer sharp fluctuations in supply. The miner's activity fluctuates notoriously. In recent years dry farming has extended itself over new areas but it also has received reverses. Technological advances in drought-resistant seeds, in pest control and in mechanised cultivation have increased the flexibility of farming operations, and we now are in a period of paradoxical government effort. On one hand the nations where nomadism prevails are seeking to put a halt to it: 'sedentarisation' is the term commonly applied to all sorts of measures to curb, fetter or eliminate the nomad. On the other hand, farmers armed with tractors and Land Rovers are becoming increasingly mobile, moving both crops and livestock in and out of hazardous areas as the weather dictates, and city dwellers in motor-cars are carving out new caravan routes and territories for exploitation. Petroleum and gas exploitation also presents a picture of quick invasions and slow retreats that leave only a small rearguard to tend the operating fields.

In this regard, the arid lands after a long time of relatively gradual social change, are moving across a threshold into a period of rapid alterations. Caravans have disappeared, the caravanserais have lost their function, nomadic pastoralism is suffering rapid contraction, and revolutions in transport have shifted the trade and cultural relations of farmers and towns alike. Cities of the arid zone such as Cairo and Karachi have grown at a high rate. Political regimes, too, continue to display instability. The magnitude of change has increased while the time span has shortened. Can we think any longer, as Brunhes did for his Saharan oasis, of a desert people having reached a state of 'absolute perfection' in their adjustment to natural conditions [5]?

Environmental complex

These diverse patterns of population and livelihood correspond to climatic patterns in some areas but not in others, and it is important to recognise the profound differences in combinations of climate, landform, soil, vegetation and hydrology that are found within the limits of the arid lands. For any area lying within the arid zone it is possible to group lands according to particular characteristics that are significant for a given purpose. For example, in working out a programme for development of the agricultural resources of Tunisia a team of government and FAO scientists found it suitable to divide the country into six regions according to the combinations of conditions affecting development opportunities and methods [6, pp. 149–50]. These were (1) the Khroumirie and Mogod uplands with rough, forested slopes and some cultivation in valleys, (2) the Medjerda plains with irrigation in poorly drained lowlands and rain-fed dry farming of other soils, (3) the hills and low mountains of the Tell with their forests and soils for fruit trees in the foothills, (4) the steppes with their broad grassy uplands devoted to grazing, (5) the Sahel with its intensive cultivation of olives and market crops, and (6) the subdesert south of Gafsa where there is only sporadic grazing of camels based on oases. This same kind of subdivision can be made elsewhere, but will change in Tunisia as technology and development needs evolve. It must be seen from the outset, however, that no two square miles of desert surface embrace precisely the same natural features. Even in the gently sloping, almost flat surface of the west Egyptian desert where the eye seems in places to scan a hazy, uniform skyline in all directions, there are significant variations in soil, vegetation and water availability. Moreover, the several natural features do not vary in the same fashion.

The same kind of heterogeneity is everywhere present in the arid zone, and must be assumed whenever broader generalisations are made.

There has been no generally satisfactory subdivision of the arid lands according to these environmental complexes. We have maps of vegetation, soils and lithology, and of certain physical features such as closed basins and sandy desert, but not of the resulting complexes. The geomorphologist may classify lands by the erosion forms or their place in a presumed cycle of erosion, and he also may distinguish the major types of surface (see Chapter IV). The botanist may distinguish associations and larger formations (see Chapter VIII) without arriving at the same boundaries as the soil expert or the hydrologist.

A special aspect of the environment deserving separate note is the character of water supply. The climate indicates the total amount of moisture that reaches the ground and the extent to which it meets or exceeds the natural demands of potential evapotranspiration, but this resulting surplus may be differently distributed in time and space according to local physical conditions, and it may be transported out of the area or supplemented from the outside by imports in streams or aquifers. Here it is important to distinguish between potential supply and available supply. By potential supply is meant the total volume of water that falls as precipitation or flows into an area in streams or aquifers. By available supply is meant the amount that is available for human use.

Six major types of water supply situation may be distinguished.

1. Extremely arid. In these the potential supply rarely exceeds 50–100 mm annually. This would apply to the great ergs and the north Chilean coast.
2. Sparse localised. In areas with a mean annual water surplus of 100–400 mm, the potential supply is small but collects or may be collected in small amounts in localised places – pans or valleys – from which it evaporates or charges the groundwater.
3. Semi-arid. Here the mean annual water surplus is sufficient to support crop growth in some but not all years, and to supply scattered perennial streams and aquifers, as in the plains of the USSR or the United States.
4. Exogenous surface. Here the potential supply is fixed by water flowing in streams fed in more humid areas, as in the case of the Colorado, Nile and Euphrates rivers.
5. Exogenous ground. Similarly, some aquifers carry massive amounts of water from intake areas at a distance, as with the Nubian sandstone in northern Africa and the Dakota sandstone in the north central United States.
6. Localised exogenous. Where sharp gradients in relief and rainfall are found over short distances, there are substantial but erratic stream flows, and occasional replenishment of aquifers in restricted areas, as on the flanks of the Tien Shan and Elburz mountains.

According to these water supply conditions the opportunities for agricultural development and the hazards of water use vary. In general, the exogenous surface and ground supplies are the most reliable. Ground supplies are particularly susceptible to exhaustion through overdraft, and the streams of localised exogenous supplies present special problems of silting and channel shifting. Highest risk and uncertainty are found in the extremely arid, localised sparse, and semi-arid situations.

Equilibrium

The sparse resources of water, soil and vegetation of arid lands are in delicate equilibrium without the intervention of man, and under his occupation they are subject both to gross deterioration and to substantial improvement. A slight change in the grassy vegetation of a semi-arid steppe or in the thin upper horizon of its soil may have a profound effect upon the area's productivity as well as upon quality and movement of water. At any time the environmental complex at a given place may be thought of as having been drastically disturbed, as slowly moving away from an equilibrium condition, as slowly returning to it, or as showing little change.

Extremely arid lands account for much of the arid zone that displays little change from precarious equilibrium conditions. Deserts of low intensity use such as the Saharan ergs, the Kalahari, the Atacama and central western Australia and their borders have been altered little by the hand of man. Even when grazing may be severe in times of unusual drought there generally is opportunity for natural recovery of plant cover, and the amount of soil erosion which is generated may not be extreme. There are exceptions, as will be noted, on the periphery zones of cities and oases.

Most striking are the lands where man has caused marked deterioration of soil, vegetation and water resources from what is believed to have been the state before extensive human use. Determining the earlier state or the natural processes that have been affected is by no means easy, and much of the knowledge of human modifications is largely speculative. For example, the degree to which the grasslands on the southern fringe of the Sahara are natural or are the product of repeated burning and grazing by man in his shifting occupance is still controversial: evidence is scanty, and some theories are conflicting. The record of human destruction is clear, however, in many instances. Thus, denudation of soil cover from steeper slopes in the uplands of south-west Pakistan and in the Rif of Morocco has been pronounced, and bare hillsides testify to overgrazing and wood removal. It is common to find wide belts around towns and cities, such as Khartoum, in which fuel gatherers, first on foot and donkeys and now in lorries, have removed virtually every stick of woody vegetation, and a desert aspect indeed is induced. Wherever large-scale cereal farming has attacked the natural grasslands the vegetative cover has been changed massively, as in the Syrian plains. Drifting sand from such operations continues to encroach upon dry farming lands in the Sind and the Punjab. Use of groundwater has exhausted some aquifers, as in the San Joaquin valley of California, and has in numerous situations permitted intrusion of saline waters that then permanently pollute the supply, as in the Tripoli Quadrangle on the coast of Libya. Curiously enough, some of the more damaging changes have taken place in irrigated areas where waterlogging from over-application and inadequate drainage of irrigation water, or salt accumulation resulting from improper application, have rendered great plots of heavy soil useless, as in the Indus valley.

A drastic disturbance of the equilibrium also may build up soil, vegetation and water so as to increase productive capacity, and this is often the case in those irrigated areas where farmers have been wise enough or fortunate enough to avoid land destruction. Altogether, about 140 million hectares of land now are under irrigation, and new lands are being watered at an increasing rate under stimulation of heavy government public works. In the four chief countries carrying on large-scale water management the conversion of desert to cropland by irrigation is proceeding with vigour: the Chinese People's Republic is reported to have raised its area from 20,000,000 to 53,000,000 hectares in 1949–58; India expanded its acreage by at least 40 per cent between 1950 and 1961; the USSR acreage doubled from 1940 to 1960, and the United States expanded its acreage by one-third from 1944 to 1959. From these totals, of course, must be subtracted the lands in the previous category which are being lost each year, and these must be well above 80,000 hectares annually. Certain grazing lands also have been reclaimed for use by installation of reliable and more closely spaced watering points, as in the Sudan borders of the Sahara, and in Kazakhstan and Turkmenia in the USSR.

Intermediate between the lands that are largely undisturbed and those that have been altered drastically for either good or bad, are the lands where significant but gradual trends in equilibrium may be observed. Thorough destruction has not yet resulted, and neither has there been dramatic recovery. With some exceptions, the whole belt of grazing lands between Morocco and India is literally and figuratively losing ground every year. Likewise, the combination of vegetation removal, wind erosion, and sheet and gully erosion places a very large part of the semi-arid farming and grazing lands of the Americas, Asia and Africa in the category of slow deterioration. As population increases in these areas and in the African lands south of the Sahara, the rate of destruction is tending to accelerate.

In a few countries strenuous efforts are being made to rebuild and improve the resource base where gradual degradation had been the rule. Much of the programme devised by FAO in collaboration with countries of the Mediterranean for agricultural development has this as its aim [6]. The USSR has made massive efforts to transform soil, vegetation, water and microclimate by means of changes in cultivation (ploughing and fallowing), strip cropping, trenching, forest planting, and various types of small-scale irrigation [7]. The United States in the semi-arid Great Plains has undertaken widespread works of contour ploughing, strip cropping, mulching and revegetation to stabilise the soil and vegetative cover. In the USSR, the USA and Australia, government measures for improvement of grazing lands include drilling of new wells, development of watering places, reseeding of range, weed control and, perhaps most important, the management of amounts and season of grazing.

On balance, much greater areas of the arid zone are deteriorating than are improving in stability. Only relatively small sections of the world's grazing lands now promise early increases in bearing capacity. Acreage of presently expanding land under irrigation is mounting, but in some regions it barely exceeds the current losses. Area, however, is a crude measure of change, and more importance should be attached to shifts in production. This is harder to gauge because there are no accurate records of livestock and crop returns for arid portions of countries including humid lands. A few general observations may nevertheless be ventured. In the semi-arid fringes where power machinery has made possible wholesale cultivation of cereals, the gains in returns have been large in recent decades. The virgin lands of the USSR, the Euphrates plain of the Syrian sector and the northern Great Plains of the USA have all enlarged returns, albeit not without some failures and disappointments in unfavourable crop years and some doubts as to the efficiency of those lands by comparison with humid lands. The net gains in irrigation no doubt more than offset the losses in returns from grazing lands.

Political and economic integration

Each of these patterns has been discussed without detailed consideration of the relations between phenomena inside the arid zone and those in adjoining areas. This ignores the elaborate and in some places crucial linkages that affect the character of arid land-use. To enumerate a few of them is to suggest their extent and importance. The spread of recreational use in southern California is directly linked to urban growth in the adjacent coastal strip having a Mediterranean type of climate. The irrigation developments in the south Peruvian desert are affected on the one hand by farming activities of Indians in the nearby humid highlands, and on the other hand by commercial growth in the coastal cities. Efforts at construction of new irrigation projects in the dry valley of the Sabi River in Southern Rhodesia were first inspired by the search for land for Africans crowded by European plantations off of the nearby rainy uplands. The life of most Saharan oases is strongly tied not only by trade but by ownership and tribal organisation with distant cities and agricultural centres.

It would be a gross error to think of the human use of the arid zone as being self-contained or largely organised within the zone. With the exception of the Kalahari, the north central Australian desert, and a few other areas, the arid lands must be seen as intricately linked with more humid areas.

Certain of the arid land problems stand out most clearly in nations which lie wholly within the arid zone. These face the integration of grazing and irrigated lands, and of dry lands with other sectors of the economy. They cannot rely on humid sectors for support of new investments, or for supplementing the range of products, and their foreign trade must provide the needed complementarity of activities. Only about seven nations (Egypt, Eritrea, Iraq, Libya, Saudi Arabia, Republic of the Sudan and Somalia), not counting the Persian Gulf and Trucial states of the Arabian Peninsula, are wholly arid, and these must be regarded as a special, extreme category. Here there can be no choice of attempting development of the arid lands, but there can and should be earnest searching as to alternatives in use of those resources.

More than thirty-seven nations having some territory within the arid zone but some territory outside it account for the larger part of the zone in land area. The typical situation is one of arid lands beside sub-humid and humid lands within the national boundaries, and national strategies for development must consider the needs and opportunities of the two groups together.

Problems

Upon the complex mosaic of physical environment and culture which the arid zone thus comprehends, several problems persistently present themselves to the geographer. Without solutions to them we can neither fully understand the past course of resource management in the zone nor build the groundwork for intelligent planning of further action.

Are there sensitivity points in the use of resources which trigger accelerated deterioration of the resource? We know, for example, rates of water pumpage for any given aquifer which exceed the safe yield. Can similar critical rates be recognised for use of vegetation and soil? Or for the application of irrigation water? To the extent these can be determined for a particular natural land type we can begin to perceive the physical limits of lasting land-use in the zone.

In the same fashion, it may be asked whether or not there are sensitivity points in the returns from risk-taking in agricultural use of arid lands. At what recurrence interval or frequency of crop failure in dry farming areas does the Bedouin cease to place a barley crop in favourable soils? What frequency of drought can be supported by the mechanised farmer in the semi-arid margins at a given price for wheat? We do not know how far or how often the pastoralist of earlier times was willing to venture in search of grass in very dry years, and we have only vague notions of this for many of the present-day nomads.

There is great doubt as to the extent to which further investment in improvement of arid land resources may be warranted by the resulting gains in productivity. Over many centuries elaborate works for water management have claimed heavy inputs of capital and labour, and new forms of land improvement, such as by wells and pasture management, as well as the traditional, are receiving attention in national development plans. Irrigation is extending its green valleys further into the sunburnt lands of the arid zone, but it may already be overextended. Although the laboriously constructed high terraces of the Roman occupation of Tunisia are not being duplicated today, even more extensive alterations in water management in the field are being made by tractor terracing and furrowing. Range improvement measures now are under way with benefit of tractor and aeroplane over large expanses that previously had enjoyed no systematic management. Yet, the colour of the re-conquest of the desert may obscure some of the less attractive but fundamental themes. To what extent is the capital going into high concrete dams, deep wells and afforestation being wisely spent to serve the ends of national development and stability? This, of course, requires an appraisal of alternative investment opportunities for the same resources and people, and it must involve an estimation of the impacts from these efforts – present and future – upon the soil, vegetation and water and upon human livelihood.

This raises the question of what conditions of social guidance must be met in order to realise full benefits from basic alterations in resource use. Whatever the theoretical gains that might be expected from a high dam at Aswan, and however competent the design and construction of the dam itself, their realisation will hinge in large measure upon the quality of complementary measures reaching deep into the social fabric and economic process of the communities involved. Have we yet reached a point where we may specify the conditions necessary to achieving fully the intended impacts from such projects? Indeed, are we yet capable of identifying and measuring those impacts?

GILBERT F. WHITE

References

[1] RIEHL, H., *Tropical Meteorology*, New York, 1954, pp. 95–7.
[2] WALLÉN, C. C., 'Fluctuations and variability in Mexican rainfall', *The Future of Arid Lands*, Amer. Assoc. Adv. Sci., Washington, 1956, pp. 148–152.
[3] WALLÉN, C. C. and G. PERRIN DE BRICHAMBAUT, *A Study of Agroclimatology in Semi-Arid and Arid Zones of the Near East*, FAO, Unesco/WMO, Rome, 1962, pp. 29–30.
[4] IBRD, *Economic Development of Syria*, Baltimore, 1955.
[5] BRUNHES, J., *Human Geography*, Chicago, 1920, p. 451.
[6] FAO, Mediterranean Development Project, *Integrated Development of Mediterranean Agriculture and Forestry in Relation to Economic Growth: A Study and Proposals for Action*, FAO, Rome, 1959.
[7] KOVDA, V. A., 'Land use development in the arid regions of the Russian Plain, the Caucasus and Central Asia', *History of Land Use in Arid Regions*, Unesco, Paris, 1961, pp. 175–214.

ADDITIONAL REFERENCE

MEIGS, P., *Geography of Coastal Deserts*, Unesco, Paris, 1966 (Arid Zone Research XXVIII).

CHAPTER III

Arid Zone Meteorology

There are three main regions in the world where climatic circumstances are dominant in limiting the activity of man: in the polar regions the cold and long dark winters prevent permanent human settlement based on local productivity, in the tropical rain forests the high temperature and humidity lower human activity, and in the arid and semi-arid zones strong insolation and scarcity of water place a limit on the growth of permanent settlement.

It is essential for the proper development of these regions that the basic factors affecting living conditions should be thoroughly known, and the study of weather phenomena and climate is fundamental for this.

Climatic causes of arid zones

MACROCLIMATIC INFLUENCES

The first questions we have to answer are, what are the causes of aridity, and how should we define an arid or a semi-arid region? We shall return to the question of how to define the arid regions later, and shall confine ourselves here to stating that a semi-arid or arid region must reveal in some way or another scarcity of water; in other words the amount of water available through rain, soil moisture and groundwater is not sufficient to balance the loss caused by run-off, evaporation and transpiration by plants. The income side of the water balance consists almost entirely of precipitation; only a very small part stems from dew and other condensation phenomena such as fog-drip. On the other hand certain areas showing deficiency of rainfall receive water sufficient to prevent aridity through surface or underground water entering them. The earth's arid regions are, however, basically caused by deficiency of rainfall in relation to evaporation.

General circulation of the atmosphere

The amount of precipitation which comes to a certain area is fundamentally related to the general circulation of the atmosphere, and the location of the area in question in relation to this. Roughly and schematically the circulation of the earth's atmosphere may be described as occurring in three different systems. The first system is found in the tropics on both sides of the equator. In this area winds at the ground are fairly stable from the north-east in the northern and from the south-east in the southern hemisphere. In the equatorial zone the air is almost constantly rising to higher levels, causing, through convection and adiabatic cooling, clouds and heavy rain all the year round, and creating tropical rain forests. At upper levels the air generally flows towards higher latitudes and above latitude 30° N and S becomes approximately westerly and very strong, forming the so-called subtropical 'jet-streams' due to interaction between the rotation of the earth and the dynamics of the air. Below these strong westerly winds the atmosphere is very stable and the air is subsiding, which means that it is warmed adiabatically, clouds disappear and practically no rain occurs. Permanent anticyclones are formed at the ground and large desert zones are created.

Beyond latitude 30° the circulation of the atmosphere is dominated by the mid-latitude westerlies occurring both at the surface and at upper levels. Cyclones and fronts are formed in these westerlies and rain is a common phenomenon in all regions covered by this circulation system.

In the polar regions, low-level anticyclones are formed due to long-lasting snow-covers or ice caps, and because of this and the low moisture content of the cold air, the amount of precipitation is comparatively low.

We see then that, by and large, there are in each hemisphere two main zones of comparatively low rainfall. The dry parts of the polar regions may, however, hardly be described as arid, because evaporation is very low and therefore only small areas actually show a negative moisture balance. The important arid zones of the world are found around latitude 30° N and S, where large areas are dominated by high-pressure systems and subsiding air all the year round. Most of the large deserts of the earth are situated in these latitudes, such as the Sahara in North Africa (Biskra), Arabia and the Syrian desert, in west Asia (Palmyra, Baghdad), Death Valley in North America (Phoenix), the Atacama Desert in South America (Antofagasta), the Kalahari Desert (Windhoek) in southern Africa, and the Australian deserts (Alice Springs).[1]

In the borderline zones between the high-pressure systems around latitude 30° and the circulation systems situated to the north in the northern hemisphere and to the south in the southern, conditions are more complicated and create regions of transition which to a large extent may be called semi-arid. According to studies carried out jointly by Unesco, FAO and WMO in the semi-arid and arid zones of the Near East [1] it may be advisable to draw the limit between the semi-arid and arid conditions, where regular dry-land farming is no longer possible due to lack of water. In areas on the low latitude side of the truly arid regions, where rain comes mainly through convection during the summer

---

[1] For data concerning towns named in parentheses see Table III – I, page 37.

season when evapotranspiration is great, the limit for regular dry land farming lies at an annual amount of rainfall of at least 500 mm (cf. Monterrey). Local factors, however, may create great deviations from this figure.

On the higher latitude side of the high-pressure systems which create the true arid regions the situation is quite different (cf. Sacramento, Palmyra, Baghdad, Teheran, Zaragoza, Marrakesh, Merredin). In winter the transition zones on this side are under the influence of the mid-latitude westerlies, implying that considerable amounts of rain are received quite regularly in connection with cyclones and fronts. However, due to the fact that rain occurs in winter when water loss is comparatively low, not so much water is needed to allow for regular dry-land farming as on the lower latitude sides. With the definition applied here 250–300 mm of annual rainfall could in most areas be accepted as the limit between arid and semi-arid zones. However, a strong reservation should again be given for deviations incurred by local factors influencing the general circulation of the atmosphere and the evaporation conditions.

Influence of topography and sea

Several other factors are also important in the creation of arid or semi-arid regions. In the interior of the Asian continent, for instance, vast areas are found at comparatively northern latitudes, which should be classified as arid or semi-arid. The Takla-Makan of east Turkestan and the Gobi of Mongolia are deserts located as far north as latitude 40° to 50° N (cf. Kashgar). Even the Kirgis steppe in west Asia is very dry, showing at least semi-arid conditions, although situated at latitude 50° N. Similar conditions at comparatively high latitudes are encountered in North America, where large parts of the central Plateau east of the Rocky Mountains or on the plains in this mountain chain are at least semi-arid at a latitude of around 40° N. In South America, Patagonia east of the Cordillera is arid to semi-arid at a latitude of 40° S.

In all these cases the decisive factor is either distance from the sea or topographic barriers which prevent the moisture carried by the winds from the sea from reaching the areas in question. Most of the cyclones created in the Atlantic and carried with the westerlies over Europe into Asia lose their strength and moisture before reaching the interior of Asia. In a similar way the moisture carried by the monsoon winds from the Indian Ocean or the Pacific is precipitated over the huge mountain barriers south and east of the interior of Asia. In the case of North and South America, the Rockies and the Cordillera create similar conditions on the great plains and in Patagonia. It goes without saying that when these topographical factors are added to those generated by the general atmospheric circulation, the effects are greatly increased. The most severe desert conditions in the world are found under such circumstances, for instance in Death Valley in the subtropical regions of western USA.

Along the coast from the Atacama Desert of northern Chile (Antofagasta) to northern Peru, extreme aridity is caused by the combined effects of the circulation of the atmosphere and the cold Humboldt current, which runs from the Antarctic along the west coast of South America. A cold sea-current, bending out, as in this case, from a coast, causes relatively cold bottom water to well up along the coast, adding an adiabatic cooling effect in the rising water to its own coldness. This acts as a strong stabilising factor on the air above. At the Atacama and Peruvian coasts, rain is practically non-existent due to the stability of the air. At the same time the air is often quite humid and in many of the coastal deserts fog is a common phenomenon, at least in winter and at night. The humidity of the air at this latitude is, however, not sufficient for the growth of trees or important vegetation, in the absence of rain.

It is obvious that when the effect of the cold sea is added to other factors creating aridity the consequences become extreme; this can be said to be true locally in many localities on the Eastern Cordillera of South America. Other regions where weaker cooling effects of the sea are operative occur along the Californian coast and on the East African coasts (Windhoek) in both the northern and southern hemispheres.

Climatic elements to be observed

In order to understand thoroughly the effects of those factors mentioned in the preceding section it is essential to study very carefully the climatic elements that are the consequences both of the general atmospheric circulation and of local factors. A fundamental condition for such studies is well-distributed networks of meteorological stations taking regular observations at least of temperature, humidity, precipitation and wind, which are the simplest climatic elements to measure. Measurements of radiation from sun and sky, and the soil-moisture conditions are also extremely important although they are more difficult to measure continuously, due to the need for more delicate instruments and more skilled technicians.

It is a regrettable fact that in the arid and semi-arid zones of the world sparse population, low production and other economic factors often prevent the establishment of a sufficiently effective network of stations. There are of course great differences between the various arid regions, but it is generally accepted that observational difficulties are greater in such regions than in any other parts of the world, except the polar regions. In fact, in many deserts, hundreds of kilometres separate the stations. The Sahara, covering as great an area as all Europe, has only one-hundredth the number of meteorological stations that Europe has. It is obvious that in this situation our knowledge of the climate in semi-arid and arid zones is in many cases extremely limited and superficial.

Radiation

The overall basic climatic element is the radiation received from the sun and sky (insolation) and the balance between the insolation which is absorbed by the earth, and the reflected and outgoing radiation from the earth. This *net radiation* governs the energy available for the origination of winds, for the dynamics of rainfall, for evaporation and other processes. It is unfortunate that practically no radiation measurements were made in extreme deserts until the last few years. Nevertheless, by means of interpolation between some few observation stations Black has made an estimate of the incoming radiation from sun and sky all over the world [2], and estimates have been made of the outgoing radiation also. Those arid areas of the world which are situated around the tropics of Cancer and Capricorn generally show maximum incoming radiation over the year for the whole world, mainly due to low cloudiness. For similar reasons the outgoing radiation is, however, also great and the annual net radiation therefore is some-what lower than around the equator. Generally speaking, the main arid zones of the world are located at latitudes where the annual amount of incoming radiation is greater than the outgoing, i.e. they have a positive radiation balance. In the border zone towards higher latitudes the surplus gradually changes into a deficit which is counter-balanced by advection of warm air from the areas of surplus by the circulation of the atmosphere.

Interpolated values of total incoming radiation from sun and sky taken from Black's maps are given in Table III-1 for January and July for a number of stations in semi-arid and arid regions. Monthly maximum values are obtained in July in the Syrian and Arabian deserts with around 750 cal/cm². For summer rainfall climates values in July come down considerably as shown for a monsoonal district (Jodhpur in India) where it is only 490. Winter values are low in dry lands of winter rain but may reach as high as 520 in Khartoum where there is a summer rainfall regime.

The net radiation at the earth's surface is partitioned into three parts: that part which heats the soil, that which heats the air by convection through contact with the soil surface, and that which is used in evaporation.

At this stage we are interested in that part of the energy which heats the air and the soil because this heating introduces the next climatic element to be measured, i.e. temperature. This element to a considerable degree integrates the heating conditions by radiation and may to a certain extent give estimates of the available energy when no direct measurements of that element are available.

Temperature of the air

Temperature of the air usually is measured, as are most other elements, in a screened space where the thermometers are protected from direct sunlight. Generally the screen is erected so as to give the temperature of the air 1·5–2 metres above the earth in a grassed area away from the influence of buildings and trees. From such measurements, more or less standardised all over the world, temperature conditions in the arid and semi-arid zones may be compared with those in other regions.

Maximum temperatures in deserts and semi-arid areas are high, particularly so in summer, but even in winter, temperatures during daytime are very high compared with other regions. As seen from Table III-I the mean maximum temperature in summer may reach as high as +43·6° in Baghdad, +41·2° in Biskra (Sahara) and +40·0° in Phoenix (USA). In arid regions influenced by cool ocean currents mean maximum temperatures in summer are considerably lower, as shown by Antofagasta (Chile) +24·4 and Windhoek (South Africa) +29·4°. Winter mean maxima are low in cold deserts in Asia as seen by Kashgar, where in January it is only 0·6° but may reach as high as around 30° in the Sudan and Sahara (Khartoum, Bilma).

Temperature is not only an indication of energy conditions: it is, as such, critical at certain values. In cold arid regions, of course, low temperatures may be critical. Certain of the arid regions particularly in Asia are affected by severe frosts due to strong outgoing radiation from bare soil at night. Studies of frost hazard and measurement and prediction of minimum temperatures therefore are essential, especially in those semi-arid parts where agriculture is possible.

However, in the majority of arid lands it is the high temperatures that create difficulties. The reason is twofold. As far as temperature gives an indication of energy conditions, high values during long periods mean that the energy available for evapotranspiration is so high that the potential water loss is greater than the amount of water available. On the other hand there are chemical processes in the plants which are intensified at high temperatures, so that the plant may die. Extremely high temperatures may thus lead to plant disaster, and they also lead to stress both for animals and human beings.

In the study of the distribution of arid regions temperature therefore has to be taken into consideration, both in relation to the moisture conditions and as far as extremely high values are concerned.

Soil temperature

The radiation energy available at the earth's surface not only heats the air but penetrates into the soil, and gives rise to fluctuations in the soil temperature which it is essential to measure and consider in studies of the complete energy and water balance at the earth's surface.

Precipitation

The radiation balance which is the driving force of the general circulation of the atmosphere is also closely coupled with precipitation. Our interest lies in the causes of precipitation as well as its amount, dependability and intensity. The causes of rainfall in any area must be studied through synoptic and theoretical considerations related to the atmospheric circulation. (By synoptic studies are meant the consideration of observations made simultaneously at different places.) As examples could be mentioned the dynamics of the easterly circulation and

TABLE III-1 *Climatic data at various arid and semi-arid stations*

	Lat.	Long.	January Mean	Mean max.	Mean min.	Rain-fall	Inc. rad.	July Mean	Mean max.	Mean min.	Rain-fall	Inc. rad.	Year Rain-fall	Period
1. Sacramento, N. Am.	38°31′ N	121 30 W	+7·5	+11·1	+3·9	96·5	190	+23·3	+32·2	+14·4	<2·5	650	472	1850–1949
2. Phoenix, N. Am.	33 26 N	112 01 W	+11·1	+18·3	+3·9	20·3	290	+32·5	+40·0	+25·0	25·4	610	191	1896–1949
3. Zaragoza, Eur.	41 39 N	00 53 W	+5·7	+9·7	+1·8	14·0	170	+23·5	+30·5	+16·6	20·0	680	377	1951–60
4. Marrakesh, Afr.	31 37 N	08 02 W	+11·5	+17·3	+5·7	28·0	290	+28·2	+36·9	+19·5	2·0	670	242	1951–60
5. Biskra, Afr.	34 48 N	05 44 E	+11·4	+16·1	+6·7	17·8	290	+34·2	+41·7	+26·7	2·5	690	158	1913–50
6. Kashgar, As.	39 24 N	76 07 E	−5·3	+0·6	−11·1	15·2	190	+26·7	+33·3	+20·0	10·2	650	81	1934–44
7. Palmyra, As.	34 33 N	38 17 E	+7·9	+13·1	+2·6	24·0	240	+29·8	+38·3	+21·2	0	740	153	1951–60
8. Baghdad, As.	33 20 N	44 24 E	+10·6	+16·7	+4·5	25·0	300	+34·6	+43·6	+25·6	0	730	151	1951–60
9. Teheran, As.	35 41 N	51 19 E	+4·0	+8·9	−1·0	37·0	230	+28·6	+35·5	+21·6	0	700	213	1951–60
10. Jodhpur, As.	26 18 N	73 01 E	+16·7	+24·4	+8·9	2·5	370	+31·4	+36·1	+26·7	101·6	490	356	1891–1943
11. Cairo/Helwan, Afr.	29 52 N	31 20 E	+13·3	+18·3	+8·3	5·1	300	+28·4	+35·6	+21·1	0	730	28	1904–45
12. La Paz, N. Am.	24 10 N	110 18 W	+18·0	+22·2	+13·9	5·0	380	+19·4	+35·0	+23·9	10·1	560	145	1917–42
13. Monterrey, N. Am.	25 40 N	100 18 W	+14·4	+20·0	+8·9	15·2	250	+26·9	+32·2	+21·7	58·4	600	579	1910–45
14. Bilma, Afr.	18 41 N	12 55 E	+17·2	+27·2	+7·2	0	480	+33·0	+42·2	+23·9	2·5	630	23	1931–40, 1949–55
15. Khartoum, Afr.	15 36 N	32 33 E	+23·6	+32·2	+15·0	<2·5	520	+31·7	+38·3	+25·0	53·3	600	157	1900–45
16. Antofagasta, S. Am.	23 28 S	70 26 W	+20·8	+24·4	+17·2	0	700	+13·8	+17·2	+10·5	5·8	340	13	1904–42
17. Windhoek, S. Afr.	22 34 S	17 06 E	+23·3	+29·4	+17·2	76·2	420	+13·0	+20·0	+6·1	<2·5	340	363	1891–1950
18. Alice Springs, Aus.	23 38 S	133 53 E	+28·6	+36·1	+21·1	43·1	690	+11·7	+19·4	+3·9	7·6	390	251	1921–50
19. Merredin, Aus.	30 46 S	121 27 E	+26·0	+34·3	+17·6	18·0	690	+10·3	+15·7	+4·9	49·0	250	300	1951–60
20. Mendoza, S. Am.	32 53 S	68 49 W	+23·9	+32·2	+15·5	22·8	680	+8·4	+15·0	+1·7	5·0	250	191	1921–50

Temperatures in °C   Rainfall in mm   Radiation in cal/cm²/day

convective storms which cause summer rainfall on the low latitude side of the true desert areas, as well as the synoptic characteristics of disturbances in the mid-latitude westerlies which cause the main part of the rain on the high latitude side. From recent investigations we know for instance that waves in the easterlies and the position of the subtropical 'jet-stream' at high levels have a considerable bearing on the causes and amount of rainfall in the semi-arid lands of the tropics. We can even better understand and follow the cyclonic disturbances created by outbreaks of cold polar air in the mid-latitude westerlies and their bearing on the rainfall mechanism in the semi-arid zones of winter rainfall.

In defining aridity the amount of precipitation annually or by seasons may be used. In Table III-1 the mean annual rainfall as well as the average amounts in January and July are given for a number of stations in arid and semi-arid conditions. The relation between winter and summer rainfall is of particular importance for a determination of the type of rainfall regime at various places. It should be emphasised, however, that for practical reasons, i.e. for plant growth and survival and for agricultural possibilities, not only the amount but also the variability is of fundamental importance.

What really counts in practice is the probability of exceeding the minimum amount of rain required for vegetation or crops to develop. This minimum value should, for instance, be exceeded in at least eight years out of ten to allow for regular dry-land farming of crops. As mentioned earlier, it may be useful to adopt this limit for a typical plant as the limit between semi-arid and arid conditions. In the Near East it has been found that this limit is situated where an annual rainfall of 180–220 mm is obtained in eight years out of ten. The dependability of annual or seasonal rainfall can be studied by various methods. It has proved quite useful in areas of low rainfall to adopt the relative interannual variability as a suitable method. This measure is calculated by summing the difference in rainfall from one year to the next, disregarding the sign, and dividing the sum by the number of years of record minus one. This value is then put in relation to the average rainfall, on a percentage basis. In the Near East the limit for dry-land farming was found to be at a mean annual rainfall of 250 mm with an interannual variability of around 40 per cent. Variability normally increases with decrease in rainfall, so that with true desert conditions it may reach 100 per cent of the rainfall and considerably higher values for monthly or seasonal rainfall.

Deserts and arid regions are often subject to very heavy showers of short duration. At such times normally dry valleys may become fast-flowing streams, and flooding may occur on flat land (Fig. III-1). Such rainstorms are important in relation to geomorphology (Chapter IV) and also for life in the deserts (Chapter V).

As was pointed out in the discussion of the causes of aridity, some areas are humid and foggy although they receive very little rain. This implies that the humidity conditions of the air are important, and from the macrometeorological point of view, relative humidity is the parameter commonly studied. There are several examples in low latitudes, where a high relative humidity, usually due to dominant onshore winds from the sea, may to a certain degree compensate for lack of rainfall. In the arid or semi-arid zones along the Persian Gulf, for instance, the vegetation is much better developed than would be expected with the rainfall,

Fig III-1. An Algerian oasis after a flash storm flooded the area. Arab inhabitants quickly took to makeshift rafts. (*Copyright STEEP – Jean Guglielini*)

and even includes several forms characteristic of a tropical flora. Permanent onshore winds give rise to a relative humidity in this area of 60–90 per cent instead of 20–40 per cent in the interior of the Iranian plateau.

Hot winds

Strong hot winds, which are generally associated with sandstorms (Fig III-2), are connected either with large-scale disturbances in the general circulation, or are of a local character. Well-known hot winds related to cyclonic disturbances of a large scale are, for instance, the 'khamsin' in North Africa and the Near East, the 'harmattan' on the south side of the Sahara, and the 'sumum' in Iran and Pakistan. The 'khamsin' is connected with cyclonic disturbances created in North Africa south of the Atlas mountains and moving through North Africa towards the Near East. Hot, strong winds with dust or sandstorms occur in front of the depression and often last for some days. The 'harmattan' is a more constant hot wind on the southern fringe of the Sahara occurring in periods when the north-east trade winds dominate the area. The 'sumum' is a fairly constant north-westerly summer wind.

Fig III-2. Sandstorm approaching Mildura, northern Victoria, Australia. (*Photo by courtesy Dr U. Radok*)

MICROCLIMATIC INFLUENCES

Although the general climatic conditions are important for the understanding of the causes of aridity, the actual local conditions at a certain place may be of greater significance for practical reasons. In the following section we shall deal with the details of climate which influence plant life more directly.

Water balance

To most plants the essential factor for life is the availability of water to allow for transpiration. The study of the availability of water for plant growth calls therefore for a thorough knowledge of the annual fluctuations in the water balance, i.e. between income of water on the one hand (precipitation, dew, moisture carried to the region as groundwater), and loss of water (infiltration into the ground, run-off and moisture transfer from the ground and vegetation) on the other.

The analysis of the water balance is particularly essential in semi-arid regions, situated as they are in a transition zone to regions where sufficient water for useful plant growth is not available. In true arid regions the evaluation is important for the practical purpose of determining the needs of irrigation for particular crops. Among the factors mentioned above, precipitation and run-off in streams are measured on a routine basis. The other factors are often more difficult to measure or evaluate and not until recent years have strong efforts been made to develop instruments and methods to measure soil moisture, groundwater fluctuations, dew conditions and evaporation.

Evaporation and evapotranspiration

Moisture losses from the ground and from vegetation, often referred to as *evapotranspiration*, may be used as an important characteristic of climate, and are of great significance in semi-arid and arid regions. In order to study this factor macroclimatological measurements carried out at 2 metres above the ground are generally not sufficient. Analysis of the evaporation and transpiration is a matter calling for detailed measurements of various climatic elements at different levels in the air layer close to the ground.

An approximation of the evaporation from a certain area may be obtained by the use of the hydrological equation, where evaporation comes out as a residual between precipitation on to the area and the discharge from the same area. Infiltration into the ground, and the soil-moisture conditions, are then disregarded. The method may be used on an annual basis when soil and ground conditions make it probable that in the course of a year, changes in soil-moisture and ground-water conditions even out. In other situations this method may lead to serious errors, and in arid and semi-arid climates is generally not acceptable.

For years attempts have been made to determine evaporation from 'artificial lakes', i.e. tanks or pans of various dimensions. Unfortunately, experience shows that several sources of error make such measurements somewhat inaccurate. Apart from difficulties due to winds, animals and accidents which may cause great errors in the measurements, the basic difficulty exists that a tank, because of different boundary conditions, heating effects due to the material of the tank, and other factors, will never act exactly as a natural water surface as far as evaporation is concerned. Last but not least, it should be emphasised that transpiration from plants cannot be very closely approximated by measurement of tank evaporation. The method may be used, however, to gauge approximately evaporation from a free water surface.

In order to avoid the above-mentioned difficulties, several attempts have been made to calculate evaporation and transpiration from theoretical principles, using measurements of various meteorological factors in the air layer close to the ground.

Moisture transfer into the atmosphere is governed by turbulent diffusion and may be determined from the theoretical principles of this mechanism. In calculating evaporation it is also possible to apply the approach of the energy balance.

The problem of determining natural evaporation from turbulence theory implies the representation of evaporation as a matter of turbulent transfer, assuming that the coefficient governing the transfer of momentum and temperature is identical with the one for moisture content. From measurements of the wind and temperature profile this coefficient may be calculated and applied to the moisture profile. This basic assumption may be accepted under special circumstances, but it has been shown that it may sometimes lead to considerable errors.

In the energy budget approach, evaporation may be determined if accurate knowledge is available of all other factors contributing to the heat balance at the evaporating surface. These factors are short-wave radiation, long-wave radiation, outgoing radiation from the surface, reflected short-wave radiation and sensible heat transfer upwards and downwards from the surface. As, however, both the determination of the outgoing radiation and of the sensible heat transfer are subject to great difficulties, this approach also does not generally lead to quite satisfactory results.

The most promising method for the determination of evaporation, developed in recent years, is the method of measuring with great precision the moisture content at two levels, and the vertical velocity of the air in between. It is then possible to calculate evaporation directly. It is to be hoped that this method may lead to much more accurate results on evaporation from a particular spot than it has hitherto been possible to achieve.

In arid zone research, however, it is not usually sufficient to know the water need or supply at a certain point. For agricultural and irrigation purposes we need to know the water balance conditions over large areas. Due to high instrumental costs it is usually too expensive to use the methods described above at a large number of stations, and climatological approaches based on empirical formulae for calculating evaporation and water need from simple measurements of common meteorological elements must therefore be adopted.

In arid regions we are mainly concerned, for practical purposes, with evapotranspiration and soil moisture. Actual evapotranspiration (ETA) in dry regions is difficult to determine, as it is considerably influenced by the moisture status and characteristics of the soil. This has led to the introduction of the concept of 'potential evapotranspiration' (ETP) defined as the 'rate of evaporation from an extended surface of short green crop, actively growing, completely shading the ground, of uniform height, and not short of water'. Thornthwaite [2], Blaney-Criddle [3], Penman [4], Haude [5], Turc [6] and others have independently derived methods for calculating the potential evapotranspiration from macrometeorological data. It is not possible here to deal with these methods in detail, so we shall confine ourselves to those most commonly used and discussed.

Thornthwaite's method is based on comparisons made in the US between the measured potential evapotranspiration and data on temperature and day length. An empirical formula was developed by which it is possible to calculate ETP using only the last-mentioned elements, and which is supposed to be valid in all climates. Penman has derived a more complicated formula, for which knowledge is necessary of radiation conditions, moisture content of the air, cloudiness and wind. For arid and semi-arid regions with a summer rainfall regime, Thornthwaite's formula – being derived from data in such climates – may be applicable, but for other parts of the world (for instance in the Mediterranean semi-arid regions) Penman's formula is recommended for more careful analysis.[1]

In Fig III-3 two diagrams are given to show how the calculation of evapotranspiration (ETA or ETP) in connection with measurements of precipitation

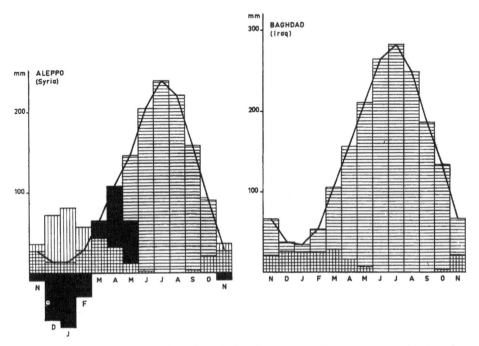

Fig III-3. Diagram to show how the calculated or measured evapotranspiration together with values of precipitation and soil-moisture content may be used for studies of the annual course of the water balance

and soil moisture may be used for studies of the fluctuations of the water balance at a semi-arid and an arid station in the Near East. The horizontally filled columns represent the evapotranspiration by months. The vertically filled columns represent the precipitation. In months of surplus of water a black column below the base line represents the amount of water stored in the soil. In the ideal case we know this amount by measurements, otherwise we have to assume that up to a certain limit the surplus of water is stored in the soil. In such cases consideration should be given to the type of soil and its storage capacity. Whenever the evapotranspiration is greater than the precipitation in a certain month, part or all of the soil moisture is used. This amount is shown by a black part at the top of the precipitation column. Studying water supply

[1] With some approximation Thornthwaite's formula may be accepted for a classification of the arid and semi-arid regions of the world (see below).

conditions during the growing period for a certain plant implies then to follow month by month the development of the balance between the factors – evapotranspiration, soil moisture and precipitation. When precipitation and soil moisture do not balance the evapotranspiration, no water is available for plant growth.

In studies of the conditions required by a certain plant, the actual evapotranspiration is of course the factor that should preferably be used in these balance studies. As said earlier, however, the ETA is difficult to determine accurately and when this is not possible the ETP may be applied. As ETA is never greater than ETP we will obtain, by using ETP, a lowest time limit for the soil running out of water, which may be used for comparative studies between various places. Whenever the available water calculated in this way is sufficient to cover the complete period of growth for a certain crop one may conclude that dry-land farming of that crop is possible. An approximate limit for dry-land farming of the key crop may in this way be established for a certain region – a limit which could be accepted as the borderline between semi-arid and arid regions.

To establish the true conditions at a certain place there is again a need for good measurements of the actual evapotranspiration from the field. In the absence of such measurements certain generally established facts regarding the relation between ETA and ETP may be used. It is quite well known that $ETA < ETP$ during the first part of the growing season, when the plants do not completely cover the fields. When the plants cover the fields and are in active growth, $ETA = ETP$ according to definition and provided water is available. At a later stage ETA may again be $<ETP$ because, particularly in very hot climates, where ETP is very great, plants may, for physiological reasons, not be able to transpire at the very intense rate that the theoretical calculations of ETP may give.

An important practical concern in arid climates is the amount of water needed for irrigation during various parts of the year (see for instance Hallaire [7]). To establish true conditions it is, of course, necessary to know the ETA, but here again the ETP may be used to calculate the maximum need for water. This amount is usually smaller than the amount applied based on experience.

It should finally be emphasised that there is for practical purposes little advantage in calculating water need on an annual basis. In winter rainfall climates ETP is at its minimum in winter when cultivated plants are grown under dry-land farming conditions. The annual value, however, is completely dominated by summer ETP and therefore does not tell anything about the conditions during the growing season for these plants. It may only give some indications of the need for irrigation to allow for cultivation of summer crops.

It should be concluded that improved methods of measuring soil moisture and evapotranspiration are the most important needs for water balance research in semi-arid and arid climates.

CLASSIFICATION OF DRY CLIMATES

The best-known attempt to classify dry climates was made by Köppen [8]. In his last world classification of climate he gave on the basis of studies of natural vegetation the following definition of a semi-arid climate (steppe climate) in regions of winter rainfall:

$$P \leqslant 20\,T$$

where P is annual precipitation in mm and T is the annual mean temperature in °Celsius. The true arid climate (desert climate) was arbitrarily defined as

$$P \leqslant 10\,T$$

i.e. the annual precipitation should be half of the amount at the border of the dry climates. In summer rainfall areas the corresponding formulas are

$$P \leqslant 20\,(T + 14)$$

and
$$P \leqslant 10\,(T + 14)$$

Where there is no clearly defined period of rainfall the Köppen limits are as follows:

For semi-arid regions $P \leqslant 20\,(T + 7)$
For arid regions $\qquad P \leqslant 10\,(T + 7)$

It is not surprising that it should be possible to find a suitable relationship between precipitation and temperature to define aridity in view of the facts that we have discussed in the preceding sections. The water balance is a function of precipitation, soil moisture and evapotranspiration and, to an approximation, temperature may give some indication of evapotranspiration. As the water supply conditions are, however, very complicated, a relationship of the above-mentioned type is not likely to work out in detail, and therefore must not be applied otherwise than in outlining the distribution of dry climates on the world map scale. In small regions it is apt to show large deviations from the true conditions.

Nevertheless several other attempts have been made to find relationships between temperature and precipitation, which would give a better and more general definition of dry climates. De Martonne [9] conceived his so-called 'aridity index' to study the degree of aridity in all parts of the world. It is defined as

$$I = \frac{n \cdot p}{t + 10}$$

where I is the index, n is number of rainy days, p the mean precipitation per day and t the mean temperature in the selected period, which may be a few days, a month or a season. It has the advantage of being applicable to shorter periods than a year and, being in accord with the observed distribution of areas of internal drainage over the earth's surface, gives fairly good indications of aridity. It still is quite empirical and does not tell anything about the physical relationship between precipitation and evapotranspiration. In de Martonne's classification of the world's climates an index of $I < 30$, calculated on an annual basis, means a semi-arid climate and $I < 20$ an arid climate.

Plant ecologists have often applied a similar approach to that of Köppen and de Martonne in their studies of the relation between limits for the adaptation of plants and the climate. Gaussen [10], from studies of the natural vegetation, has developed a system based on the number of arid months in a year. A certain month is classified as arid if

$$p < 2t$$

where p is the mean monthly precipitation in cm and t is the mean monthly temperature. The system appears to work rather satisfactorily in the Mediterranean climates but seems to give considerable errors when applied on a world-wide basis as attempted by Walter and Lieth [11]. Emberger [12], relating extremes of weather to the critical conditions for plant survival, has developed a system similar to the one developed by de Martonne, essentially based on the rhythm of seasonal climatic variations – the existence or non-existence of rainfall and extreme temperature and humidity conditions.

Thornthwaite [2] has tried to develop a system for classifying world climates which considers the factors important in arid regions. By establishing indices based on the relation between precipitation and evapotranspiration, his system has a sound physical basis and is in principle suitable for application in dry climates. In practice it is not applicable to all dry climates – as was mentioned earlier – due to the fact that his simple method of calculating potential evapotranspiration from only temperature and day length is derived in a summer rainfall climate and seems not to work out in other types of dry climates. However, it certainly is possible by his method to obtain a fairly good idea of the relation between precipitation and evapotranspiration on a world scale basis. Meigs [13] therefore adopted the Thornthwaite system refined by certain definitions from Emberger for the very arid regions, in drawing for Unesco a basic map of the distribution of semi-arid and arid zones in the world. This map is given on the endpapers of this book. Everything indicates that this map gives a representative picture of the distribution of dry lands. This has been proved for instance by a comparison with the maps prepared by Köppen and de Martonne. Only small differences are found between the outcome of the various systems on a world-wide scale. This, of course, does not mean that the differences may not be rather great when applied in more detail to smaller regions, where it is much to be preferred to establish aridity on the basis of the water balance.

Effects of human interference

The practical implications of climate for human activity in the dry parts of the world will be thoroughly discussed in other chapters of this book. There is just one aspect of the interrelationship between climate and man which needs some discussion in this chapter, namely the changes of climate and other natural conditions which may occur due to unintentional and voluntary influences by man.

Many of the areas where we are now trying to extend settlement and agriculture were fertile and used for extensive agriculture in ancient times. So far we are not able to say what has been the ultimate cause of this change. Although it is obvious that in certain parts of the world changes of the plant-ecological conditions must have been due to large-scale fluctuations in the general circulation, i.e. to a real change of climate, it is clear that in areas of rather dense population in semi-arid regions one must not neglect the influence of man and his animals on microclimate, soil and vegetation conditions through activities such as deforestation, soil erosion, urbanisation, irrigation and grazing. The more generally accepted idea at present is that the deterioration of the agricultural conditions in arid and semi-arid regions of the world since the time of flourishing culture has been mainly due to the activity of man, but the problem of how this secondary influence on climate has occurred in detail remains unsolved. Many investigators maintain that vast deforestation in semi-humid and semi-arid lands has not only increased soil erosion and thereby extended desert conditions, but should also have increased as air humidity and precipitation may have diminished as a consequence of a decrease of evapotranspiration. Surface run-off certainly also increases by the cutting down of the forests, thus allowing less water to percolate into the soil and to become available for plants. Others argue that trees generally do use a lot of water and that therefore deforestation would increase the amount of soil moisture available for cultivated plants. The replacement of forests by grasslands and agricultural lands has, in fact, been a major effect in the opening up of vast areas in all continents. The argument for cutting forests in developed communities relates particularly to an attempt to increase stream flow by reducing the drain on subsurface water made by plants with deep water systems, which reach shallow groundwater. These are the *phreatophytes*, such as the tamarisk and salt cedar of western USA.

Deforestation has direct effects on microclimates by removing the protective cover of foliage, by reducing fog-drip from the leaves, and by reducing the effective elevation of hills, which in fact reduces precipitation, especially with trees 30–100 feet or more in height. Irrigation, by changing the air humidity conditions, has a clear influence upon the local microclimate, but, if applied on a large scale, may also have some consequences for the macroclimate. Grazing affects the vegetation and therefore has to be taken into account whenever we try to establish climatic conditions by plant-ecological methods. It is most likely that changes in the agricultural conditions in semi-arid regions are due to the combined effects of human activities and fluctuations in climatic elements such as precipitation and temperature. Leopold [14] gives an interesting example for the western United States where the cattle industry reached its apex in the 1880's at the same time as extended dry periods were punctuated with individual years of very heavy precipitation. Overgrazing in combination with extreme intensity of individual rainstorms caused a period of intense erosion and an extension of semi-arid conditions.

c

Rain-making and evaporation reduction

Another possibility of man influencing climate is by his direct intervention in meteorological physics. Attempts have been made for instance to induce precipitation by artificial means and to prevent evaporation likewise.

Artificial rainfall may be induced only from already existing clouds and only if certain conditions are fulfilled (Fig III-4). Methods are based mainly on the knowledge that ice crystals, when available, are rapidly growing at the cost of the water droplets in supercooled water clouds due to a lower saturation pressure over ice crystals than over supercooled droplets. The difference is greatest at a temperature of $-12°$ and leads to a diffusion of water from the droplets to the ice particles. If one ice particle is available for 1,000 droplets, the diffusion causes the ice particle to become ten times as big as each of the droplets.

The basic problem in artificial rain-making from existing clouds is to increase the number of ice crystals available. In clouds which do not by themselves reach a level where their temperature is below $-10°$, the number of ice crystals is not sufficient to cause significant precipitation. It has therefore been attempted to lower the temperature in them to values considerably below $-35°$, when ice crystals start to form spontaneously. This has been done by seeding the clouds by carbon dioxide crystals ('dry ice') by which means temperature may be reduced to $-60°$ or $-80°$. In recent years sublimation nuclei for ice crystals have been created by seeding the cloud, a contribution of sublimation nuclei having a hexagonal crystal form similar to that of ice. Experiments with other chemical compounds having an appropriate crystal structure are also being made.

Theoretical considerations show that there are rather few types of clouds where the chances of obtaining significant results by using the above-mentioned methods are really good. Most convective clouds, which are dominant in semi-arid and arid regions, reach by themselves the ice nuclei level at which no seeding is needed to start precipitation. Stratiform clouds usually are colloidally stable, i.e. contain only water droplets and cannot be made unstable by seeding with 'dry ice'. It has therefore been mainly in mountainous regions where orographic clouds occur that experiments with 'dry ice' or silver iodide have sometimes proved to be successful. A recent development has shown, however, that even in 'warm' clouds (above 0°) a precipitation mechanism occurs by coalescence of droplets, and attempts have been made to stimulate this mechanism by introducing a great amount of large hygroscopic nuclei such as common salt particles into the cloud. In this case also only limited success has been attained.

It may be stated that although it may be possible to obtain successful results in areas favourably located as regards topography and atmospheric circulation, we cannot hope that artificially induced rainfall will serve as a solution to the problem of water shortage in arid lands, where clouds suitable for seeding very rarely appear. Only a major interference with the atmospheric circulation could

Fig III-4. Effects of cloud seeding, shown by successive photographs of the same cloud before and after seeding with silver iodide, near Wagga, NSW, Australia. (*Photo by CSIRO Radiophysics Laboratory*)

alter this basic deficiency of arid areas. This possibility, although remote, cannot be ruled out.

A basic approach to the water supply problem is to make the best possible use of the available water. One way of doing this is to reduce or prevent evaporation by artificial means. Attempts have been made in recent years in Australia, East Africa and India to prevent water molecules from leaving a water surface and thereby to reduce evaporation from reservoirs by means of a film of cetyl alcohol spread over the water. These experiments have generally been successful when applied to small reservoirs, but with larger reservoirs or lakes difficulties have been encountered due to the breaking-up of the film by wind-induced waves. New methods of spreading the cetyl alcohol may overcome these difficulties sufficiently to make the cost of application so low that the economics of water storage are greatly improved, as against the construction of larger dams.

C. C. WALLÉN

References

[1] PERRIN DE BRICHAMBAUT, G. and C. C. WALLÉN, 'A study of agroclimatology in the Near East', *WMO Technical Note* no. 56, Geneva 1963.

[2] BLACK, N. J., 'The distribution of solar radiation over the earth's surface', *Arch. f. Met. Geoph. und Bioklim.*, Ser. B. Bd 7, 1956, p. 165.

[3] THORNTHWAITE, W., 'An approach towards a rational classification of climate', *Geogr. Rev.*, Vol. 38, 1948, pp. 55–94.

[4] BLANEY, H. C. and W. D. CRIDDLE, 'Determining water requirements from climatological and irrigation data', *U.S. Dept. Agric. S.C.*, S-T-D No. 96, 1950.

[5] PENMAN, H. L., 'Natural evaporation from open water, bare soil and grass', *Proc. Roy. Soc.* (A), Vol. 193, 1948, pp. 120–45.

[6] HAUDE, W., 'Über die Verwendung verschiedene Klimafaktoren zur Berechnung potentiellen Evaporation und Evapotranspiration', *Met. Rundschau*, Vol. 11, 1958, pp. 96–9.

[7] TURC, L., 'Evaluation des besoins en eau d'irrigation, evapotranspiration potentielle', *Ann. Agronomiques*, INRA, Vol. 12, 1961, pp. 13–49.

[8] HALLAIRE, M., 'Irrigation et utilisation des réserves naturelles', *Ann. Agronomiques*, INRA, Vol. 12, 1961, pp. 87–97.

[9] KÖPPEN, W., *Die Klimate der Erde*, Berlin, 1931.

[10] DE MARTONNE, E., 'Une nouvelle fonction climatologique: l'indice d'aridité', *La Météorologie*, 1926, p. 449.

[11] GAUSSEN, H., 'Théorie et classification des climats et des microclimats', *8th Congr. Internat. Bot.*, Paris, 1954, Section 748 (Rapports et communications parvenues avant le congrès).

[12] WALTER, H. and H. LIETH, *Klimadiagramm-Welt Atlas*, Jena, 1961.

[13] EMBERGER, L., 'Une classification biogéographique des climats', *Rec. des traveaux. Fac. Sc. de l'Univ. de Montpellier*, 1955, pp. 3–43.

[14] MEIGS, P., 'World distribution of arid and semi-arid homoclimates', *Reviews of Research on Arid Zone Hydrology*, Unesco, Paris, 1953, pp. 203–9.

[15] LEOPOLD, L., 'Water and arid zones of the United States', *Proc. Paris Symp. on Problems of the Arid Zone*, Unesco, 1962.

ADDITIONAL REFERENCE

UNESCO/FAO, *Bioclimatic Map of the Mediterranean Zone*, and *Explanatory Note*, Unesco, Paris, 1963 (Arid Zone Research XXI).

CHAPTER IV

Geomorphology

Landscapes in all parts of the world, including deserts, are not permanent but ever changing. Climatically induced processes of weathering, erosion and deposition are constantly at work on the basic geological structures, destroying old and creating new landforms, so that at any one time the landscape includes relics of former land surfaces, together with the newly created landforms.

Arid lands are hot regions, characterised not only by a low average annual rainfall but also by the incidence of heavy, if rare and spasmodic, downpours. Temperatures are high during the day but may be low at night, so that there is a large diurnal temperature range. Vegetation is either sparse or absent, and the conditions of climate and vegetation allow wind to assume a significance in moulding the land surface that is unknown elsewhere except for the snow-covered polar regions.

It is the physiographic processes induced by these climatic conditions that give the deserts their geomorphic unity, for the processes act upon rocks and structures that are no different from those found beyond the confines of the arid lands. Deserts do not correspond with any special geological environment, and as with all other land areas the gross relief of their major landforms is due to large-scale crustal movements of up- or down-warping, folding or faulting, with the additive effects of vulcanicity in certain areas. Local relief is determined by the disposition of rock masses, their lithology or rock-types, and the local structures such as bedding or jointing within them, as they are dissected by the agents of weathering and erosion. Deposition of the detrital products of erosion also contributes to the landscape features such as alluvial fans, salt-lake beds, and, in the true deserts, sand dunes of various kinds. Due to the lack of vegetation and soil on steep slopes, geological structures are particularly well exposed in arid regions (see Figs IV-2, 4).

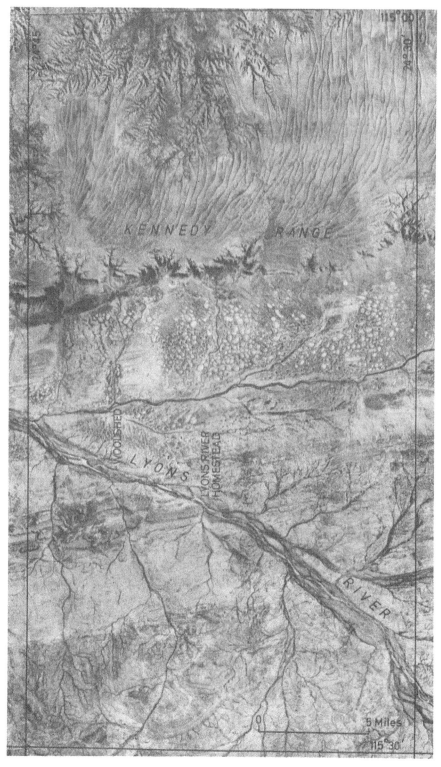

Fig IV-1. An aerial view of an arid landscape at Lyons River, Western Australia. See page 76 for explanation. (*Photo mosaic by courtesy of the Division of National Mapping*)

Structure of arid lands

BLOCK TECTONICS

Large areas of south-western USA including the arid regions exhibit block tectonics. The crust is divided into blocks that are separated by faults or by sharp flexures. The lines of faulting and flexuring are of some antiquity, and there have been recurrent movements along them, both lateral and vertical, up to the Quaternary, and some are still active. As a result of the differential block

Fig IV-2. Dissected anticline, Zagros Mountains, west Iran. (*Photo by Hunting Surveys Ltd associated with Aerofilms Ltd – copyright reserved*)

movements, there is a basin and range topography with abrupt escarpments between uplands and lowlands [1]. Fault scarps range in height from a few feet to several thousands of feet: the San Gabriel scarp rises abruptly 6,500 feet above the alluvial fans at its base, and the great scarp fronting the Sierra Nevada on the east towers 10,500 feet above the Owens Valley. Scarplets are common even in the unconsolidated alluvial fans that front the ranges, and they are found also in valley floors, as in the Mojave Desert. Fault-line scarps, which are due to the differential erosion of rocks having contrasting resistance, and which have been brought into juxtaposition by faulting, also are common in the deserts of the American south-west. So also are landforms due to transverse faulting, in which the relative movement of the blocks takes place principally in a horizontal direction. Among the minor effects of such faulting are offset stream courses, one example from southern California having a lateral displacement of $2\frac{1}{2}$ km.

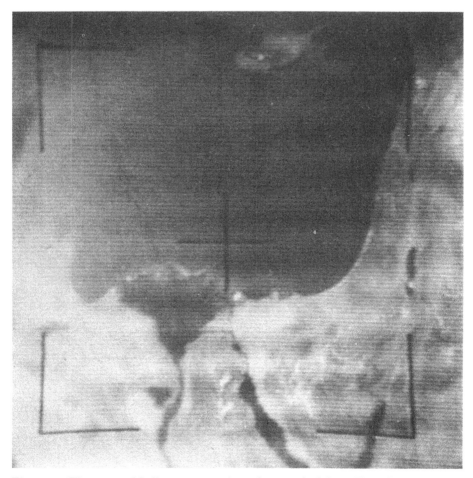

Fig IV-3. The eastern Mediterranean region, photographed from Tiros V, 1962. Note the Nile Valley and Delta, the Faiyum, the Gulf of Suez and the Gulf of Aqaba, and the Dead Sea and Lake Tiberias in the Palestinian rift valley. (*Orbit 30R/O29, 21 June, height 659 km. Courtesy of the United States of America*)

The raised fault-blocks display upland surfaces which cut across the structures within the blocks, indicating past periods of planation in the region before the faulting. The basins of the down-faulted areas are commonly regions of interior drainage. They are gradually filled in by detritus brought from the adjacent highlands by streams, past and present, giving rise to many depositional features.

The American south-west provides the best-known example of 'basin-and-range' topography but similar structural landforms are found elsewhere, as for instance in some of the ranges of the Gobi Desert, and the Little Kharas mountains of South-West Africa.

Rift valleys are major elongate fault troughs which transect plateau regions. They dominate the landscape where they are present, as with the Dead Sea Rift (see Fig IV-3), a branch of the vast Red Sea and east African rift system.

SHIELDS AND PLATFORMS

Not all deserts are underlain by diastrophically disturbed bedrock. The Sahara, for instance, is a typical continental platform, a region in which ancient crystalline or strongly contorted basement rocks, mostly Pre-Cambrian, are exposed over considerable tracts, but covered elsewhere by little-disturbed sedimentary rocks of Palaeozoic, Mesozoic or Tertiary age.

Arabia is in many respects similar to the Sahara. There are rugged mountains developed on the exposed Pre-Cambrian crystalline basement in the west and north-west, similar to the mountainous eastern desert of Egypt, with younger sediments dipping off this domed basement to the east and south to produce a magnificent series of cuestas sweeping around in approximately concentric arcs [2]. The Thar Desert of north-western India consists largely of a sand plain overlying a peneplain eroded across the basement of the Deccan Shield. Its surface is interrupted only by occasional low outcrops and ridges, outliers of the Arvalli Hills that border the desert of the east.

The Gobi, although it has its high mountains, as mentioned above, also possesses platform characteristics, for there are large areas of undeformed Mesozoic sediments in the Gobi basin.

The Australian Shield consists of Pre-Cambrian basement and is bordered by basins of younger rocks such as the Barkly (Cambrian), Canning (Palaeozoic and Mesozoic) and Eucla (Cretaceous and Tertiary) basins. The deserts of central and western Australia extend across sedimentary basins and Pre-Cambrian basement alike, and the Simpson Desert extends beyond the confines of the Shield on to the great Lake Eyre basin, one of the massive depressions bounding the Shield to the east.

The Australian deserts are mainly regions of low relief, although in altitude they extend from 500 feet to 2,000 feet on the plains, with mountains up to 5,000 feet. There are many east–west trending ranges such as the Macdonnell Ranges near Alice Springs, and the series that includes the Musgrave, Everard, Mann and Birksgate ranges and which extends for over 500 miles from the Lake Eyre depression westward to the Western Australian border and beyond.

The Macdonnells and some other Australian desert mountain ranges such as the Willourans in the north of the Flinders Ranges (South Australia) are examples of landforms which follow rock structures picked out by differential erosion of stratified rocks. The ridges form long lines of hills, cuestas or hogbacks, and such landforms are common in arid regions underlain by folded sedimentary rocks (Fig IV-4).

Fig IV-4. Vertical aerial photograph of part of the North Flinders Ranges, South Australia, showing the admirable way in which geological structure, with folded and faulted sedimentary formations, is revealed in an arid region. (*RAAF photograph, reproduced with permission*)

VULCANICITY

At the present time active volcanoes are rare in arid regions, but extinct volcanoes are important in the landscapes of several deserts. In northern Kenya and Abyssinia for instance there are large areas covered by volcanic rock, which in the absence of much weathering or erosion seems to retain its original structures almost indefinitely. Even tiny wrinkles in the lava skin are still preserved.

There are volcanic mountains and lava flows of Recent age in the Sahara, for instance in the Hoggar and Tibesti massifs. Volcanic landforms are widespread in Anatolia, Transjordan and Aden, and considerable vulcanicity is associated with the African rift valleys, giving rise to prominent volcanic mountains as in Abyssinia and Kenya. In the American deserts vulcanicity has occurred throughout the Tertiary and Quaternary up to the present. In the neighbourhood of Death Valley, for instance, there are numerous small lava flows and cinder cones.

WEATHERING IN THE DESERTS

In the deserts, as elsewhere, rocks at the earth's surface are changed by weathering, which may be defined as the disintegration of rocks *in situ*. Weathering processes are either chemical, when alteration of some of the constituent particles is involved, or mechanical, when there is merely the physical breaking apart and fragmentation of rocks. Which process will dominate depends primarily on the mineralogy and texture of the rock and the local climate, but several individual processes usually work together to the common end of rock disintegration.

The great diurnal changes in temperature of deserts have long been supposed to be responsible for the disintegration of rocks, either by the differential heating of the various rock-forming minerals or by differential heating between the outer and inner parts of rock masses. However, both field observations and laboratory experiments have led to a reassessment of the importance of insolation in desert weathering. Almost half a century ago Barton remarked that the buried parts of some of the ancient monuments in Egypt were more weathered than were those parts fully exposed to the sun's rays, and attributed this to the effects of hydration below the ground surface [3]. Laboratory experiments have shown that rocks subjected to many cycles of large temperature oscillations (larger than those experienced in nature) display no evidence of fissuring or fragmentation as a result; but when marked fluctuations of temperature occur in moist conditions small rock fragments quickly form [4]. Blackwelder [5] cites the case of a granite statue in Egypt which fell on its side and which, even after 3,000 years, exhibits no sign of weathering on the surface exposed to the hot sun, yet has disintegrated and shed skins on the underside where the granite was in contact with the ground, where there is some moisture. On the other hand some shattered pebbles and boulders attributed to thermal changes consist of flint and quartzite, both homogeneous and almost inert chemically. Flints split by straight and parallel fractures are common in Egypt, quartzite pebbles with 'onion-skin' weathering have been reported from Peru, and blocky disintegration of pure quartzite has been noted in central Australia. Insolation may not be as important as was formerly supposed but it contributes materially to rock weathering in the arid lands.

The expansive action of crystallising salts is often alleged to exert sufficient force to disintegrate rocks. Few would dispute that this mechanism is capable of disrupting fissile or well-cleaved rocks or rocks already weakened by other weathering agencies; wood is splintered, terracotta tiles disintegrated and clays disturbed by the mechanism, but its importance when acting upon fresh and cohesive crystalline rocks remains uncertain.

Weathering achieves more than the disintegration of rocks, though this is its most important geomorphic effect. It causes specific landforms to develop. Many boulders possess a superficial induration of iron oxide and/or silica, which substances have migrated in solution from the inside of the block towards the surface. Not only is the exterior thus case-hardened but the depleted interior is left friable. When weathering penetrates the shell the inside is rapidly attacked and only the indurated zone remains to give hollowed or 'tortoiseshell' rocks.

Fig IV-5. Duricrust of ancient laterite, showing the indurated capping and underlying pallid zone, central Australia. (*Photo by courtesy CSIRO Land Research and Regional Survey Section*)

Another superficial layer, the precise nature of which is little understood, is the well-known desert varnish or patina, a shiny coat on the surface of rocks and pebbles and characteristic of arid environments. Some varnishes are colourless, others light brown, yet others so dark a brown as to be virtually black. Its origin is unknown but is significant, for it has been suggested that the varnish grows darker with the passage of time; obviously before such a criterion could be used with confidence as a chronological tool its origin must be known with precision. Its formation is so slow that in Egypt for example it has been estimated that a light brown coating requires between 2,000 and 5,000 years to develop, a fully formed blackish veneer between 20,000 and 50,000.

The development of relatively impermeable soil horizons that are sub-sequently exposed at the surface because of erosion of once overlying friable materials, and which thus become surface crusts, is widespread in arid regions, although it is also known outside the deserts, and indeed many of the examples in arid lands probably originated in former periods of humid climate. The crusts prevent the waters of occasional torrential downpours from penetrating deeply into the soil, and thus they contribute to the rapid run-off associated with desert storms. Also, after erosion has cut through the crust and exposed underlying soil layers, the hard layer forms a resistant capping (duricrust) on plateaux and mesas, such as are common in many parts of arid and semi-arid Australia (Fig IV-5). This happens mostly with siliceous (silcrete) and ferruginous (lateri-tic) cappings. Calcareous crusts (calcrete; *croûte calcaire*) are generally thinner and less resistant, and they mantle the countryside following the undulations of the present topography.

Some duricrust layers have been used as time markers in geomorphic and stratigraphic histories. The necessary conditions for this are that the crust form fairly rapidly, and that it be sufficiently distinct in appearance to preclude the possibility of confusion with other crusts formed at other times. The Barrilaco calcrete of Mexico for instance is believed to date from about 7000 B.C. [6]. The main silcrete of the northern districts of South Australia is believed to date from the Lower Miocene, the laterite of northern Australia to be of the Lower or Middle Miocene age.

SUB-SURFACE WEATHERING

In the last few years it has been found that in deserts, as in non-desert regions, the depth of weathered material may go down hundreds of feet. This is parti-cularly commonly observed on granite. Erosion at the present time strips away the weathered rock easily, but unweathered rock is naturally more resistant. It is found that with granite (and perhaps with other rocks too) there is usually, though not always, an abrupt knife-edge contact between hard rock and com-pletely weathered rock, which is called the 'weathering front'. This is an irregular surface with many hollows and towers, and gives rise to an equally irregular topography when exposed. Many exposed weathering fronts have now been discovered, often littered with corestones of fresh rock. These features were first described from deserts, before it was realised that deep weathering had a very wide distribution. Many landforms due to exhumation from a deep regolith, although often attributed to desert weathering and erosion, occur as well in areas of fluvial erosion.

Landforms due to weathering are significant in the morphology of the desert lands, but unquestionably the most important contribution of the various weathering agencies is in the preparation of materials for transportation and erosion.

Rivers and streams in arid regions

RAINFALL

No desert is completely waterless, although parts of the Atacama and the Sahara bid fair to warrant this description. There are extended periods of many years without rain, but at intervals of the order of a decade or even a century there are heavy downpours. These achieve more in the way of erosion and deposition than would a long period of more evenly distributed but less intense rainfall, and it is

Fig IV-6. Dry river bed near Jebel Ram, Jordan, with sandy alluvium in valleys and bare rock exposures of jointed sandstones resting on granite. (*Photo by Hunting Surveys Ltd associated with Aerofilms Ltd – copyright reserved*)

now generally accepted that most major erosional forms found in desert landscapes result from fluvial action. Even in temperate regions a large part of the work of rivers is achieved by periodic floods, but in deserts the effectiveness of occasional storms is enhanced by the sparseness of vegetation, the effectiveness of raindrop impact on bare surfaces and by the presence of pans or crusts that cause heavy and rapid run-off. No doubt many erosional and depositional forms now present in the deserts and of fluviatile origin are, at least in part, relict features deriving from past periods of higher rainfall; but many are attributable to the occasional heavy downpours of the present climatic regime.

There are good accounts in the literature of such torrential downpours and resulting floods. Hume [7] records them for Upper Egypt, and for the deserts of north-west Peru, which are said to be swept by violent floods only two or three times a century, Bosworth [8] reports that:

The flood of 1891 was brought about by daily torrential rains in February and March. The water poured down the deep mountain valleys and immediately overflowed the quebradas [deep, trench-like canyons cut across a plateau below the mountains] uniting as a great sheet, which rolled on to the sea. Many drainage channels were changed, and an immense load of stones was swept from the mountains to the plain below. For two years afterwards the country was moist and green, so that herds of cattle and goats were introduced, and also cotton planted. In the next two years the country dried up again completely, the vegetation died, and the cotton gin became a curio.

Clearly the possibility and potential of such floods must be borne in mind when landforms are interpreted even in the true deserts, and particularly when landforms are used as evidence of climatic change. In arid regions where rain, though not plentiful, is nevertheless normal, it is of course to be expected that landforms of fluvial origin will also be prevalent, with some modifications due to aeolian action.

WATERCOURSES

A few great through-flowing or exogenous rivers derive their waters from well-watered country beyond the deserts and have sufficient volume to survive the journey across them. Most rivers that enter deserts, however, gradually shrink and are lost by evaporation and downward percolation of the water. Deserts are characteristically areas of interior drainage, as the streams that discharge into them or which rise within them following falls of rain lose their water and deposit their sediment within the confines of the desert. Desert basins contain shallow lakes which are dry for most of the time. River valleys in mountainous arid regions are steep-sided with V-shaped cross-sections and sharp interfluves. In plains and plateaux the streams carve wadis which are steep-sided and flat-floored. Their beds are covered with alluvial debris. The wadis are usually widely spaced with considerable tracts of unbroken plateau between. Where soft rock occurs numerous steep-sided valleys consume the hills or plateaux to give badland topography – a maze of ravines separated by sharp crests and pinnacles. On the plains where the rivers are depositing their debris they flow in wide shallow courses that are very changeable, branching and rejoining repeatedly in braided or anastomosing patterns (Fig IV-7).

ALLUVIAL FANS, PLAYAS AND SALINAS

The streams of desert uplands collect immense amounts of debris from the bare unvegetated hillslopes, and most of the stream load is deposited close to the upland border or mountain front, where the streams lose gradient and also lose water through seepage into the unconsolidated sediments of the basin lowlands, and by evaporation. In exceptional cases detritus may be spread considerable distances from the uplands, as in the Atacama Desert, where alluvial fans extend

Fig IV-7. Neale River in flood, central Australia. A braided stream in an alluvium-filled valley. Note rock outcrops top right. (*Photo by courtesy of the Division of National Mapping*)

as much as 60 km from the ranges, and are actively infilling the great bolson or basin. Cones or fans of alluvial material that coalesce with one another to form continuous aprons fronting the uplands are characteristic of arid and semi-arid areas of bold relief. Alluvial fans are built of gravels, sands and muds, and there is a rapid decrease in size of average particle from apex to base of a fan.

Further from the mountains fine detritus forms the alluvial plains of the basins. In the lower parts of the depressions there are lakes which are usually dry, and whose beds form clay plains called *playas*. Areas of internal drainage accumulate salts, and the playa sites may be occupied by salt flats or salinas. Sodium chloride is most common in the evaporite deposits, but there are lakes with gypsum, sodium carbonate and other salts.

The mountains of the Australian arid regions are neither extensive nor high and the alluvial fans are not as well developed as in other deserts. There are, however, vast salinas such as Lakes Eyre, Amadeus, Frome and Gairdner. Such salinas and playas are by no means confined to intermontane valleys; indeed so common are they in parts of the West Australian plateau that they give their name to a physiographic region – Salinaland. As in other deserts there is evidence that the playas and salinas were once more extensive. Lake Woods in the Northern Territory is now a mere 70 sq. miles in area but old shore lines show that it once extended over some 1,100 sq. miles and that, moreover, it attained this size after the sand ridges of the surrounding desert had formed. In Salinaland of Western Australia the lakes are relics of a once extensive river system.

In north-west Peru the principal accumulation formed in the desert is the Amotape Breccia Fan, composed of angular quartzite and slate, and produced by spasmodic floods. The oldest cone, of early Quaternary age, forms the bulk of the deposit, but this has been dissected and later breccias have been laid down at lower levels.

The basins or bolsons of the western USA are regions of interior drainage, with numerous playas and salinas which display clear evidence of former greater extent. Large Pleistocene lakes such as Lakes Bonneville and Lahontan occupied some of these basin depressions. The sedimentary sequences of the lakes offer possibilities for the elucidation of the recent climatic history of the region. The basins are gradually being filled by detritus, and alluvial fans cover large areas of the deserts of the western USA, fully 60 per cent of the eastern Mojave Desert for instance being occupied by fans and playa basins.

Similarly in the Gobi Desert the lower parts of the basins are occupied by playas and swamps, and, as elsewhere, there are incontrovertible signs of once more extensive lakes. Berkey and Morris [9] noted at least seven old beach lines associated with a former lake, one strand line being 29 feet higher than the 1922 level of the modern shrunken lake. In Arabia also there are extensive alluvial fans associated with the mountain ranges. It is believed that the bulk of the deposits date from a pluvial period which terminated some 5,000 years ago.

DESERT PLAINS

Plains feature prominently in desert landscapes in some areas, but not all desert plains are of the same nature. Some are constructional, being built up through the gradual accumulation of detrital materials; others are relics of former landscapes now undergoing erosion and destruction. All are affected by wind action, and deflation leaves an accumulation.of stones (gibbers). But many desert plains are cut in solid bedrock, and these are now generally considered to be the work of water, although the precise mechanism remains obscure. In arid regions the junction between hill and rock-plain is in many places extraordinarily abrupt, so much so that it is quite possible to stand astride the upland–lowland boundary.

Such a planate bedrock surface with a sharp break of slope (called a piedmont angle) between it and the backing scarp is called a pediment, and on this there

Fig IV-8. Ayers Rock and The Olgas, inselbergs in central Australia. Ayers Rock, in the foreground, reaches an elevation of 2,820 ft and stands about 1,100 ft above the surrounding plain. Twenty miles away are 'The Olgas', 3,419 ft high and 1,700 ft above the plain. Note vertical bedding in the indurated greywackes of Ayers Rock, and the line of caverns above the present plain. (*Photo by courtesy Australian National Travel Association*)

may or may not be a veneer of alluvial-colluvial debris. Coalescence of pediments gives a pediplain.

The origin of the pediment, and as a corollary the piedmont angle, has given rise to considerable discussion from which it is fair to say no agreement has emerged. A number of factors and mechanisms have been invoked including lateral corrasion by intermittent streams, the action of sheet-wash, and parallel scarp retreat.

1. Wind erosion, being most effective close to the ground, would tend to erode the base of slopes, which could be steepened, and the junction of scarp and plain would be made angular. In some cases, however, detailed study of small-

scale weathering phenomena has indicated that wind erosion at piedmont angles is of little or no significance.

2. Lateral erosion by intermittent rivers has been suggested as a possible means whereby the piedmont slope can develop, but slopes remote from such stream courses still display the abrupt junction.

3. Different types of water flow have been invoked. Sheet-wash might pare off thin laminae of weathered bedrock from the pediment, while gullying occurs on the steeper back slope. The different types of water flow, however, only come into existence if the break of slope is already present.

4. It is claimed that in any desert area of diverse relief the hillslopes or scarps tend to be at or about the same angle of inclination, though there are variations attributable to lithological differences from place to place. This suggests that at any stage of landscape development the slopes will all be at the same angle, and the idea is now prevalent that steep slope elements retreat parallel to themselves. A retreating slope such as an escarpment leaves behind a planate slope as it retreats; this newly generated slope which is of low inclination and cuts across bedrock is the pediment.

RESIDUAL REMNANTS

Steep-sided residuals (inselbergs) that stand abruptly from the plains in arid regions are often spectacular in appearance (Fig IV-8). They are in many cases monolithic remnants of circumdenudation and are regarded by some as merely a late stage in the cycle of pedimentation and parallel scarp retreat. Many of them, however, are the result of deep sub-surface weathering (see p. 61) followed by erosion of the soft weathered rock.

The work of wind

Where the surface soils or weathered rocks are desiccated, and where the cover of protecting vegetation is sparse or non-existent, wind becomes important as an agent of transportation, erosion and deposition. The winds of the desert are no stronger than elsewhere, but their effect is facilitated by their uninterrupted passage across the sparsely or non-vegetated plains. Wind sorts the debris on the desert floor, blowing away the fine material and concentrating the coarse. Sand is built by the wind into landforms, and wind armed with sand is an agent of erosion. A wind-modified landscape can therefore have many forms.

DEFLATION AND WIND EROSION

Dust and fine sand picked up and transported by the wind are commonly transported from the desert regions to be deposited far beyond their borders. Removal of these fine materials as 'dust' results in a general loss from the desert known as 'deflation'. It is estimated that some 850 million tons of dust are carried 1,440 miles each year by wind in the western USA, and that 8 feet of material was eroded by the wind from the surface of the Nile delta in 2,600 years. In summer, storms of red dust are common in Western Australia, South Australia and Victoria, as well as the western parts of New South Wales, and considerable quantities of bright red dust are at times brought down by rainstorms ('red rain')

Fig IV-9. Desert pavement with gibbers, Coober Pedy, South Australia.
(*Photo C. D. Ollier*)

in these areas. The best-known deposit of dust is loess, which attains great thicknesses in many parts of China, Europe and North America. Much of this was derived, however, from regions of ice and frost, and in part from periglacial regions of the Pleistocene ice ages, and not from the hot deserts. The loess of northern China is derived from the Gobi Desert of central Asia, where winters are severe and where there is intense frost action.

Where wind acts on loose soil containing grains of different sizes, the finer material is blown away, leaving gravel and stones behind. These accumulate until they form a continuous carpet of stones on the desert floor, after which deflation is much reduced, for the stone layer protects the underlying material from further erosion. This layer is aptly called 'desert armour' in North America, but the same feature occurs in many deserts, and is called 'reg' in the Sahara, 'serir' in Libya and 'gibber plains' in Australia (Fig IV-9). The desert pavement itself is prone to wind abrasion. The upper surfaces of the stones are sand-blasted off

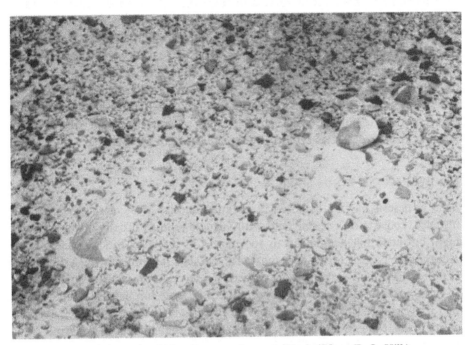

Fig IV-10. Wind-faceted pebbles (limestone), north Sinai. (*Photo E. S. Hills*)

until the fragments come to have a flat upper surface, when the desert pavement looks like a mosaic. Wind-faceted stones are themselves frequently diagnostic of a desert environment (Fig IV-10).

In the past it was commonly assumed that wind erosion was responsible for most erosional forms in the hot deserts. Even planation of continental proportions was attributed to this agency. Wind armed with sand is an effective eroding agent, and sand-blasting is particularly strong close to the ground, but most landforms due to such erosion are in soft sediments. Peculiarly shaped boulders, mushroom rocks and similar strange forms usually owe their formation to the sand-blasting of soft strata interbedded with hard. Wind erosion on soft rocks can give rise to sharp ridges called yardangs, separated by round-bottomed chutes or troughs aligned to the wind direction. Yardangs normally range in size from about a metre to 10 metres in height, though some reported from Iran stand 200 metres above the adjacent depressions. Similar forms eroded in limestone have been reported from Dahran in Arabia.

Characteristic of many desert areas underlain by soft rocks are depressions up to 100 km or even more across. Tectonic effects are important in some of these hollows as may be solution where the bedrock is calcareous, but deflation is either partly or wholly responsible. The great depressions in which lie the Kharga oasis of Libya and P'an Kiang and Tsagan Nor of the Gobi appear wholly attributable to wind action. Excavation is limited eventually by the presence of groundwater, for the wind cannot easily pick up moist debris. Also, a depression will tend to be filled with debris carried in from the sides during occasional rainstorms. In many depressions the water table is close to the surface and these become oases. All the Egyptian oasis depressions are bounded by scarps on the northern side and some such as the Baharia oasis are completely surrounded by hard rock. As the floor consists of the same soft sediments found at the base of the escarpments, the depression cannot have been formed tectonically.

Where wind-swept desert is underlain by hard rock, depression forms are not easily produced, but the whole area is wind abraded and kept free of sand, forming a rocky desert as typified by the *hammada* of the Sahara.

LANDFORMS OF SAND

Sand in deserts may be disposed in irregular sheets. Most of the Kalahari, for instance, consists of a sea of sand with vague swellings usually about 5 metres high, but attaining heights up to 100 metres in a few localities. There are no recognisable regular forms over most of the Kalahari. Similarly the Selima sand sheet on the Sudan–Egypt border extends for several thousands of square miles. It has a gradient of only 2–3 feet per mile, is of table-top flatness, and although uniformly covered by sand, this appears to be only about a foot thick over the entire area. It is extraordinarily smooth and has somehow been veneered by an extremely even thickness of sand.

There are sand sheets also in other desert regions, but the most typical form resulting from wind deposition and moulding is the sand dune. This is by no means the only element of desert landform assemblage but it is a characteristic one. Sand dunes are known outside the deserts, in littoral areas and in association with dry stream beds in such semi-arid areas as the Riverina of Australia, but only in the true deserts do dunes have an extensive development. In the Sahara, for instance, there is an unbroken field of dunes extending over an area larger than France.

Although perhaps appearing on the ground chaotic and disorderly, from the air dunes and dune-fields present a picture of repetition and orderliness: 'instead of finding chaos and disorder, the observer never fails to be amazed at the simplicity of form, an exactitude of repetition and a geometric order unknown in nature on a scale larger than that of crystalline structure' [10].

Several types of dune are known, occurring either in isolation or in fields in which there is contact between adjacent dunes. Some of the dunes are fixed, others mobile. Several distinctive types have been given technical names distinguishing them by shape, which is related to the mode of formation although this is not always known. Since the technical terms, which come from a variety of languages, had non-technical meanings originally, there is danger of some confusion, and furthermore where the same landform is found in different regions, several terms may be current for one type of dune.

Perhaps the simplest type of fixed dune is a drift of sand piled up against an obstacle, often a tree. These consist of a variety of wind-blown material and obviously are stationary. Nebka, rebdou and tamarisk mounds are some of the

Fig IV-11. Small barchan on the playa (takyr) of the ancient delta of the Amu Darya River, USSR. (*Photo by courtesy Professor M. P. Petrov*)

names applied to such forms. A rather peculiar form, apparently trapped by vegetation, is the conical *sable mamelonne* of some of the central Asian deserts. They occur only in valleys and on alluvial fans where the water table is close to the surface. Another type of fixed drift occurs downwind of gaps between hills, where the wind drops its sand in an irregular elongated dune. Sand shadows are dunes formed in the lee of obstacles; they are of varied shape and size, depending on the nature of the obstacle, and are, of course, stationary.

The best-known mobile dune is the barchan. This is crescentic in plan, with tapering horns pointing downwind. The windward slope is gentle, the crest abrupt, and the lee slope is steep, at the angle of rest of the sand which is blown up the gentler slope and over the crest, whence it slips down and comes to rest (Fig IV-11). At the same time sand is swept around the main body of the dune and into the horns. These mechanisms cause the dune to move downwind in the course of time, although it retains the same form. The sand is usually very well sorted in barchans, for it is constantly re-worked as the dune 'marches'. The marching also causes cross-bedding inside the barchan, with a dip parallel to the sand-fall face. Barchans are found in many parts of the Sahara, Arabia, the American deserts, the Gobi, and Kazahkstan; they are not known in either the Kalahari or Australia. There are all sizes up to about 100 metres high, and giant barchans with other minor barchans on their windward slopes have been reported from the Takla Makan Desert of the USSR. Small barchans tend to move faster than large ones, and while moving the small ones grow bigger. When they attain the maximum size they become virtually immobile and further evolution is into other dune forms.

A fairly constant wind direction is necessary for the formation of the most perfect barchans – side winds tend to make sickle-shaped or irregular dunes. In the Gobi, for instance, Berkey and Morris report barchans in the area of strong westerlies related to the monsoonal circulation; in areas that are subject to violent easterlies the barchans are sinuous and irregular. In Arabia barchans occur only in the north-east of the peninsula where the wind is essentially unidirectional; elsewhere seif dunes are present, except that dome dunes occur in some places where winds are both variable and strong.

Barchans are best developed where the sand supply is limited, so that they march over a bedrock floor. In Mauretania there are barchans where there is only a thin veneer of sand overlying a gneissic platform.

Where sand is too plentiful there is mutual interference between adjacent dunes and with complete sand cover an irregular sand sea is liable to develop, or possibly another regular form such as the transverse dune. Occasionally in an area of plentiful sand great mounds will be formed with windows of bedrock exposed between; barchans are found in these areas although they could not form in the sand sea. This situation is found in the dune area east of the Salton Basin, in southern California.

Another type of crescentic dune which in some respects is similar to the barchan occurs in semi-arid and coastal areas. They are, however, immobile and are oriented with the horns of the crescent pointing upwind, contrary to the situation in barchans. Like barchans they possess a long gentle windward slope and a short sand-fall face leeward, but they are generally less clearly defined morphologically than barchans, and once formed are immobile. These are parabolic or U-dunes and they result from the blow-out of a more or less transverse sand ridge, the ends of which are fixed.

Lunettes are yet another type of crescentic landform. Lunettes are widely distributed across southern Australia and are also known from North Africa. They occur on one side of lakes or swamps and determine the shore line, which follows the foot of the steep windward face of the lunette. The crest is broadly rounded or humped, and there is a long and gentle lee slope downwind. Lunettes are most commonly built of silt which appears to be derived from the neighbouring lake floor, and was probably trapped amongst lakeside grasses.

Transverse dunes occupy large areas of the North African, Persian and Soviet deserts. Their crests are normal to the prevalent wind direction, they have a gentle windward slope, and a steep sand-fall slope on the lee, and move slowly downwind. If for some reason the sand supply diminishes, transverse dunes break up into barchans, but with abundant sand they aggregate into dune complexes.

Dome dunes have been reported only from Arabia, though morphologically similar (but fixed) forms occur in the western USA. They are roughly symmetrical and of course dome-shaped. They are believed to develop under the influence of strong winds blowing from various quarters; however, in Arabia they have a tendency to move generally from west to east.

Fig IV-12. Part of a field of longitudinal dunes, Simpson Desert, South Australia. Scale 1 : 100,000. (*Photo by courtesy Division of National Mapping*)

Longitudinal dunes are elongated sand ridges, most commonly forming a series of parallel or gently curving and branching dunes. Rarely, an isolated dune-ridge is formed, as for example near Cairo, where the Gebel Asfar is a high sand peak on such a single long ridge. Price [11] recognises three sub-types as follows: (i) Seif ridges, originally identified in Egypt and Libya, but also known in the Kalahari. These are almost bare of vegetation and have local peaks and crests which in places are due to barchan-like forms crossing the main ridge transversely. Many seifs are winding in plan, and merge with or diverge from each other. (ii) The draa of the western Sahara are bare ridges attaining considerable heights, with two or more lines of sand-fall faces, mostly aligned longitudinally. Many minor transverse dunes may be located on the broad ridge, giving the draa a rough surface of ridge and trough. (iii) The sand ridges of the Australian and Thar deserts, which where well developed, are remarkably straight, long and regularly spaced, although branching and linking is also common (Fig IV-12).

The Australian longitudinal dunes have sharp asymmetrical ridges on their crests, but these are formed of a small amount of mobile sand on the top of a fixed and vegetated plinth. The mobile ridge-tops are formed and maintained by side winds, and on any one longitudinal dune they have a consistent asymmetry.

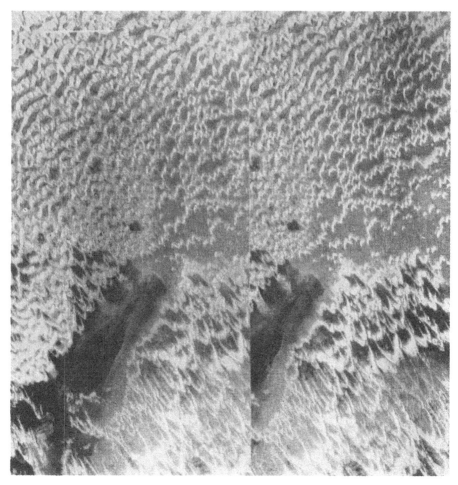

Fig. IV-13. Transverse dunes in the top left, giving way to barchans, which are converted into U-dunes beyond an area of moisture and vegetation (dark area). Otero County, New Mexico. (*Stereophotos by Soil Conservation Service, US Department of Agriculture*)

Longitudinal dunes are best developed where there is an abundant supply of sand, especially in large alluvial basins. This applies, for instance, to the Thar, the Kara Kum, the Simpson and the west Sahara deserts. The sand in longitudinal dunes is not very well sorted, except for the mobile sand in the crests.

Longitudinal dunes are parallel to the dominant sand-shifting wind, but a second marked component in the windrose also appears to affect them, mainly perhaps with respect to their asymmetry or to superimposed smaller dunes. The pronounced variations in trend are connected with the wind systems, although the trend varies markedly in the vicinity of uplands and ranges, where the winds are affected by local topography.

DESERT SAND

Desert sand, especially that of the dunes, has a number of features which are characteristic and help to distinguish it from sand of other sources.

Mechanical breakdown is important in debris transported by wind. Hard grains such as quartz become rounded and roughened into 'millet seed' sand grains, and mica (a common component of many types of bedrock), which is a soft mineral, is worn to destruction and removed by deflation during wind transport. Aeolian sand deposits, therefore, are usually deficient in mica.

The source of dune sands is more complex than at one time seemed to be the case. Earlier workers, such as Gautier, believed that the deserts were developed from great tracts of pre-existing sandy alluvium, and this is very probably true in many instances, such as the Simpson Desert of Australia and the Kara Kum east of the Caspian Sea. In West Africa, the Thar, and in Peru, some of the sands were brought in during low stands of the sea when expanses of beach sand became available. Bagnold says of the Libyan Desert, 'The place of origin of the sand is usually fairly obvious, an escarpment or a series of depressions where wind erosion is actively taking place,' and in many other deserts, including large areas in Australia, the source has been shown to be in local rocks. In the Kalahari the sand is largely derived from the Stormberg Sandstone. In Arabia the source is mainly the granitic basement, with additions from younger sandstones.

E. S. HILLS
C. D. OLLIER
C. R. TWIDALE

References

[1] JAHNS, R. H., 'Geology of Southern California', *Calif. Dept. Nat. Resources, Div. of Mines*, Bull. 170, 1954, Chaps. IV, V.

[2] BROWN, G. F., 'Geomorphology of western and central Saudi Arabia', *Rept. 21st Int. Geol. Congr. Copenhagen*, 1960, Pt 21, pp. 150–9.

[3] BARTON, D. C., 'Notes on the disintegration of granite in Egypt', *Journ. Geol.*, Vol. 24, 1916, pp. 382–93.

[4] GRIGGS, D. T., 'The factor of fatigue in rock exfoliation', *Journ. Geol.*, Vol. 44, 1936, pp. 781–96.

[5] BLACKWELDER, E., 'Geomorphic processes in the desert', Pt 2 of Chap. V in Jahns (reference 1).

[6] ARELLANO, A. R. V., 'Barrilaco pedocal, a stratigraphic marker ca. 5000 B.C. and its climatic significance', *Rept. 19th Int. Geol. Congr. Algiers, 1952* (1953), Pt vii, pp. 53–76.

[7] HUME, W. F., *Geology of Egypt*, Vol. I, Cairo, 1925, particularly pp. 82–6.

[8] BOSWORTH, T. O., *Geology and Palaeontology of North-West Peru*, London, 1922.

[9] BERKEY, C. P. and F. K. MORRIS, *Geology of Mongolia*, New York, 1927.

[10] BAGNOLD, R. A., *Physics of Blown Sand and Desert Dunes*, London, 1941.

[11] PRICE, W. A., 'Saharan sand dunes and the origin of the longitudinal dunes: a Review', *Amer. Geogr. Soc.*, Vol. 40, 1950, pp. 462–5.

ADDITIONAL REFERENCES

CHRISTIAN, C. S., J. N. JENNINGS and C. R. TWIDALE, 'Geomorphology', Chap. 5 in *Guide Book to Research Data for Arid Zone Development*, Unesco, Paris, 1957.

Int. Geol. Congr., Rept. 19th Session, Algiers, 1952 (1953), Fasc. VII, Deserts Actuels et Anciens.

PEEL, R. F., 'Some aspects of desert geomorphology', *Geography*, Vol. 45, 1960, pp. 241–62.

TRICART, J. and A. CAILLEUX, *Les Models des Régions Sèches*, Centre de Documentation Universitaire, Paris, 1962.

EXPLANATION OF FIG IV-I

An aerial view of an arid landscape at Lyons River, Western Australia.

The geological basement rocks are stratified, folded and faulted, and in the eastern section of the photo parallel strike-ridges have been formed by differential erosion.

The basement is overlain by a sequence of flat-lying sediments, preserved in the Kennedy Range. A series of longitudinal dunes, varying in direction, as presumably did the winds which moulded them, and exhibiting complicated forking and anastomosing, extends westerly from the Kennedy Range. These are fixed, fossil dunes and are being eroded.

In the west, erosion by many ephemeral streamlets with a fine drainage texture has formed a badlands type of scenery. In the east, the Lyons River has a braided course, and it and its tributaries have excavated below the level of the Kennedy Range exposing the basement complex, covered in part with alluvium. Here a new generation of longitudinal dunes has formed, smaller than and not parallel to the old set on the Kennedy Range. These dunes include a complicated pattern of clay pans (or salt lakes) many of which are almost circular.

CHAPTER V

Water Supply, Use and Management

In the arid regions even more than elsewhere, water is truly the basis of life for man, for beast and for plants. Moreover, it is not only a matter of quantity, for unless the water is also of suitable quality it is useless, and in some arid areas the problem is in fact more one of quality than of quantity.

Fig v-1. Dry river bed, Karamoja, north-eastern Uganda. Water for the cattle is obtained from a deep hole dug in the river bed. (*Photo C. D. Ollier*)

Few experiences can give a more powerful impression of the supreme importance of water in arid lands than to see women and children crowding around a well or water-hole and awaiting their turn, often for hours, to obtain their meagre daily supply; or to see great herds, often in country much overgrazed, patiently waiting while the tribesmen labour continuously at their task of raising water from shallow wells already in need of replenishment from new rainfall (Fig v-1).

In most arid lands there is great need to find new sources of water and to develop them; but it is equally important that such development should be carried out efficiently and economically, not only for current use, but also on a well-planned long-term pattern, so that water may continue to be available for future years, and to ensure that the land should not itself be rendered valueless by salinisation and waterlogging.

Although in the following paragraphs surface waters and groundwaters will be described separately, they are in fact intimately associated and are but different phases of the hydrologic cycle (Fig v-2). In the course of this cycle water is evaporated from the oceans and other water bodies and is precipitated on land[1] in the form of rain, snow, hail or sleet, or forms as dew; some of the precipitation flows back to the sea in streams, some is evaporated or transpired by plants, while the remainder infiltrates into the land and percolates downwards to augment the soil and groundwaters. Ultimately, some groundwaters return to the surface as springs or as increments to stream flow, and after a long dry spell the flow of streams may in fact derive entirely from the outflow of groundwater. The management of water resources can be improved by making use of this interrelation, for example by using as a combined resource the water of a stream and that in hydraulic continuity with it within the adjacent alluvium. Groundwater reserves may be used as a buffer to tide over dry periods, or underground aquifers may be recharged with excess surface water during the wet season.

The relationship of the salinity of surface to groundwaters is also significant, in that surface waters concentrated by evaporation may, if they pass underground, increase the salinity of groundwaters, or fresh run-off may be contaminated by emerging saline groundwater, or emerging groundwater of good quality may give rise, by evaporation, to shallow saline lakes, as in the *chotts* of northern Africa.

Action aimed at increasing the benefits of water to mankind may be taken at any point in the hydrologic cycle, and because this involves not only the water itself and the rain from which it comes but also soils, rocks, plants, animals and human beings, the range of possible action is very great. This is dealt with in the paragraphs that follow and in other chapters devoted to special topics in this book, in which reference is made to the collection of basic data and to various lines of fundamental and applied research.

It must, however, be said that in addition to these studies and their application there is often a simple need not to waste the water that is already available. In the Iraqi section of the Tigris–Euphrates basin, for example, the mean annual loss of water by seepage and waste in the irrigation enterprises is of the order of five or six times the amount which could possibly be used by the cultivated crops in their annual growth. Where artesian supplies are drawn on, they must be controlled so that they do not run to waste. Sometimes where water is locally available in reasonable quality and quantity, as in an oasis, its utilisation is limited by

---

[1] Precipitation at sea does not enter into consideration in the cycle.

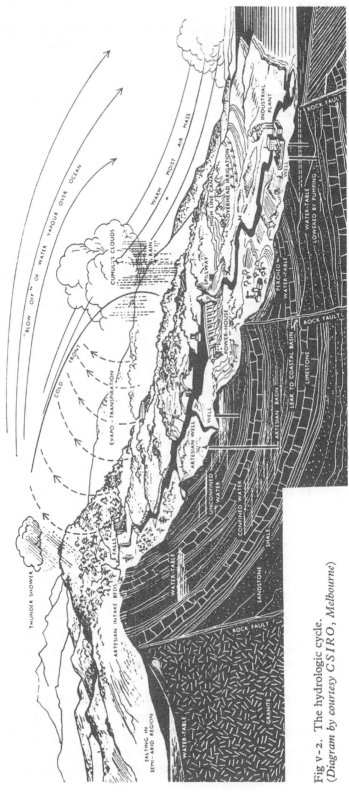

Fig v-2. The hydrologic cycle.
(*Diagram by courtesy CSIRO, Melbourne*)

social and economic factors – as in questions of ownership, and ownership of the surrounding land, and the competing demands for its use, say for drinking purposes for man and beast, for subsistence or cash agriculture, or for industry, as in mining.

The question of the ownership of sparse water supplies is of great importance in pastoral areas, leading not infrequently to tribal strife in times of unusual shortage, or to disputes as to whether a given run-off should be used to fill small dams (Arab, *hafirs*) or to augment the groundwater of wells lower downstream.

Rivers and other surface waters

SURFACE WATERS IN RELATION TO CLIMATE, GEOLOGY, TOPOGRAPHY AND VEGETATION

The amount and availability of the water supplies of an area are dependent upon many different factors, including particularly climate, geology, topography and vegetation. Climate affects water supplies through rainfall, temperature, evaporation and, in relation to plant cover, through transpiration. Moreover, the distribution of rainfall in time is often as important as the annual amount; many areas are 'arid' not because of low total rainfall but because it falls within a short period of a few weeks or months and leaves the land for the rest of the year rainless and parched. Furthermore, minor climatic variations such as would pass almost unnoticed in more humid regions may give rise in arid regions to disastrous droughts, which in fact do occur in irregular, and unfortunately unpredictable, cyclic patterns of a few years to a few score years or more. In many lands a succession of relatively humid years has induced agricultural settlement that has failed disastrously when drier conditions have returned.

Surface water resources, like groundwater resources, are much dependent on the geology of an area, for the relative proportion of the rainfall that sinks into the ground is dependent upon the nature of the rocks and the overlying soils that are largely derived from them.

Limestone terrains in arid regions usually form karst lands which readily absorb the rainfall and provide little or no run-off, and the absorbed water follows underground courses or reappears as strong springs. Fine-textured argillaceous rocks such as clays and shales, when dry, can absorb large quantities of water, but when saturated they become impervious and yield high run-off. A moderately porous sandstone often possesses the requisite degree of porosity and permeability to maintain streams in more or less continuous flow, and jointed igneous rocks also trap large quantities of groundwater, which may emerge as springs or seepages.

In more arid areas springs and seepages are generally rare, except for short periods after rain, and the continuity of stream flow from run-off is largely dependent, apart from evaporation, on the absorptive capacity of the rocks exposed along the stream beds. On reaching a particular porous or fissured formation the whole of the stream flow may sink into it and augment the groundwater below. In general, the drier the climate the greater the tendency to give rise to large or small closed drainage basins, in which sediments accumulate and saline lakes occur.

A close study of these and other geomorphological features, which have been discussed in Chapter IV, assists in understanding the current hydrological regime of an area and that of the recent past. For example, the geomorphology of a river basin will throw much light on the nature, intensity, extent, frequency, volume and depth of the floods to which the area is or has been subject: such studies will trace also the past and probable future movements of sand dunes, which may block water courses and form temporary lakes or diversions, or submerge springs and marshes.

The relation of vegetation to the hydrologic cycle is an important question to which much attention has been given, and, while some aspects require further research, much is already known. In many circumstances the protection of the natural vegetation cover, particularly forest, will maintain streams in flow, and control erosion and sedimentation, while the indiscriminate destruction of such vegetation will lead to rapid run-off, soil erosion, and replacement of steady stream flow by intermittent flow and torrential floods, and to the disappearance of springs; but the hydrological effects of the replacement of forest by grassland or cultivation, or vice versa, are highly variable according to the climate, topographic and geological conditions, and the nature of the vegetation itself. Reference has been made in Chapter III to the harmful activities of man in producing 'deserts' by the destruction of forests. While true deserts arise solely from climatic factors, the balance of the natural vegetation is so precarious in the semi-arid lands fringing the climatic deserts that, once destroyed by man and succeeded by loss of soil and soil moisture, it cannot well be restored. In such circumstances the whole appearance of the country changes, and it assumes a far more arid aspect than is justified by the purely climatic factors, for the streams become intermittent and torrential or even disappear altogether, and locally lands that once were cultivated or grazed may be affected by drifting sands. Such changes have taken place in most of the Mediterranean lands, as for example in Greece, Algeria, Tunisia, Turkey, Syria and Lebanon.

STREAMS AND LAKES

The streams of the arid regions include local and ephemeral streams arising from local storms, those that flow seasonally from surrounding uplands for longer or shorter periods before becoming lost by evaporation and by absorption into their channels, and, rarely, perennial rivers from distant humid areas that flow across the arid lands. Among the latter may be mentioned the Nile, the Euphrates, the Indus, the Rio Grande, the Amu-Daria, and the Murray–Darling system. Such streams afford natural irrigation to the arid lands they traverse and have for centuries or even millennia supported permanent agricultural settlement along them, or have provided water for nomadic tribes.

Such benefits have been extended by various engineering works, but the complete control of flooding, or the impounding of storage water to provide supplies during dry seasons, is a major and costly undertaking. The economic considerations are indeed complex, as they not only include the costs of the dam and reservoir and of the supply of water by canals or pipelines, but also the direct and indirect benefits to a community of the maintenance of irrigation. Water from permanent rivers may be brought by canal from great distances for irrigation purposes; for example the great Fergansky canal in the USSR is 350 km long, and the first section only of the Karakum canal is 400 km.

Ephemeral streams

Arid countries are characterised by innumerable dry sandy river beds which are in flood only rarely. The streams may flow only once in years, or annually, or several times a year, and the individual flows may last for hours, days or weeks,

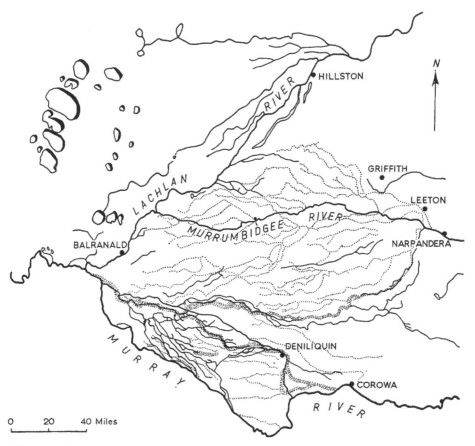

Fig v-3. Part of the Murray River Basin in New South Wales, with present streams shown by continuous lines, prior streams by dotted lines (or wider dotted areas for a prior stream complex). Lunettes are shown by solid black areas on lake shores. (*After Pels, 1962*)

and a river may flow for a time along a particular part of its course and not elsewhere. Some streams flow fairly regularly for some months after seasonal rains, and the point of termination of flow then recedes slowly upstream.

Long after such rivers have ceased to flow, pools remain in their beds, often maintained by sub-surface flow from farther upstream. These occur wherever the regime of the river or the nature of its bed have permitted it to scour out its channel below average depth, as at sharp bends in its course, or at the foot of waterfalls, or again, debris deposited by a tributary stream may serve as a dam. Many pools lie at intervals along the fossil stream courses of former humid time, as on the arid plain of northern Kenya. On the plains of the Murray–Darling system in south-eastern Australia (the Riverina district), which are the results of long-continued deposition by wandering streams, a sequence of now buried stream deposits in which the courses of 'prior' streams may be identified as well as buried levee banks and flood plains, has been recognised [1]. The prior stream courses constitute collecting channels for underground water, and the whole intricate pattern of ancient stream deposits greatly influences both underground water development and irrigation and drainage practices in the area, which is today traversed by the Murray River, its tributaries and a complex system of river anabranches and effluents (Fig V-3).

The more or less circular pans of the arid and semi-arid plains of southern Africa also probably derive in part from past humid cycles, but some may have been formed and maintained by wind action or by the trampling of game or stock. The *chotts* and *sebkhas* of northern Africa are in some cases fed from artesian groundwaters, while others are eroded in part by wind action. These various lakes and ponds, even when small and short-lived, are of great value in enabling flocks and herds to graze in distant and scattered pastures that could not otherwise be reached.

The control of surface waters

STORAGE IN RESERVOIRS, TANKS AND CISTERNS

In arid lands even very small and temporary reserves of fresh water may be vitally important, and people living close to nature are usually aware of plants from the roots or shoots of which water may be obtained in quantities sufficient to sustain life. The Bushmen of the Kalahari store quantities of water melons for use when no other water supplies are available, and they also get small quantities of drinking water from wet sand by sucking through a hollow stem, and collecting the drops in a container [2]. But from very early times people in arid lands have also been accustomed to store rainwater, storm water run-off, or water carried from wells or soaks, for domestic use. Tanks, cisterns, jars, or even, as in the Sudan, hollow baobab trees may be used, and cisterns large enough to supply towns have been constructed. These essentially local or domestic storages are more fully treated in Chapter x.

In many countries, where the subsoil is sufficiently impervious, excavated tanks are extensively employed to store the water of surface run-off or of ephemeral streams; they are particularly well developed in the Sudan, in Australia and in India for agricultural and pastoral use. Because there may be several on one farm, the practice of using them is sometimes called 'water harvesting'. Some of the tanks are elaborately constructed, being provided with sediment traps, wire fencing to keep the stock away from the embankments, and oil-driven pumps or windmills to raise the water to drinking troughs. The use of modern earth-moving equipment has made it possible to build tanks of larger capacity than hitherto and in shorter time, and has obviated the difficulties attendant upon assembling large numbers of labourers in isolated places.

Dams are, of course, extensively used to impound surplus water, and they range from small stock dams across minor streams to great earth-fill or concrete structures intended to yield permanent supplies. Such dams have the great disadvantage, as compared with those of more humid lands, of being exposed to greater evaporation, greater sedimentation, and greater uncertainty of supply. There is an economical limit to their size owing to a general increase of evaporation losses as the volume and particularly the surface area of stored water increases. It is sometimes necessary, as has been found in Israel, to store as much as three times the amount of water actually required.

The economic and technical success of any project will depend upon the ability to control the amount of sediment carried by the rivers and consequently the rate of silting-up of the reservoirs, and also on the control and management of water use. It is, however, rarely possible to make complete use of the extremely variable run-off of arid lands, and the best that can be hoped for is to provide storage for a minimum regular demand and to plan for some limited use of the remaining fluctuating supply [3]. Ways of estimating the potential storage capacity and proper rate and amount of water use are currently receiving much attention [4], as well as the social and economic aspects of large-scale river control and irrigation projects [5].

Much experimental work is being done on various methods of reducing evaporation from reservoirs (see pp. 50, 441), but losses of water stored in reservoirs and tanks by downward percolation may also be very great, where the soil and sub-surface rocks are pervious to water. Small tanks may be treated to reduce such losses, even by laying a sheet of plastic material over the whole floor; but with large reservoirs this is not possible, and the water soaks down until the porous subsoil is saturated beneath and around the reservoir, often for some miles. The water table is raised, and this may lead to deleterious effects if the sub-surface water, raised near to the zone of plant roots, is saline.

QUALITY OF SURFACE WATERS

As compared with those of more humid regions, the surface waters of arid lands, both flowing and stored, tend to be (i) more saline, owing to higher evaporation rates and to salts already present in the soil; (ii) more laden with silt and matter in suspension, owing to poorer plant cover and more rapid soil erosion, often accentuated by overgrazing and improper methods of agriculture; and (iii) fouled by plant and animal organic matter. They are none the less commonly consumed by man and stock under these conditions.

Surface waters, while they are usually less saline and more potable on this account as compared with local groundwaters, are silt-laden and often highly polluted; where they are required for community use, as when stored in reservoirs, it is necessary to protect them as much as possible from human and animal pollution and from harmful vegetable matter, and from the outflow from cultivated lands, but even then they generally require purification before use.

The extent to which saline waters can be used for various purposes is described on page 98.

Underground water

INTRODUCTION

In view of the great scarcity of surface supplies, or even their complete absence throughout much of the year, groundwater assumes a vital importance in those arid lands to which water cannot be brought by channels. In most areas the water is required essentially for drinking purposes for man and beast and to a greater or less extent for agriculture or for community and industrial purposes including mining. In addition, groundwaters are normally more reliable than surface waters in that they can be drawn upon throughout the year, and are usually less affected by a relatively dry period of two or three years.

The inhabitants of certain arid lands have long been accustomed to sinking wells by hand in suitable places, even to depths of as much as 400 feet (Fig v-4), but greatly improved supplies may often be obtained by means of bore-holes, which can be constructed under difficult conditions of hard rock or loose sand, and which can be drilled to great depths, beyond the resources of the local people.

To one unaccustomed to arid conditions it is astonishing to see the abundance of groundwaters in certain favoured localities, and to note the vast herds that are watered from them and feed upon the sparse grazing of the surrounding country.

The beef cattle industry of the interior plains of Queensland and New South Wales in Australia is completely dependent upon artesian wells, drilled to various depths according to the geological structure, from a few hundred to over 4,000 feet in the Great Artesian Basin of Australia, which covers an area three times that of France and crosses the boundaries of three Australian states and the Northern Territory (Fig v-5).

It is indeed common enough for large basins to be crossed by National or State boundaries, and accordingly a large measure of intergovernmental collaboration is required in their study and in controlling the use of the artesian water, in order to ensure that the supplies will be properly and equitably developed. In northern Africa there is one great project that can be attacked satisfactorily only on an international basis, and that is the survey of the artesian basins, which together comprise the world's greatest related group of such structures.

Fig v-4. Open well, with balance bar, in date-palm oasis, Algerian Sahara. (*p. 223 of 'Algeria' by H. Innard. Translation by O. C. Warden, Nicholas Kaye Ltd, London*)

Since underground waters are replenished directly or indirectly by rainfall, and since, of course, arid countries owe their aridity to sparse rainfall, it might be expected that underground waters in arid countries would be small in volume. Often, however, they percolate laterally from some distant humid region at a higher elevation where the 'intake' occurs.

Groundwater

Underground waters that occur in porous beds or fractured hard rocks without an overlying impervious layer, and which are fed by local rain or stream flow,

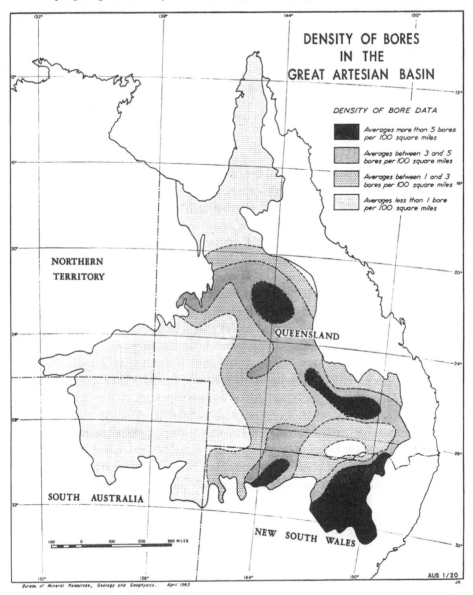

DENSITY OF BORES
IN THE
GREAT ARTESIAN BASIN

DENSITY OF BORE DATA

Averages more than 5 bores per 100 square miles

Averages between 3 and 5 bores per 100 square miles

Averages between 1 and 3 bores per 100 square miles

Averages less than 1 bore per 100 square miles

NORTHERN TERRITORY

QUEENSLAND

SOUTH AUSTRALIA

NEW SOUTH WALES

Bureau of Mineral Resources, Geology and Geophysics. April 1963

AUS 1/20

Fig v-5. Density of bores in the Great Artesian Basin, eastern Australia. (*After Hahn & Fisher, Fig. 2, p. 170, 'Water Resources Use and Management', Melbourne, 1964*)

are known as 'phreatic' or 'free' groundwaters; they lie within a zone of saturation at moderate depths beneath the surface, depending upon meteorological and geological conditions, and yield water to wells and bore-holes. The upper surface of the zone of saturation, when unconfined, is known as the *water table*.

Where nomads have roamed over arid lands for many generations, they have usually discovered, by long-continued observation, trial and error, the places at which the free groundwaters lie within reach of open wells excavated by hand, and usually it is difficult even with modern methods of exploration to improve on such well-field sites. Modern techniques can, however, make the wells more stable and more productive where required, and enable waters to be reached in hard rocks and at much greater depths than were possible by older well-sinking methods. A truly sophisticated technique was, however, developed long ago in Iran, whence it has spread widely. This is known as the *foggara*, *karez* or *qanat*.

Qanats

The physical foundation of qanats is the combination of mountains receiving a considerable and quite stable amount of rainfall and, therefore, groundwater recharge, and nearby plain-land with good agricultural soil but an inadequate supply of water, or salinity problems of water available locally. This complex is frequently found on the border of the central Iranian basins and the mountain ranges fringing them. The groundwater horizon becomes accessible at moderate depth – down to 20 metres, but in some places 100 metres and even more – either on alluvial fans issuing from valleys at the mountain border or below the plain nearest the mountains. Here a vertical shaft is dug to the groundwater horizon. From there a tunnel is excavated at a very slight gradient, sometimes less than that of the groundwater horizon. This type of conduit solves the power problem, permitting gravity flow of water and obviating pumping; at the same time it prevents evaporation losses, though losses by seepage may be considerable.

The upstream part of the qanat serves as an infiltration gallery and has often feeders branching out to make more water available. A qanat may be a few hundred metres or several kilometres in length, and qanats of 25 km have been mapped; even larger ones, perhaps up to 160 km, may exist, and the yields are commonly more than 500 gallons per minute. The excavation of a qanat is done from vertical shafts, working in both directions. The distance apart of these shafts is 8–50 metres, depending on the overburden and the depth below surface of the tunnel. Through these shafts all the excavated material is removed. Consequently, an elevated ring of earth and debris surrounds their opening and at the same time prevents material and impurities from being washed into them from the surface. Sometimes these manholes are even covered. These manholes are often the first indications of a qanat to the traveller; their long and regularly spaced lines make them very conspicuous from the air (Fig v-6). If the material in which the qanat has to be dug is not strong enough, these shafts require a revetment or lining, often of stones. Where the overburden of a qanat is too heavy to be borne, the walls of the tunnel are lined – in Iran with tiles.

In countries where qanats have been particularly developed their numbers are astounding. Cressy considers for Iran a figure of at least 2,000 as substantiated,

Fig v-6. Lines of pits and mounds along the course of a qanat, Pakistan. (*Photo E. S. Hills*)

and mentions estimates up to 40,000 (Fig v-7). Whatever the correct figure, the work, effort and cost invested in them over generations is stupendous. With the economic conditions, including the price of labour, still persisting in these countries it often still pays to dig qanats regardless of their high cost, as their

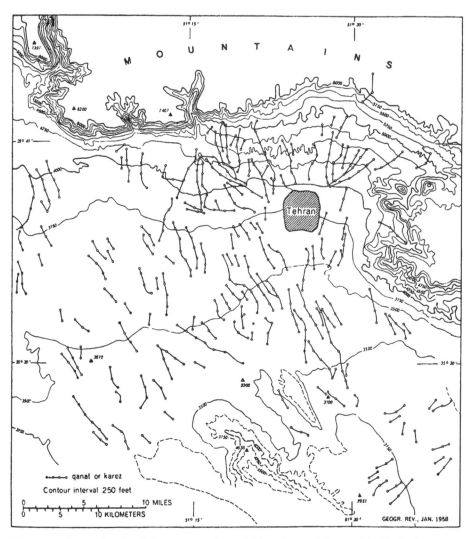

Fig v-7. Qanats in the Teheran area, Iran. (*After· Cressy, 'Geographic Review',* Vol. 48, 1958)

operation is much cheaper than those of any water delivery system requiring power-operated pumps.

According to Cressy at least 36 qanats still operate in Teheran. 'Where a qanat passes beneath a residential area, some houses have a summer living room alongside the flowing stream, several tens of feet underground.'

Qanats have a wide distribution over the arid zone from western China through

Afghanistan [6, 7] and Iran to the Atlas countries, where ancient foggaras, carefully maintained, provide part of the water supply of Marrakesh. The Spaniards introduced them into South America: Isaiah Bowman described them from Pica in the Atacama Desert of Chile. Those *socavónes* of Pica which are still in full operation derive their water from an alluvial fan at the foot of the

Fig v-8. Man-holes (lumbreras) of qanat at Pica in the northern desert of Chile. (*Photo D. H. K. Amiran*)

Andes. They number about fifteen and they augment the water supplied by Pica's springs and wells. Their length is up to 1½ km and the spacing of the *lumbreras*, the square manholes, is about 50 metres; their opening is 180 by 180 cm (Fig v-8). The height of their tunnel is about the size of a man. The water delivered by socavónes has a salinity of about 550 mg per litre, more than the approximate 200 mg per litre of the water of the springs.

Sub-surface flow in dry valleys – underflow

Dry river beds (variously known as wadis, oueds, arroyos or dry gulches) contain quantities of coarse and fine sediment. After surface flow ceases, water continues to percolate downstream through the interstices of the sediments, constituting a sub-surface or *underflow*. When the underflow ceases, small bodies of water remain in stream-bed hollows beneath the sand and gravel. Such supplies of water may persist far into the dry season and when reached by shallow wells and water-holes they form a major source of supply, sufficient even for large townships, as at Kano in northern Nigeria and Alice Springs in central Australia.

Local people may recover water with great difficulty from a depth of 30 feet or more in loose, running sand, using funnel-shaped excavations. In such cases supplies are more readily obtained from perforated tube wells driven into the sand, from wells or bore-holes on the stream banks, or from infiltration works driven along the base of the sediments.

Sub-surface dams

In stream beds of the kind described it may be possible to hold up the underflow by means of a sub-surface dam, and thus produce a sub-surface reservoir which has the great advantage of being well protected from evaporation. The town of Massawa on the Red Sea coast is supplied in this way, and throughout the drier parts of Africa, and of South America, where conditions permit, such dams are occasionally built. They are, however, not nearly so generally applicable as is commonly supposed, since to be successful they require rather special site conditions, and where large structures are envisaged and a strong underflow persists in coarse pebble beds at the base, the difficulty and cost of construction are apt to be greater than the conditions warrant. Also, if it is to reach to the surface of the sand, the upper part of the structure has to be very strongly built so as to withstand the pressure of the mass of drifting sand when the stream is in flood.

The main difficulty, however, is usually to find a suitable site, of which the first essential is impermeability of the bottom and sides of the infilled stream channel; this demands that the sediments rest in a channel eroded in solid rock or in deep clay, conditions rarely realised, since the dry rivers concerned are usually aggrading their beds and their channels are cut through more or less permeable sediments. But sometimes the required conditions are fulfilled, and a sub-surface dam of clay, concrete, driven piles or other available materials can be well bonded to the bottom and sides, as exposed in a trench cut through the sediments, and a valuable water supply thereby secured. Even when not completely watertight the structure will at least retard the underflow and cause the water to bank up behind the dam.

Seepage water

While in humid regions the water table normally declines towards the floor of a valley, where it feeds the stream, in arid regions it is often inclined downward when traced away from a river course, and represents a loss of water from the river. Thus along the lower Nile the water table as seen in wells is found to lie at increasing depths when traced away from the river. Nile water has been shown to make its way by this means into Wadi Natrun, a lake, or group of small lakes, lying in the Western Desert 65 miles west-north-west of Cairo and 22 miles from the nearest branch of the Nile; it stands about 25 feet below sea-level, and its waters have long been known for their high soda content, due to changes in composition as the water seeps through the rocks, and to high evaporation under the prevailing desert conditions.

When the Murray River, South Australia, is in flood the river supplies water by seepage into the surrounding country and raises the water table for a distance of at least seven miles from the bank, but when the river is low, water from the surrounding country drains back into it.

Artesian water

Rainwater entering a porous stratum where it outcrops in the highlands is conveyed underground by that stratum, which is called an *aquifer*, to distant parts of the plain whence it may be recovered as artesian water from bore-holes; such waters are known as 'artesian', 'pressure' or 'confined' waters because the water of the porous stratum is confined within that stratum by impervious beds above and below. Where the stratum reaches a level lower than the intake, the water is held under pressure which may cause it, in a well or bore-hole, to rise to a level

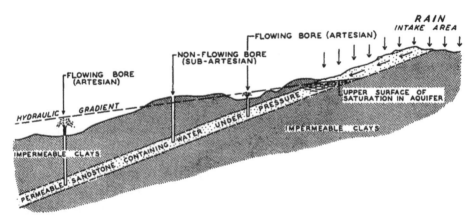

Fig V-9. Diagrammatic cross-section of the edge of an artesian basin. (*After South Australian Dept of Mines 'Groundwater Handbook', 1959*)

below, at, or above the surface of the ground at that point (Fig V-9). Water that is under insufficient pressure to reach ground level and overflow is sometimes referred to as 'sub-artesian' water.

An imaginary surface joining the levels to which the water would rise in pipes extending above the ground is called the *piezometric surface*. This surface slopes away from the highlands where the intake area of the aquifer outcrops, and the slope of the surface is the *hydraulic gradient*. The slope may be represented by lines joining places of equal pressure, known as *iso-piestic lines* (Fig V-10).

The water pressure in an artesian bore is found to decrease as time goes on, the first flows often being very much greater than those obtainable after some years have elapsed. These first 'flush' flows, as they are called, result from the so-called *elastic effect* of the aquifer. If we imagine this to be a rubber tube, the pressure of water due to hydraulic head would expand the tube. In rocks, the grains are forced apart into an open-packed condition. A bore lets the water

escape, and the grains settle down into closer packing, the rocks subsiding somewhat as the collapse goes on. When the grains have settled into their close-packed
condition no further collapse can occur, the pores are reduced in size, and all
later flow is due to the normal passage of water through the pores towards the
bore. The artesian pressure is now lower than formerly, and a *steady state* is
established. The pressure may be so low that bores once artesian become sub-
artesian. This has occurred in parts of the Great Artesian Basin of Australia. The

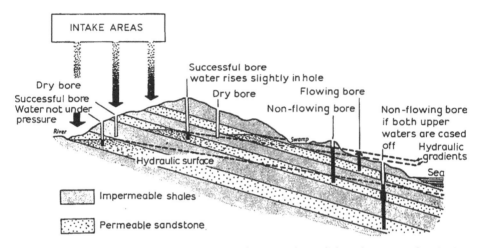

Fig v-10. Diagrammatic cross-section showing varied conditions in an artesian basin.
(*After South Australian Dept of Mines 'Groundwater Handbook', 1959*)

economic effects are considerable, as pumping must be resorted to under sub-
artesian conditions to bring the water to the surface. Control of the flush flows
prolongs the period of decline of pressure, and is accordingly recommended in
all artesian basins.

Important artesian systems are well known to occur in large parts of Australia,
North Africa and the USSR; but, as already indicated, such systems occur only
under special geological conditions, and there are numerous and extensive arid
lands in which it is clear from geological examination that artesian conditions
do not exist.

In most of the artesian basins of the arid lands there is room for an increase
in the number of bore-holes and corresponding improvement in the economic
prospects of these areas; in Algeria, for example, the aquifers seem to be capable
of providing safely more than twenty times as much as at present, and those in
Morocco, Libya and elsewhere could also yield greater supplies.

Fossil water

Certain groundwaters of semi-arid regions represent bodies that accumulated in
a former more humid period and are no longer subject to annual replenishment;
they are accordingly termed fossil groundwaters. While they are of little value

for long-term development, since they are exhaustible, they could sometimes be 'mined' with advantage for short-term purposes, or kept in reserve to supplement surface supplies in time of drought. In the arid south-western United States some home sites are self-supplied with wells that tap groundwaters emplaced during Late Pleistocene time, and for which recharge is now negligible.

In many places water is used with little concern for some future time when the pumping lift increases beyond the limit of economic feasibility, or when the supply might be exhausted. Also it would appear that much artesian water in northern Africa may be fossil water, as the intake areas of the Sahara have very little rain today.

The recharge of groundwater

The recharge or replenishment of groundwater bodies normally takes place by infiltration of rainfall or by absorption from surface-water bodies or both.

The direct artificial recharge of groundwater is assuming greater importance with the growth of population and the increased demands of agriculture and industry. In arid countries it is particularly valuable because of the scarcity of surface water and also because groundwater storage is not subject to as much evaporation as surface storage. In Algeria and Morocco recharge has long been practised by leading floodwaters away from streams to the surrounding land; after use for agriculture and absorption into the ground, a portion becomes available for groundwater development and pump irrigation. In arid countries the main difficulty in using floodwaters for recharge is often their suspended load of silt which, when deposited, soon reduces the infiltration capacity of the ground. In the United States, especially California, and in European countries where recharge methods are highly developed, water is often treated before being led into wells and infiltration basins of various kinds, suitably situated on highly permeable formations overlying the aquifers to be recharged. In Greece it is proposed to use surplus water from springs, when not required for irrigation, for recharging groundwater through wells.

The use and misuse of underground waters

Owing to the great increase in human and animal populations, and to the need for economic expansion, there has been a growing demand for new water supplies in most of the arid countries of the world, and the governments concerned have accordingly made the requisite investigations and developed many new groundwater supplies by means of wells and bore-holes for pastoral, agricultural and industrial needs.

In the Algerian Sahara, for example, apart from the important artesian supplies that have made possible the development of the new oil and gas fields, the government has installed on the steppe grazing grounds thousands of watering points, each comprising a well or bore-hole, and a pump, worked by hand, windmill or internal-combustion engine according to circumstances, a tank, drinking-troughs, etc., and in addition numerous new oases have been established for the cultivation and export of dates and other fruit. The artesian and groundwaters of the geological formation known as the *Continental Intercalaire*, largely of Cretaceous age, have been tapped by tunnels (*foggaras*), driven tube wells, and by bore-holes; some of the artesian bores, recently constructed, penetrate to depths of 1,000 to 2,000 metres and yield 100 to 300 litres of water per second (Fig V-11a, b). Extensive groundwater development along roughly similar lines has been in progress also in other parts of northern Africa, and in South Africa, the Middle East, the south-western United States, the Karakum and other deserts of USSR, and in Australia.

Groundwater as a natural resource may, however, be misused in various ways, as by the over-pumping of bore-holes, i.e by extracting water from them over a period of years at a rate exceeding the average rate of annual recharge by accretions of percolating rainfall; this leads to a gradual lowering of the water table in the vicinity of the bore-hole and the reduction or eventual exhaustion of the supply.

Also, in cases where the fresh-water body is in contact with salt or brackish water, as in many coastal and some inland areas, over-pumping leads to the rise of salt water into the wells, with consequent pollution of the supply, rendering it unfit for agricultural or domestic use. This has actually occurred in many coastal areas, particularly in the Mediterreanean, northern Europe and in California, with disastrous results in the fruit-growing and other communities.

Groundwater bodies may be polluted also by the discharge into wells of oil-field or industrial waste, as has happened extensively in southern California. Great care has to be taken also to avoid the pollution of groundwaters by atomic waste.

THE CHEMISTRY OF GROUNDWATER IN RELATION TO USE

The content of mineral matter in solution in groundwater is normally higher than that in surface waters, and with increasing aridity of climate groundwater tends to become more and more saline. But within any one arid area a great range of water types may be found: fresh, brackish, salt, or even super-saline, and the physical conditions may be such that these different types may occur in close proximity; thus in an area of saline groundwaters relatively fresh groundwater tends to occur in the vicinity of stream channels owing to the flushing action of storm waters, and elsewhere new precipitation will percolate into the ground and form a layer of fresh water resting upon brackish or saline water. In the coastal areas of arid lands, as in humid regions, fresh water not infrequently rests upon sea water that has percolated inland through permeable formations. In discharge-less basins, the salinity of groundwaters increases from the edges towards the centre and in general in arid regions the salinity of groundwater tends to increase with depth; but some deep artesian waters may be of lower salinity than waters

Fig v-11a. Artesian bore-hole, Guerrara, Algerian Sahara, under test. Flow from Albanian artesian formation; depth of bore-hole 1,000 metres, yield 150 litres per second. To supply new date-palm oases. (*Photograph No. 15, 'Atlas Photographique d'Algérie', XIX, International Geological Congress, Algiers, 1952*)

Fig v-11b. Artesian well at the same locality, under full control

lying above them as in Australia, although the temperature increases with depth and some water emerges nearly boiling. Thus in most arid regions groundwater problems tend to become problems of salinity, and in the search for low-salinity waters reliance must be placed on a thorough hydrological investigation of the occurrence of the saline and fresh waters so that safe prediction of the relatively rare occurrences of fresh water may be made.

In view of their general salinity any assessment of the quality of arid zone groundwaters must take into account the purpose for which they will be used; thus water that would be satisfactory for stock would be unfit for human beings, and water suitable for drinking purposes may be too saline for agricultural use. Groundwaters are, however, normally free of organic pollution, whereas surface waters in arid countries, particularly in pools and around wells and springs, are usually highly polluted with organic matter, but the human and stock populations concerned tend to regard this as normal and do not appear to find it objectionable.

In humid countries, such as England and the United States, 570 parts per million (ppm) has been regarded as the extreme limit of mineral content for drinking waters, but in arid countries people commonly accept 2,500 ppm and may tolerate 4,000 ppm, or more. Much depends, however, on the nature and proportion of the salts present in the water. In certain areas of North Africa it is customary for the local population to drink water containing as much as 3,000 ppm of chlorides, chiefly sodium chloride.

There is much greater latitude in the standard of water quality for irrigation and the watering of stock than for domestic use. For irrigation, where highly mineralised water must be developed as the only available source, the concentration of the minerals must be examined with care. Among the more important of these substances are the chlorides, sulphates and nitrates of sodium, potassium, magnesium and calcium. Experimental agricultural research on actual use and results is advisable, for there are no generally accepted standards of irrigation water quality, since so much depends also upon the nature of the soils and crops considered. An important part of the problem is to find or to develop plants, such as the date palm, that are highly tolerant of salts. Often, however, the groundwaters of arid countries are unsuitable for irrigation both in amount and quality; exceptions occur where there is strong annual recharge, as in aquifers tapped by qanats, or in some artesian supplies, as in Libya.

The limits for animals are similarly variable; in Victoria, Australia, up to 3,000 ppm of dissolved salts has been considered safe for working horses, dairy cattle and pigs; higher salt contents are acceptable for sheep and grazing cattle, provided concentrations of over 7,000 ppm are used only for emergencies. For comparison, it may be stated that sea water contains 35,000 ppm, mostly as sodium chloride.

The various methods of desalination recently developed are likely to have in the near future considerable bearing on the availability of usable waters in arid lands, especially for domestic purposes and for stock. In Algeria, for example, solar stills are used for very small communities, and resin ion-exchangers, in their modern form of resin-packed polythene containers regenerated on reaching good water, are used by mobile teams forced to supply themselves from brackish wells and springs for limited periods; but the only procedure at present practicable on any scale is that of selective membrane electrolysis. When, in the near future,

Fig V-12. Water carriers, Luxor, Egypt. (*Photo E. S. Hills*)

cheap power becomes available, it is likely that demineralisation of local saline waters will cost no more than long-distance pumping from outside sources. In South Africa where an important mine converts saline groundwater for township use, experiments are being made with a view to the conversion of small local supplies, as from bore-holes, required for stock and other purposes. At Kuwait, the conversion of sea water from the Persian Gulf is being carried out on a large scale for township purposes, and at Eilat on the Gulf of Aqaba a plant using a new process developed in Israel has recently been established.

GROUNDWATER USAGE

The use of groundwater raised from wells is closely bound up with the ways of life in small communities, which often exist at a low level of productivity in arid regions (Fig V-12). These are treated in Chapter x.

F. DIXEY

References

[1] PELS, S., 'Geological investigations in the Murray–Darling Basin in New South Wales', *M.Sc. Thesis*, University of Melbourne, 1962.

[2] VAN DER POST, L., *The Lost World of the Kalahari*, London, 1958.

[3] LEOPOLD, L. B., 'Water and the arid zone of the United States', *Proc. Paris Symp. on Problems of the Arid Zone*, Unesco, Paris, 1962, pp. 395–9 (Arid Zone Research XVIII).

[4] LANGBEIN, W. R., 'Surface water (including sedimentation)', ibid., pp. 3–22.

[5] *Aust. Acad. Sci. Nat. Symp. on Water Resources, Use and Management*, 1963.

[6] CRESSY, G. B., 'Qanats, karez, and foggaras', *Geogr. Rev.*, Vol. 48, 1958, pp. 27–44. Cressy gives a list of the many names applied to qanats in various countries.

[7] HUMLUM, J., *La géographie de l'Afghanistan*, Copenhagen, 1959, pp. 205, 212–14, 226–8, and detailed map (1 : 2,100) of Kushki-i-Nakhud oasis. Humlum quotes an estimate by Wilber that 13 per cent (10,400 sq. km) of the cultivated area of Afghanistan is irrigated by qanats.

ADDITIONAL REFERENCES

The author acknowledges the assistance he has received from the following publications, in which will be found additional useful references.

ASKCHENSKY, A. N., 'Basic trends and methods of water control in the arid zones of the Soviet Union', *Proc. Paris Symp. on Problems of the Arid Zone*, Unesco, Paris, 1962 (Arid Zone Research XVIII).

AWAD, HASSAN, 'Groundwater and human geography: ways of life in desert regions', *Proc. Ankara Symp. on Arid Zone Hydrology*, Unesco, Paris, 1953, pp. 235–8 (Arid Zone Research II). (Existe aussi en français.)

BAARS, J. K., 'Pollution of ground water', *International Association of Scientific Hydrology, General Assembly of Toronto*, 1957, Vol. II, pp. 279–89 (Publication No. 44).

BOGOMOLOV, G. V., 'Conditions of formation of fresh waters under pressure in certain desert zones of North Africa, the USSR and south Asia', *Salinity Problems of the Arid Zones, Proc. Teheran Symp.*, Unesco, Paris, 1961, pp. 37–41 (Arid Zone Research). (Existe aussi en français.)

— and A. I. SILIN-BEKCHURIN, *Specific Hydrogeology*, Moscow, 1955.

CASTANY, G., *Traité pratique des eaux souterraines*, Paris, 1963.

— *Méthodes d'études et de recherches des nappes aquifères*, Bur. Res. Geol. et Min., Paris, 1963.

DIXEY, F., *Practical Handbook of Water Supply* (2nd ed., 1950), Murby, London.

— 'Some recent studies in ground-water problems', *Colonial Geological and Mineral Resources*, Vol. 4, No. 2, London, H.M. Stationery Office, 1953; also in *Proc. Ankara Symp. on Arid Zone Hydrology*, Unesco, Paris, 1953, pp. 43–59 (Arid Zone Research II). (Existe aussi en français.)

— 'Variability and predictability of water supply', G. White (Ed.), *The Future of Arid Lands*, Washington, D.C., 1956 (American Association for the Advancement of Science Publication No. 43).

DIXEY, F., 'The search for water beneath the desert', *New Scientist*, Vol. II, No. 244, 1961, p. 164.

— 'Geology and Geomorphology, and Groundwater Hydrology', *Proc. Paris Symp. on Problems of the Arid Zone*, Unesco, Paris, 1962 (Arid Zone Research XVIII).

— and SHAW, 'Introductory paper: Hydrology with respect to salinity', *Salinity Problems in the Arid Regions, Proc. Teheran Symp.*, 1961.

DHIR, R. D., 'Hydrological research in the arid and semi-arid regions of India and Pakistan', *Reviews of Research on Arid Zone Hydrology*, Unesco, Paris, 1953, pp. 96–127 (Arid Zone Research I). (Existe aussi en français.)

— 'The hydrological balance and influence of the utilization of underground water upon it', *Proc. Ankara Symp. on Arid Zone Hydrology*, Unesco, Paris, 1953, pp. 111–14 (Arid Zone Research II). (Existe aussi en français.)

DROUHIN, G., 'The problem of water resources in north-west Africa', *Reviews of Research on Arid Zone Hydrology*, Unesco, Paris, 1953, pp. 9–41 (Arid Zone Research I). (Existe aussi en français.)

— 'The possibility of utilizing minor water resources in Algeria', *Proc. Paris Symp. on Problems of the Arid Zone*, Unesco, Paris, 1962 (Arid Zone Research XVIII).

FOURMARIER, P., *Hydrogéologie*, 2nd ed., 1958, Paris.

FRENKIEL, J., *Evaporation Reduction, Physical Principles and Review of Experiments*, Unesco, Paris, 1965 (Arid Zone Research XXVII).

FROMMURZE, R. F., 'Hydrological research in arid and semi-arid areas in the Union of South Africa and Angola, *Reviews of Research on Arid Zone Hydrology*, Unesco, Paris, 1953, pp. 58–77 (Arid Zone Research I). (Existe aussi en français.)

HILLS, E. S., 'The hydrology of arid and sub-arid Australia with special reference to underground water', *Reviews of Research on Arid Zone Hydrology*, Unesco, Paris, 1953, pp. 179–202 (Arid Zone Research I). (Existe aussi en français.)

International Association of Scientific Hydrology, 'A legend for hydrological maps', *Bull. I.A.S.H.*, Ann. 7, No. 3, 1962, pp. 5–32.

LINSLEY, R. K., 'A report on the hydrologic problems of the arid and semi-arid zones of the United States and Canada', *Reviews of Research on Arid Zone Hydrology*, Unesco, Paris, 1953, pp. 128–52 (Arid Zone Research I). (Existe aussi en français.)

MARGAT, J., 'Établissement des cartes hydrogéologiques', *Association internationale d'hydrologie scientifique, Assemblée générale de Toronto*, 1957, Vol. II, pp. 36–43 (Publication No. 44).

SCHOELLER, H., *Arid Zone Hydrology: Recent Developments*, Unesco, Paris, 1959 (Arid Zone Research XII). (Existe aussi en français.)

SHOTTON, F. W., 'The availability of underground water in hot deserts', *Biology of Deserts*, London Institute of Biology, 1954, pp. 13–17.

SILIN-BEKCHURIN, A. I., 'Quelques règles de la formation des eaux souterraines dans les régions arides du globe', Moscou, Académie des sciences de l'URSS. Institut de recherches géologiques. (Personal communication.)

— 'Conditions of formation of saline waters in arid zones', *Salinity Problems in the Arid Zones, Proc. Teheran Symp.*, Unesco, Paris, 1961, pp. 43–7.

SIMAIKA, Y. M., 'Research in hydrology and fluid mechanics and development of underground water in the arid regions of north-east Africa', *Reviews of Research* on *Arid Zone Hydrology*, Unesco, Paris, 1953, pp. 42–57 (Arid Zone Research 1). (Existe aussi en français.)

SIMAIKA, Y. M., 'Alternative uses of limited water supplies in the Egyptian Region of the United Arab Republic', *Proc. Paris Symp. on Problems of the Arid Zone*, Unesco, Paris, 1962 (Arid Zone Research XVIII).

STAMP, L. DUDLEY (Ed.), *A History of Land Use in Arid Regions*, Unesco, Paris, 1962 (Arid Zone Research XVII).

TISON, L. J., 'Salinité des eaux artésiennes en Belgique due nord', *Association internationale d'hydrologie scientifique, Assemblée générale de Toronto*, 1957, Vol. II, pp. 290–9 (Publication No. 44).

TIXERONT, J., 'Les ressources en eau dans les régions arides', *Ann. Ponts Chauss.*, 1956, Vol. 126, No. 2, pp. 956–97.

— 'Water resources in arid regions', G. White (Ed.), *The Future of Arid Lands*, Washington, D.C., 1956 (American Association for the Advancement of Science Publication No. 43).

UNESCO, *Proc. Ankara Symp. on Arid Zone Hydrology*, Unesco, Paris, 1953 (Arid Zone Research II). (Existe aussi en français.)

— *Reviews of Research on Arid Zone Hydrology*, Unesco, Paris, 1953 (Arid Zone Research I). (Existe aussi en français.)

— 'Salinity problems in the arid zones', *Proc. Teheran Symp.*, 1961 (Arid Zone Research XIV).

— 'The problems of the arid zone', *Proc. Paris Symp.*, 1962 (Arid Zone Research XVIII).

UNITED NATIONS, 'Large scale ground-water development', *U.N. Publication No. 60*, UN, New York, 1960, II. B. 3.

WHITE, G. F. (Ed.), *The Future of Arid Lands, Papers and recommendations from the International Arid Lands Meetings*, Washington, D.C., 1956 (American Association for the Advancement of Science Publication No. 43).

— *Science and the Future of Arid Lands*, Unesco, Paris, 1960.

— 'Alternate uses of limited water supplies'. *Proc. Paris Symp. on Problems of the Arid Zone*, Unesco, Paris, 1962 (Arid Zone Research XVIII).

CHAPTER VI

Soils of Arid Lands

Introduction

Arid land soils are defined for the purpose of this discussion as those of regions sufficiently arid to make both rain-fed cropping and grazing potentially impossible or marginal. The climate is responsible for their most important character, which is that they are soils in which the soluble products of weathering are accumulated within the upper part of the soil profile, and are present as calcium carbonate and soluble salts. This accumulation of both carbonate and soluble salts tends to separate them from the well-known grassland soils of rather wetter regions, where usually the soluble salts do not accumulate in the upper part of the soil column though the carbonates do so. The texture of arid soils varies widely from heavy clay to coarse sands, reflecting the influence of parent material, although most have one horizon of heavy texture.

The present use of soils in deserts is mostly for extensive or dispersed grazing and short-period cropping, both of course depending upon the precipitation. Cropping is frequently practised using special water-conservation measures, such as run-off-preventing bunds, and fallowing in alternate years.

Both grazing and cropping tend to be concentrated in areas with specially favourable moisture conditions and soils. When cropping takes the form of using lands inundated by seasonal streams, irrigation is in fact being practised, and the development of this into controlled irrigation is the natural development of desert lands. Much of the interest in desert soils is directed to the assessment of their value under irrigation, and the best methods of irrigation and cultivation.

Soil genesis – an outline

The special features of the soils of arid lands are best understood in relation to the general processes of soil formation, an outline of which follows.

The formation of soil from rock or rock-like deposits is the consequence of the breakdown, both physical and chemical, of rock under the influence of climatic forces and biological action. The chemical and biological processes are immensely accelerated when they are preceded by physical disintegration of the rock, since this greatly increases the area exposed to change while also providing an environment in which moisture lodges. There are many forces active in breaking down solid rocks. Large diurnal changes of temperature causing cracking and fissuring, a process particularly active under typical desert climates with extreme daytime insolation followed by cold nights. Plant roots forcing their way into fissures lead to fractures and water freezing in fissures also does this. In deserts wind-blown sand scores and abrades exposed rock surfaces, setting free particles of sand size or smaller.

Erosion by glaciers, especially the enlarged glaciers and ice caps of the Pleistocene period, has been responsible for the production of great quantities of comminuted rock powder in some cases of texture as fine as clay. After the melting of the glaciers such material is the parent material of later soil formation, either *in situ* or after transport by wind or water to other sites. Transport of rock material by rapidly flowing water is accompanied by grinding as they move along the stream bed, a process particularly rapid and effective in the torrential waters of desert rivers.

The main process of chemical weathering is the reaction of water with rock minerals in hydration and hydrolysis, the normal products being soluble material in the form of salts, resistant residues such as the oxides and hydroxides of iron, aluminium and silicon, with clay minerals partly as decomposition residues and partly as synthesised minerals. The clay minerals have a basic structure of silicon, aluminium and oxygen atoms, with associated mobile atoms such as those of sodium, calcium, potassium and hydrogen. The number of sites on the clay mineral particles available for the attachment of metallic and hydrogen atoms is constant for a particular kind of clay, but the proportion of these sites occupied by the different atom is determined by their concentration as positively charged cations in the soil solution, and changes in the composition and concentration of the soil solution are followed by changes in the proportions of the cations attached to the clay minerals. These are the exchangeable cations of the clay, and they determine some of the most important properties of soil.

The processes of chemical weathering which follow the first physical breakdown of rocks are slow in the presence of water alone, but they are greatly accelerated in the presence of carbon dioxide and organic acids, both of which are products of biological activity, carbon dioxide being present also in the atmosphere and in rainwater. The rate of chemical attack is strongly dependent upon temperature and ceases in the absence of water. It is therefore not surprising that the most extreme expressions of chemical weathering are found in the wet tropics, where moisture conditions, biological activity and temperature are all favourable.

The processes of physical and chemical weathering produce a medium in which plants and micro-organisms become established, and this is the first stage in the formation of soil. The next stage of soil formation is the redistribution within the weathering mass of the mobile products of weathering under the influence of soil water, which mainly derives from rain, plants and micro-organisms. In a humid climate rainfall is in excess of water losses due to transpiration by plants and evaporation from the soil surface, and if drainage permits, leaching is sufficient to remove progressively the soluble products of weathering. The soluble salts are lost and the exchangeable cations of the clay become dominated by the hydrogen ion and a soil with an acid reaction is formed. The leaching water may leave the soil column either by lateral flow down slopes or vertically through a porous parent material. Lateral leaching, it will be noticed, involves the local relief as a factor in soil formation. If regional drainage is not fully effective soils at the foot of slopes may receive the soluble products of weathering from higher sites and salt accumulations deleterious to plants can form. Under conditions of smaller rainfall and less effective leaching the soluble salts do not leave the soil column and a saline soil results with the exchangeable cations being mainly those of the metals, and the reaction of the soil is alkaline. Such a condition of restricted leaching is the basic cause of salinity in soils. Calcium salts rendered soluble as the bicarbonate in the root zone are precipitated deeper in the soil as the less soluble carbonate. Magnesium and sodium salts accumulate in the soil at a depth determined by the depth of water penetration. Under certain conditions both in acid and alkaline soil, clay becomes mobile as well as the soluble products of weathering and a clay-enriched layer forms within the soil column. Plants with their main roots in the upper layer of soil play several roles in soil formation. The carbon dioxide respired by their roots directly affects the process of chemical attack on mineral particles and influences the mobility of calcium. The organic matter they produce, concentrated in the surface layers, is one of the characteristics of normal soils; it provides the food supply and environment of the micro-organisms, and profoundly modifies the physical state and structure of the soil. Plants also play an important part in the movement of soluble substances within the soil, providing a mechanism which opposes the normal downward leaching process.

When the various soil-forming processes have reached a state of near equilibrium, a recognisable and often characteristic soil profile is formed. Within the profile, horizons are formed whose nature depends on the intensity of the various processes. Most soils have a surface horizon with an accumulation of organic matter, the nature and quantity of which depend on the type and intensity of vegetation, themselves controlled by the climate. Below this horizon there may be a horizon of clay accumulation, and below this a horizon dominated by the freshly weathering rock in which the processes of decomposition are active, in contrast to those in the upper horizons where the processes may be almost completed. If the profile has been formed under a low rainfall with no through leaching, the soluble products of weathering will still be present, concentrated probably in the deeper horizons. Under such conditions soil will be likely to show a distinctive horizon of calcium carbonate accumulation, the depth of which will depend on the rainfall. If on the other hand the profile has been formed under intensive rainfall with good drainage, there will be no accumulation of salt but there may be horizons with massive accumulations of iron or aluminium oxides.

The nature of the parent material influences the kind of soil profile which develops, particularly when the process has not gone on for a very long time. It is true that under conditions of very advanced weathering similar soil may develop on a range of parent materials, but even under these circumstances contrasting rocks such as sandstones and carbonates will yield different soils. The parent material of soils is not always a solid rock, it is often water or wind transported

Fig VI-I. Polygonal surface cracking on bare, very saline desert soil devoid of vegetation

material. Soils formed on such material frequently have profiles with horizons which are depositional features and not the result of the soil-forming processes discussed above. Alluvial soil, with a succession of layers or 'horizons' of differing granulometric composition or texture, are such soils commonly found in arid regions.

The soil profile can be partly described and defined in terms of the particle size analysis and the chemical nature, including that of the organic matter. It is also necessary to recognise that soils differ in their structure, that is in the arrangement of the primary particles of sand, silt and clay. Normally the primary particles are bound together into larger structures which occur in characteristic sizes and shapes. These are the units into which the soil will break or in which it occurs naturally, and the crumbs or granules characteristic of arable land in good condition are a desirable structure. In the lower horizons of the soil the units may be closely packed and separated by thin planes of weakness, and they only appear as separate aggregates when the soil is broken. In addition to these structural units the microstructure of the units is also important, in particular the porosity, and the nature of the system of pores. Movements of air and water in soils are mainly confined to the larger pores and these, to be effective, should form a continuous system. The description of soil structure is that of the various pore dimensions and their stability. The development of structure depends mainly on the bonding power of clay and organic matter. In a sand with little clay or organic matter there are no aggregates and the structure is then simply definable in terms of grain size and packing. Obviously a decline in the bonding power of either clay or organic matter will result in a decrease in the stability of the structure.

It is obvious from the sketch of soil genesis given above that the classification of soils must be complex and difficult. The stage of general agreement or finality has yet to be reached. Those interested in the subject can read an historical account and the latest thinking of United States Conservation Service in a recent publication [1].

Soil genesis in arid regions

In arid regions the rainfall is only sufficient to wet a limited depth of soil, and the wetting is usually only seasonal or even at intervals of years. The rainfall element in weathering and soil formation is therefore of restricted importance. However, since temperatures are high and for a short season the soil-moisture conditions may be favourable, too much stress should not be placed on the slowness of weathering due to low rainfall. Further, though the regional precipitation may be low, some desert landforms are particularly liable to local concentrations of run-off, and some sites are therefore much wetter than the mean rainfall would indicate. Closed basins are found under arid conditions on both small and large scales. Such sites may indeed be permanently flooded if their catchment is large, and they often have permanently high groundwater. Their existence gives warning of one of the likely consequences of irrigation, that is, a rise of the water tables.

A second means by which soils receive much more water than is provided by the local rainfall is by the seasonal or occasional flooding of alluvial plains by large through-flowing rivers which are fed by high rainfall in humid catchments. The Nile is the classic example of this. The control of natural river flooding has been one of the great tasks of irrigation engineering throughout history, as the embanked rivers of Iraq, Pakistan and China all witness, and indeed in many such situations the concept of desert soils being the simple product of the action of an arid climate on parent material can be pressed too far.

So far as the physical weathering of rock is concerned it has been indicated that these processes can be intense in arid climates. The principal mechanisms at work are rock disintegration due to the differential expansion of their mineral constituents under conditions of extreme diurnal temperature changes, the abrading effect of wind-blown sand and the grinding of rocks in torrential desert stream flows. The high intensity of precipitation in arid regions in conditions where plant growth provides little protection leads to high run-off rates and rapid removal of weathered material, so that the weathering rock never builds up a protective weathered layer of appreciable thickness. This, with wind erosion on bare surfaces, tends to prevent the formation *in situ* of soils of much thickness in

Fig VI-2. Characteristic vegetation and surface pattern of a central Iraq desert soil. The grass is *Aeluropus lagopioides*

deserts. Conversely the accumulation sites tend to have deep deposits. Thus rocky deserts with bare rock surfaces, and deep coarse sediments in depressions, constitute a well-recognised desert assemblage.

Many of the soils we are concerned with are alluvial deposits which, while forming, receive large additions of transported and partly weathered materials. If the catchment of the river is large the material laid down on the flooded plain is likely to be variable and may of course differ from that of the local sources of soil material. Most alluvial soils are stratified, with sharply distinguished depositional horizons. Profile development may take place in such deposits and one common form of this is the formation of a horizon of calcium carbonate accumulation. Generally speaking, however, alluvial soils do not show marked effects of soil-forming processes, and this is usually a consequence of youth.

Most of the large irrigated areas of the world have been developed on alluvial soil, among these being the central valley of California, the Nile valley in Egypt and areas in Iraq, Pakistan, India, Australia and China. The reason is primarily that these are precisely the areas to which water may be led by gravity, but apart from this, alluvial areas usually have soils of great depth and the irrigated column of soil is thus usually deep enough to ensure adequate water-holding capacity, at least until it is restricted by the water table rising.

Since irrigation water usually arrives at the fields carrying a considerable load of suspended matter, irrigated areas are subject to a soil-building process as this is deposited. Calculations made for certain areas irrigated by the Indus show that this addition amounts to around 0·2 inch annually.

Other large desert areas have soils of aeolian origin. The sand dunes may be composed of transported quartz grains, as they are on the Goz soils of the central Sudan, or of carbonate grains as in some of the dunes of the central basin of Iran. In Australia there are dunes formed from the reworked and wind-sorted residues of the weathering of ancient lateritic material. On the plains of Iraq there are dunes of finely granulated soil of medium texture, and in Arizona there are gypsum dunes.

The effects of plants and micro-organisms on weathering in deserts are much reduced in comparison with their effect in humid climates, but are not negligible. The typical vegetation of deserts is sparse scrub with little leaf-fall, and what there is does not usually become incorporated with the soil. When the soil has a permanent saline crust, as is the case for example in the Sebkhas of the Sahara, no vegetation survives and such saline salt flats are uninfluenced in their development by plant action. The processes of weathering and soil formation are, however, indirectly influenced by plants through their effect on the moisture regime. The condition for leaching is that annual precipitation should exceed losses by evaporation and transpiration. The moisture content of desert soils is greatly influenced by the transpiration of the natural vegetation, whose roots are typically deep and extensive (see Chapter VIII). Phreatophytes have roots which extend into the water table, even when this is beyond the reach of the roots of any crop plant except perhaps alfalfa. Such plants transpire sufficiently large quantities of water to influence the groundwater levels, and therefore the drainage conditions. It has been estimated that some phreatophytes may transpire a quantity of water equivalent to a surface depth of 3 feet a year. Transpiration at anything approaching this rate could not be sustained except under specially favourable conditions of supply, such as along springs and water courses. It is evident therefore that even the sparse vegetation of arid regions can play a dominant part in limiting the period when the soils are moist enough for weathering to take place, and even in areas with a moderately high water intake they can ensure non-leaching conditions.

Many desert plants can accumulate large quantities of sodium in their foliage.

El Gabaly [2] working in Egypt has recorded sodium concentrations in excess of 500 milliequivalents per 100 grammes of dry material. Such large accumulations of salt in desert plants can influence the salinity of the soils in which they grow. Roberts [3] has demonstrated marked differences between the salinity under shrubs and in the barren areas between them. The deep roots of *Atriplex confertifolia* absorb sodium from the lower part of the profile and concentrate it in the surface horizons through leaf droppings, with a significant change in the salt profile. The results he gives are for only slightly saline soil, and the mechanism is probably not very important in highly saline soils.

The high concentrations of sodium found in halophytes suggests their use as a means of removing important quantities of the element from saline soils. Greene [4] explored the value of *Atriplex muelleri* for this purpose in the Sudan Gezira. He found that an irrigated saltbush crop removed 580 pounds of sodium per acre, as compared with an addition of 136 pounds in the irrigation water. The removal compared with the total sodium in the top 3 feet of the soil profile was trifling, but it does appear that the process might have value in certain circumstances, particularly if a salt-removing crop of economic value could be grown when irrigation water was freely available.

Though plants and micro-organisms do not find active expression in deserts, larger organisms such as rodents and insects are very active, and must indeed be adapted for a soil habitat. The consequence is that the desert soils commonly show macroborings and cavities due to the activities of such organisms. Such borings often fill with surface soil, and where this is darker than the subsoil, channels of surface material known by the Russian word *crotovinas* are formed. The surface soil may differ in texture as well as colour from the subsoil, and there is here a mechanism which influences profile development. The infilling of the deep cracks which form during the dry season in some soils has, however, a more important soil circulating effect.

Earthworms are not a characteristic member of the fauna of desert soils; termites and ants are, however, common, and their channels have an important influence on the aeration and drainage of the soil. The harvesting species of ants play a part in compounding plant material with the soil, sometimes to an extent which seriously diminishes the erosion resistance of the vegetation cover. Harvesting ants can be sufficiently active to cause easily observable redistribution of newly sown cereal seeds, with a great concentration of seedlings growing round their mounds. Termites even when they do not form mounds are active in circulating soil and in transporting organic matter to their underground nests and fungal gardens.

It is evident from this discussion that though biological factors are not so important in deserts in weathering and soil formation as they are in humid climates, the plants do have important effects on soil genesis and particularly upon moisture conditions.

The presence of soluble salts in desert soils is an inevitable consequence of the failure of the precipitation to exceed evapotranspiration. It is mainly the salts of sodium which make up the salts; calcium, and to a lesser extent magnesium, accumulate as the insoluble carbonates, or in the case of calcium as the slightly soluble hydrated sulphate, gypsum ($CaSO_4.2H_2O$). Whatever the parent material, calcium carbonate is usually found somewhere in the profile, often but not always in a visible horizon of accumulation, where it may appear as nodules

Fig VI-3. A saline desert soil in Iraq with a groundwater at 140 cm. The surface soil is dry and very hard, and is lighter in colour than the wet soil below

or concretions, either hard or soft, or it may appear as thick continuous cemented masses. Where climatic or vegetational changes have occurred which have allowed erosion to be active, horizons of accumulation may be exposed, forming one type of desert crust. Such formations have been recorded in many deserts, including North Africa, Palestine, western USA and Australia. They are usually exposed horizons of accumulation, but it is possible that they could form directly on the surface in certain situations, as crusts of soluble salts do.

The presence of thick carbonate layers in a soil is a serious restriction on its value for irrigation development, in that it restricts the available depth of soil, and if it amounts to an impermeable layer, has serious consequences on the drainage of the soil. Similarly soils laid down or formed on gypsum beds present special problems. Such soils are found in Iraq, USSR and Spain; they have been recently discussed by Smith and Robertson [5].

The soils of deserts are not the only ones in which calcium carbonate accumulation in a distinct horizon takes place. The black chernozem grassland soils of Russia, and their American equivalents, have horizons of visible carbonate accumulations. These soils merge with increasing dryness into more arid desert groups with more intense carbonate accumulations nearer to the surface.

This discussion shows that desert soils may be expected to be very variable in depth, texture and composition, and in their moisture and drainage properties. Except where dominated by parent material of a purely siliceous nature they may be expected to contain soluble salts, to have an alkaline reaction and calcium carbonate accumulations. Weathering of the parent material is limited in degree and never seems to proceed to the final residues of the process which are formed in the humid tropics.

Salinity and alkalinity

The more spectacular manifestations of salinity usually arise when a saline groundwater approaches the surface, and this condition is often the consequence of irrigation. The extent of salt additions due to irrigation, apart from the consequences of raising the water table, can be illustrated by considering a model area irrigated for 100 years with water containing 300 parts per million (ppm) of soluble salts, the annual application being 4 feet of water. In this case irrigation would add 2·7 per cent of salt to the upper 3 feet of soil. It might well be that the precipitation in the soil of insoluble salts would reduce the addition to half this figure, but long-continued irrigation with almost any irrigation water will clearly lead to salinisation if there is no removal. Such additions are sufficient to account for much of the salinity of ancient irrigated areas without it being necessary to postulate pre-existing accumulations or weathering within the profile.

It is of some interest to examine the amount of salt which may be added to soil in rainwater. There are good reasons for believing that the origin of salt in rain is from the sea. Drischel [6] in Europe showed that the chloride content of rainwater decreased rapidly away from the coast. Drischel's work gives 6 to 13 mg of chloride per litre at a distance of 5 to 10 km from the coast. This is equivalent to an annual addition of 0·025 per cent of sodium chloride to the upper 10 cm of soil with a rainfall of 370 mm. Even with much smaller additions much further inland cyclic salt can be an important factor in soil salinity. Australian soils are in general not exceptionally saline, and against this background the addition of cyclic salt is considered to play an important part in the general salt balance. Aerial movement of salts also takes place when winds cross salt-encrusted desert surfaces. Holland and Christie [7] calculated that some 100,000 tons of salt were transported annually from the salty exposures of the Rann of Cutch near the Indus delta to Rajputana.

Generally, in the absence of a high water table, salt distributes itself in the soil with increasing quantities with depth, and the higher the rainfall the deeper and more pronounced this effect becomes. It is often obscured in very saline soils, and profiles which are extremely salty from the surface to extreme depth are common.

The salts usually found in soils are the sulphates and chlorides of sodium, magnesium and calcium. In isolated areas considerable quantities of nitrates are found and it occasionally happens in peasant agriculture that selected salt deposits are regarded with favour as a source of fertiliser. The concentration of calcium is limited in the presence of sulphates by the crystallisation of gypsum, and magnesium is then often present in greater quantities than calcium in the soil water. In the presence of these salts the reaction of the soil does not exceed pH 8·5, but if carbonates are present the soil may be very alkaline with pH as high as 10·0. Such very alkaline solutions cause the organic matter of the soil to take on a black colour, giving soils originally called 'black alkali'. High alkalinity is not always associated with the darkening of the organic matter, and it is not characteristically associated with saline soils, but rather with soils which have been leached of the bulk of their salts.

The soluble salts of the soil do not constitute the total content of metallic ions. The surface of clay minerals is the site of loosely held ions such as sodium and calcium, and these are in mobile equilibrium with the ions in the surrounding soil solution. The composition of the exchangeable cations in arid soils differs markedly from that in soils of humid areas. Sodium usually constitutes a much larger percentage and hydrogen is not normally recognised at all. The presence of a large proportion of sodium ions has damaging effects upon the physical properties of the clay: it reduces the stability of the clay particles so that they disperse into very fine particles, the favourable grain structure is destroyed and the permeability of the soil can be reduced to very low values.

If the soil is saline the evil effects of the large percentage of exchangeable sodium is masked by the flocculating effect of the high concentrations of salt, and the decrease in permeability only becomes serious when the salinity is reduced below that which causes flocculation. Thus if by irrigation or rainfall the salinity of the soil is reduced while the exchange complex of the clay contains 15–20 per cent of sodium, dispersion of the clay, loss of structure and permeability are to be expected. Since the exchangeable cations are in equilibrium with the soil solution, the proportions of, say, calcium and sodium depend on the nature of the salts present, and, in the later stages of leaching, on the composition of the leaching water. The equilibrium is favourably influenced by additions of calcium salts to the soil or water, gypsum, which is slightly soluble, being commonly used for this purpose. Many soils contain large quantities of the practically insoluble calcium carbonate, which is, however, able to supply soluble calcium in small quantities by the formation of the much more soluble bicarbonate, in the presence of carbon dioxide respired by crop roots. Carbonate soils are therefore considerably easier to leach without forming impermeable sodium-influenced clays than are other soils.

Russian scientists observed that if the leaching process occurs with the formation of a clay containing a large proportion of exchangeable sodium, the saline

Fig VI-4. A 50-cm-long prism of soil extracted unbroken from a pit. This illustrates the extraordinary hardness associated with structural deterioration

(*solonchak*) soil becomes a very alkaline non-saline soil called a *solonetz*. The process is characterised in its extreme form by the leaching not only of salts but of some of the dispersed clay, so that the clay content of the profile shows a zone of accumulation. When dried out this horizon cracks and on occasions forms extraordinarily well-developed and regular columns with rounded tops with dimensions of the order of 5 to 12 cm. Soils with similar morphology occur in western Canada and the United States, where a high proportion of magnesium and not of sodium occurs. The upper part of a solonetz may be neutral or even acid, perhaps because the soil before leaching did not contain a high proportion of sodium salts. It is the solonetz soil which is likely to present difficulties in irrigation development, and it should be recognised that solonetz soils which are also saline exist. Some of the difficulty with such soils is due to the compacted clay horizons. The exchangeable cation characteristics of a solonetz are reversible, but the movement of clay is not reversible on resalinisation. It is now recognised that not all horizons of clay accumulation are formed by the mechanism discussed above. The process of weathering within the profile in a horizon which stays wet for a long time under the high temperatures of arid regions may develop *in situ* a higher clay content than the drier horizons above, and since the increasing fineness of texture tends to maintain moist conditions the process may be self-accelerating.

The two principal limitations on the irrigated use of the soils of arid regions are salinity and drainage difficulties. The former is a soil property, the latter depends on topography and geology as well as upon the properties of the soil. Drainage forms the subject of a later section of this chapter.

The best method of expressing soil salinity is by the electrical conductivity of an extract of the saturated soil paste. Since the moisture content of such a paste is directly related to moisture contents found in the field, this method takes account of the differences in field moisture content which arise from textural differences between soils. A salinity scale based on electrical conductivity of the saturated extract, which relates salinity levels to their effect on crop growth, has been drawn up by the United States Department of Agriculture [8].

Electrical conductivity of the extract, mmhos/cm at 25°C	Effect on crops
0–2	Mostly negligible
2–4	Yields of very sensitive crops may be restricted
4–8	Yields of many crops restricted
8–16	Only tolerant crops yield satisfactorily
>16	Only a few very tolerant crops yield satisfactorily

In applying these limits to the selection of land for irrigation it must be recognised that if the irrigation has any measure of success a favourable redistribution of salts will take place. Furthermore it must be noted that acute difficulties with germination occur at salinity levels which are satisfactory for the established crop.

A soil survey provides salt profiles to a considerable depth, say 6 feet, and though salinity in the upper soil is more disabling than salinity at depth, classification should not be based solely on the upper salinity, since salts in the lower profile present a potential menace. Drainage conditions are not likely to be uniformly perfect over the whole area. There is always the possibility of local uprise of the deep salts towards the surface. If the situation is one where success depends entirely upon removal of salts, the total salt content of the profile to at least 2·5 metres should enter into the judgement of the profiles and the area. This considerable depth is chosen because, although the upper 1·5 metres may

Fig VI-5. A stratified, rather coarse-textured, desert soil of good quality. This pit is 150 cm deep

suffice for the assessment of salinity in relation to crop growth, a greater depth will eventually be desalinised and the time to do this will depend on the total salt content to a considerable depth. The safe disposal of drainage water may present a serious difficulty and the total quantity of salt to be removed has a direct bearing on this.

Reclamation of saline soils

Over a wide range of soils and climate there are many examples of successful leaching and use of saline soils. These successes have all been in situations with either a deep water table, unusually effective natural drainage or effective installed drains.

Reeve and others [9] had no difficulty in removing the bulk of the salt from highly saline silty clay loam soil in the Coachella Valley, California. They found that the passage through the profile of 2 feet of water for each foot of soil depth reduced the salt content to less than 20 per cent of its original value. This was done using Colorado River water which has a high salt content corresponding to an electrical conductivity of 1·0 millimho per cm.

Evans [10] found that leaching with a 6-foot column of water was effective in reclaiming a 4-foot column of soil in the upper Colorado River basin, and recommended an additional application of water every fourth year for leaching purposes. The land in question was suffering from salinity and high groundwater after 50–70 years of irrigation. In a typical area a heterogeneous stratified soil lies over a very good gravel-cobble aquifer in which tubewell pumping was effective in controlling the water table. Tubewells are slotted or perforated tubes of small diameter sunk into the lower aquifers, sometimes surrounded by a fine gravel layer. The water extracted is replaced by water from the upper part of the aquifer within the range of the crop roots, which is responsible for the water-logged condition of the soil. The drainage water is disposed of in shallow surface drains if it is saline, or re-used for irrigation if of acceptable quality. The soil in the experimental area was extremely saline throughout the upper 5 feet, a leaching with 2 feet of water was sufficient to reduce the electrical conductivity of the saturation extract from 18 mmhos per cm in the upper 4 feet of the profile to 9·5 mmhos per cm, and this allowed a crop of alfalfa to be established. Subsequent seasonal cropping was accompanied by further reduction in the salinity.

In leaching experiments at Dujailah near Kut in Iraq [11] leaching of salty silty clay and silty loam soils was achieved without difficulties. The passage through the soil of 3 feet of water reduced the electrical conductivity of the saturated extract in the upper foot of soil to 4 per cent of its initial value, that of the upper 2 feet to 5 per cent and that of the upper 3 feet to about 10 per cent. The Dujailah soils contained small quantities of gypsum and very large amounts of finely divided calcium carbonate. Before leaching the conductivity of the saturated extract exceeded 35 mmhos per cm everywhere in the profile, and was much higher than this in the upper foot of soil. These experiments show that in some cases even very saline soils can be freed rather easily of the bulk of their salinity by the passage of limited quantities of water.

There are other soils in which the process is slow, primarily because of slow passage of water. Such conditions may be due to a horizon of high density and very low permeability. This restricting layer may pre-exist in the profile, or it may be formed near the surface by the formation of a thin leached layer of soil with a high proportion of exchangeable sodium which prevents entry of water. When this happens it is usually necessary to use a soil amendment which increases the concentration of soluble calcium in the soil moisture, which reduces the proportion of exchangeable sodium and, by increasing the rate of entry of water, allows leaching to proceed. Where leaching has succeeded but has left a soil with a high ratio of exchangeable sodium to exchangeable calcium, it may still be advantageous to apply amendments with the object of improving the soil structure. The end product of removing soil salts depends on the quality of the leaching water, the quality being measured by the ratio of calcium to sodium in the water rather than by the absolute concentration of salts, for at this stage the composition of the exchangeable cations is being determined by the composition of the water. In the early stages of leaching the composition of the soil water is determined by that of the soluble salts, and prediction of the course of leaching is helped by a knowledge of this.

Soil amendments include gypsum, sulphur, sulphuric acid and ferrous sulphate. The first of these is a slightly soluble calcium salt, the others act as sulphuric acid and are most effective if the soil contains calcium carbonate. An example of the effect of soil amendments on the percentage of exchangeable sodium in the soil under irrigation treatment is found in the work of Overstreet, Martin and King [12] on a nearly sodium-saturated soil near Kerman in California. A compacted zone of poor permeability was first broken by a deep chiselling cultivation, the amendments were applied and the following month a pasture mixture was sown, and yields were taken at various dates during the next two years. Table VI–I below[1] shows the effect of the treatments on the exchangeable sodium content, and on yields.

TABLE VI – I *Yield of pasture plants and exchangeable sodium in the soil for the reclamation plots at Kerman, California, 1948–50*

Treatment	Exchangeable sodium		Mean yield	Fresh weight	
	Original, m.e./100 g	22/7/49, m.e./100 g	10/8/49, t./a.	3/11/49, t./a.	17/8/50, t./a.
No treatment	10·4	7·7	1·47	1·06	0·83
Sulphur, 1·86 tons/acre	8·6	8·4	1·71	1·17	0·77
Gypsum, 10 tons/acre	9·1	2·0	2·46	2·46	1·18
Sulphuric acid, 5·7 tons/acre	8·7	3·6	3·47	1·97	2·21
LSD at 5% level			0·70	0·48	0·94

The first irrigation was in January 1949, and subsequently irrigation was at weekly intervals. For about two years the yields were better with sulphuric acid than with the other treatments. Sulphur appeared to be largely ineffective compared with the gypsum and acid, but it should be noticed that irrigation alone caused a considerable improvement in the exchangeable sodium percentage, and visual comparisons at the time of the third cutting showed a general improvement in growth on all plots, including the untreated ones, over that of the first year.

[1] This table is as rearranged by Thorne and Peterson [13].

In contrast to these examples of relatively easy reclamation the experience of Snyder and others [14] is a warning that some soils defy for many years all rational attempts to reclaim them. These workers were studying alkaline bottom lands in Idaho, and the more impermeable soils after seven years' treatment were still only a little improved.

It is not always possible to predict from the chemical and physical properties of soils whether they will reclaim rapidly or not. It is therefore desirable that leaching trials should be performed in advance of development in doubtful cases. This may be expensive, but is justified if it resolves doubts as to the behaviour of the soil under irrigation, for though most soils can be brought into use, the need for a prolonged period of reclamation may have a large effect on the economics of development.

The fertility of desert soils

It is clear from the foregoing discussions that salinity is likely to be a major restriction on the fertility of soils of arid regions. This is controlled and ultimately removed in irrigated areas with effective drainage installations, provided that the irrigation water is itself of low salinity. If the available water is not of the best quality, effective irrigation can still be practised, but a large surplus over transpiration needs to be applied and the capacity of the drainage system increased over that required with better water. Given suitable soils and adequate drainage, water containing up to 4,000 ppm of soluble salts can be used, but a degree of loss of yield potential has to be accepted.

Assuming control of salinity, and irrigation water of a quality which will prevent the development of alkaline soils with a high percentage of exchangeable sodium (which requires a high ratio of soluble calcium and magnesium to sodium), crop production in arid regions can be very high. A basic reason for this is that such regions have in general favourable climates with long growing seasons. The soils chosen for irrigation are usually deep alluvial deposits which, with control of the water table, provide ample depth of soil for a wide range of crops. The productivity of irrigated soils is shown clearly by their contribution to world production. About 13 per cent of the arable lands of the world are irrigated but about 25 per cent of the total food is produced on them. The climate, depth of soil and adequate water supplies provide the physical backgrounds to high fertility. Irrigated soils, like others, require the application of adequate supplies of plant nutrients. Nearly all of them require applications of nitrogenous fertilisers and many respond to phosphate additions. Potassium is a less usual requirement, a fact related to the potassium reserves associated with an unleached soil remaining adequate for a long time under irrigation. The same is in general true of phosphate, but here availability is restricted. In general soils of desert regions have an alkaline reaction and contain large quantities of calcium carbonate and limited amounts of organic matter. These conditions have repercussions on the availability of some nutrient elements. In the soils of temperate regions a considerable proportion of the soil's phosphate is contained in organic matter and remains available to plants. Where temperatures are high it is not possible to maintain a high content of organic matter, and this reserve of phosphate is not available, which makes the problems of phosphate fixation in the soil itself more acute.

The elements boron, iron, manganese and zinc are all less readily available to plants in alkaline and calcareous soils than they are in normal soils. Deficiency of such elements is common in irrigated soils of arid regions. Boron occupies a

Fig VI-6. Irrigated land in Pakistan, abandoned because of waterlogging and salinity, mainly in this case due to canal seepage

special place among the nutrients which are required in small quantities, for the limits of deficiency are very close to those of toxicity and both conditions are known. Excessive concentrations of soluble boron are often found in saline soils, and the excess boron is much more difficult to remove by leaching than the ordinary soluble salts.

In this discussion on the fertility of desert soils the fact that in many cases they have been developed under conditions of restricted leaching has been emphasised. It must, however, be recognised that desert soils developed on material poor in plant food may exhibit deficiencies unconnected with high alkalinity. The fact that irrigated soils usually enjoy climatic conditions favourable to crop growth, and that moisture deficiency is eliminated, places added weight on the importance of providing the nutritional requirements appropriate to the high potential yields.

The drainage of irrigated soils

As is indicated in some earlier parts of this chapter and discussed in more detail in the chapter on Irrigation, a rising water table usually follows the installation of an irrigation scheme. The rise may rapidly reach a dangerous level even where the original water table was very deep. In the Kerang irrigation district in Victoria, Australia, the groundwater prior to irrigation about 1880 stood 8 metres below the surface; by 1900 it had risen 6 metres, and in 1957 it was everywhere within 1·8 metres of the surface, and under active irrigation higher than this [15].

Fig VI-7. A poor irrigated sorghum field in Pakistan. The near crop failure is due to salinity, possibly accentuated by insufficiently heavy irrigation

The United States provides many notable examples of the effect of irrigation on groundwater levels, as do both the Punjab and Sind in West Pakistan. Except near an incised river or water course, natural regional drainage is rarely good enough to remove irrigation surpluses unless the sub-stratum is of coarse sand or gravel and there is a considerable groundwater gradient. The narrow strip of land along the Nile which has continued fertile since antiquity is an example of the rare category of lands with adequate natural drainage, for there is natural drainage to the Nile at low river. The natural drainage to the river is illustrated by Fig. XII-4 for a site which is actually just outside Egypt. However, during the last century important changes have been made in Egyptian irrigation. The seasonal basin irrigation has been largely replaced by perennial irrigation, and the barrages which make this possible change the river regime, largely eliminating the low water stage at which natural drainage is possible.

The high transpiration rate of some desert plants has an important effect on drainage requirements. This removal of water by natural vegetation is of decreased importance in intensive irrigation, where only a small proportion of the land carries natural vegetation, thus accentuating the increased surpluses which arise from greater applications of water. The magnitude of removals of water by evaporation and transpiration from uncropped areas can be illustrated by calculations which have shown that in some irrigated areas in Pakistan 50 per cent of the applied water is removed by these means. It is evident that these abstractions of water from the area have a very important bearing on the required capacity of

Fig VI-8. A soil pit in Pakistan; the soil is sandy and very saline. The depositional horizons are not horizontal. The striking whiteness of the upper layer is due to the crystallisation of salt during the drying out of the profile

installed drainage works. It will be noted that both evaporation and transpiration from non-cropped land depend upon the depth to groundwater.

Some theoretical and experimental studies of water loss from a water table by evaporation from the surface have been made by Gardner and Fireman [16]. Their results show that under conditions of extreme aridity 1 mm of water a day could be removed from groundwater standing 200 cm from the surface, and as much as 0·5 mm a day when it was 400 cm below the surface. Figures comparable to these have been measured in tank experiments in Pakistan. Altovsky and Konoplyantzeva [17] in Hungary recorded losses of 1·2 mm a day from a depth of 240 cm. Such figures show that evaporation from a water table can be an important element in the water balance of arid regions. Such removals, combined with those by transpiration, will normally cause a rapid lowering of the groundwater during the dry season to depths from which the losses become negligible.

However, neither natural drainage nor evapotranspiration removals can be relied upon to control the groundwater at an acceptable level, and most considerable irrigation schemes require some degree of drainage by artificial means. Any investigation to determine the feasibility of irrigation on desert soils must therefore include an assessment of the feasibility of drainage.

There are two main methods of draining agricultural lands, open or tile drains and pumping from wells. The design of either system depends on the drainage characteristics of the soil, though more critically with the shallow systems than with wells, for which the deeper aquifer constants are decisive. Apart from the quantity of water to be drained and the depth at which the groundwater is to be maintained, the intensity or spacing of shallow drains varies approximately directly with the permeability of the soil above and for some distance below the drains. Methods of measuring this quantity are discussed by Reeve, Luthin and Donnan, in Luthin's *Drainage of Agricultural Lands* [18]. It is possible when the permeability of the soil is known to make preliminary estimates of the intensity of the drainage network which is likely to be needed. The other information required is the drainage discharge. This requires a knowledge of the water which will be applied, the cropping pattern and a preliminary knowledge of the design of the irrigation scheme. The internal drainage characteristics of the soil may be satisfactory but the disposal may involve engineering difficulties, as for instance in closed basins. It is evident that the planning of an irrigation scheme involves close co-operation of soil scientists, engineers and agriculturalists.

Important advances have been made in the last twenty years in the study of drain spacings in their relationship with discharge and the permeability of the soil. A recent graphical solution of the problem is given by Toksoz and Kirkham [19]. The following results may be given to illustrate the intensity of the drainage network required in selected circumstances. The maximum height of the water table above the drain is taken as 0·6 metre and the discharge as 0·20 litre per second per hectare, or 1·73 mm depth of water per day. The permeability is taken as 1·20 metres per day corresponding to that of a medium textured soil. Then, if the soil is homogeneous to great depth, drains are required 185 metres apart. If the water in the drains is 2 metres below the surface the water table is 1·4 metres below the surface, midway between parallel drains. If there is a substantially impermeable layer 8 metres below the surface the required spacing contracts to 118 metres. If this layer is 3 metres below the drain the spacing becomes 42 metres, and an impermeable layer at 1 metre below the drain requires a spacing of only 7 metres. The question of spacing has been illustrated at some length in order to emphasise that drainage works are likely to be required at intensities comparable with the irrigation channels, and that some soils are virtually undrainable at any economic spacing.

The minimum value of the drain discharge is determined by the leaching factor, or the proportion of the applied water which must pass through the soil column to maintain a satisfactory level of salinity in the drainage water. A reasonable figure for the electrical conductivity of drainage water is 5 millimho per cm. If that of the applied water is 0·5 millimho per cm, 10 per cent of the water applied to the fields must pass to the water table. This is minimum value of the drain discharge, because there are always involuntary excesses of water applied, and both field channels and larger canals add water by seepage to the ground-water. As has been indicated above, evapotranspiration removals reduce the necessary drain discharge. Channel seepage is influenced by permeability, so that in any practical scheme high permeability may lead to increased seepage tending to offset the wider spacings required when the soil permeability is high.

The soil properties, salinity, permeability, depth and texture, which influence the value of a soil for irrigation use must be integrated with engineering and agricultural factors to yield an estimate of the real value of the land for irrigation. The methods of doing this to yield a land classification are discussed in the chapter on Irrigation.

T. N. JEWITT

References

[1] UNITED STATES DEPARTMENT OF AGRICULTURE, *Soil Classification: A Comprehensive System*, 7th Approximation, Washington, 1960.

[2] EL GABALY, M. M., *Studies on Salt Tolerance and Specific Ion Effects on Plants*, Unesco, 1961, p. 169 (Arid Zone Research XIV).

[3] ROBERTS, RAY C., 'Chemical effects of salt-tolerant shrubs on soils', *Transactions Fourth International Congress of Soil Science*, Vol. 1, Amsterdam, 1950, p. 404.

[4] GREENE, H., 'Soil improvement in the Sudan Gezira', *Journ. Agric. Science*, Vol. 39, 1939, p. 1.

[5] SMITH, R. and V. C. ROBERTSON, 'Soil and irrigation classification of shallow soils overlying gypsum beds, northern Iraq', *Journ. Soil Science*, Vol. 13, 1962, p. 106.

[6] DRISCHEL, H., quoted by G. E. HUTCHINSON, *A Treatise on Limnology*, Vol. 1, New York, 1940, p. 544.

[7] HOLLAND, T. H. and W. A. K. CHRISTIE, 'The origin of the salt deposits of Rajputana', *Records Geol. Survey India*, Vol. 38, 1909, p. 154.

[8] UNITED STATES SALINITY LABORATORY STAFF, *Diagnosis and Improvement of Saline and Alkali Soils*, Agricultural Handbook No. 60, United States Department of Agriculture.

[9] REEVE, R. C., A. F. PILLSBURY and L. V. WILCOX, 'Reclamation of a saline and high boron soil in the Coachella Valley of California', *Hilgardia*, Vol. 24, 1955, p. 69.

[10] EVANS, N. A., 'Reclamation and drainage of saline–sodic soils in the Upper Colorado river basin', *Fourth Congress of Irrigation and Drainage*, Vol. 5, 1960, pp. 13, 113.

[11] GOVERNMENT OF IRAQ, *Reports on Dujailah Drainage Experiments*, First Technical Section, Ministry of Development, Baghdad, Iraq, 1957.

[12] OVERSTREET, R., J. C. MARTIN and H. M. KING, 'Gypsum, sulphur, and sulphuric acid for reclaiming an alkali soil of the Fresno series', *Hilgardia*, Vol. 21, 1951, p. 113.

[13] THORNE, D. W. and H. B. PETERSON, *Irrigated Soils*, New York (2nd ed., 1954).

[14] SNYDER, R. S., M. R. KULP, G. O. BADER and J. C. MARR, 'Alkali reclamation investigations', *Idaho Agr. Exp. Stat. Bull.*, 1940, 233.

[15] WEBSTER, A., 'Drainage in the riveraine plain of northern Victoria with special reference to groundwater hydrology', *Third Congress of Irrigation and Drainage*, Vol. 5, 1957, pp. 10, 33.

[16] GARDNER, W. R. and M. FIREMAN, 'Laboratory studies of evaporation from soil columns in the presence of a water table', *Soil Science*, Vol. 85, 1958, pp. 244–9.

[17] Quoted by G. Y. KOVAKS, *Internat. Comm. on Drainage and Irrigation*, Ann. Bull., 1959, p. 70.

[18] LUTHIN, J. N., *Drainage of Agricultural Lands*, American Society of Agronomy, Madison, Wis., 1957.

[19] TOKSOZ, S. and D. KIRKHAM, 'Graphical solution and interpretation of a new drain-spacing formula', *Journ. Geophysical Research*, Vol. 66, 1961.

CHAPTER VII

Deserts in the Past

The findings of historical geology are very clear that during the last 1,000 million years, for which good stratigraphical evidence is available, world climates have undergone several periods of great change, some of long duration but others much shorter. At the two extremes we have on the one hand the ice ages, when areas of continental dimensions were covered with ice sheets and mountain glaciers were greatly enlarged, and on the other, periods when tropical climates extended as far as 50–60° north and south of the equator and the polar regions were not ice-covered as they now are. In this changing scene the deserts of today were at times well-watered lands, and again, regions at present well watered were once deserts. No account of the existing arid areas can omit reference to the relatively recent climatic changes of the Quaternary era, which include several ice ages and interglacial periods, the effects of which on life and on geomorphology and soils in arid lands, as elsewhere, are still with us; but even the older events of the Tertiary era are also of direct significance, since fossil soils of Tertiary age are still preserved in the present land surface, especially in arid regions, and many major geological and geomorphological features were initiated during the Tertiary. In interpreting the record of the rocks as to climates during geological time, the conditions of present deserts afford an analogue for aridity. Reference will be made to this in later paragraphs, but the complexity of the topic is best seen by reference to the inferred changes that have affected existing arid areas, and evidence for the extension of contraction of the arid areas during recent times.

Criteria of climatic change in the recent past

LACUSTRINE CONDITIONS

Playas and salinas are characteristic of desert basins. Despite their being commonly termed 'lakes' they are dry except after the rare periods of heavy rains, but there is unequivocal evidence in the form of old beaches and other shore-line features, and of bottom deposits, that they were once, in reality, lakes. Perhaps the best known of these is Lake Bonneville in the USA [1], which in Pleistocene times extended over a maximum area of almost 20,000 sq. miles but which now is represented by the Great Salt Lake, a shrunken remnant varying in area between 1,500 and 2,200 sq. miles. Also in the Western Cordillera of the USA all that remains of Lake Lahonton, a vast Pleistocene lake, are several small lakes such as Pyramid Lake [2]. There are numerous playas of economic importance in the lower, western part of the Atacama desert basin, and here also there are indications that in times past the lake and drainage system was more extensive. In the Gobi Desert Berkey and Morris [3] noted at least seven old beach lines associated with a former lake, one shore line 29 feet higher than the 1922 level of the modern, shrunken lake. In Australia likewise there is evidence that the interior salt lakes were formerly more extensive and much deeper (see p. 65).

RIVER TERRACES

River terraces have been much used in the interpretation of climatic events. A river terrace may be arbitrarily defined as a former valley floor into which the river has incised and which is beyond the reach of present flooding. The incision in question may be caused directly by climatic change through variation in the discharge/load relationship of the stream, or indirectly through eustatic shifts of sea-level. It may, on the other hand, be initiated by tectonism. Interpretation is often complex because both climatic and eustatic factors operate simultaneously, and because climatic events in the headwater reaches of a river like the Nile may have far-reaching effects downstream. Despite these and other difficulties terraces have been widely used in the synthesis of climatic history, especially in Europe and Africa (see pp. 134–5).

PALYNOLOGY

Although most often applied in more humid environments the techniques of palynology (pollen-analysis) are applicable to the arid zone, as Pons and Quézel have shown in the Sahara [4]. Pollen is blown by the wind and settles on land as well as on water surfaces; in the first case the pollen will most likely decay, but in the second it will sink to the bottom of the lake or swamp and be incorporated in a sedimentary layer. By comparison with modern climates and associated vegetation complexes, the climates of the times can be inferred (see pp. 136–7), and climatic changes are reflected in the pollen spectra of successive sedimentary layers.

SEDIMENTS AND FOSSIL SOILS

A study of sediments and of soils may also suggest climatic change. They may be found in areas where they could not have formed under existing conditions, and are out of harmony with the modern environment. For example, there are lacustrine deposits in arid regions indicative of greater humidity at the time of their formation. In Mauretania, dunes dating from the end of the Flandrian transgression are covered with a red soil which could not have formed under desert conditions and which is at present suffering erosion [5].

DUNE FORMS

The occurrence of dunes beyond the confines of the modern deserts and in areas where (apart from coasts) dunes cannot now be supported is clear evidence of former aridity. Such dunes are now usually vegetated and fixed. In southern Australia, fixed longitudinal dunes cover great areas of Eyre Peninsula and the

Fig VII-I. Ancient longitudinal dunes in the Mallee country, north-western Victoria, now under cultivation. (*RAAF photograph, reproduced with permission*)

'Mallee' country, areas noted today for wheat and sheep production (Fig VII-I). They occur also even in the eastern suburbs of Melbourne where the climate is now cool-temperate with a 25-inch rainfall. There are dunes now fixed by vegetation in the western Great Plains of the USA, also in the Sudan and West Africa, far south of the present Sahara Desert.

DENDROCHRONOLOGY

Analysis of tree rings may be used to detect climatic changes during the very recent past. In regions with regular seasonal changes of weather trees generally produce one growth ring per annum. The thickness of growth rings varies with the age of the tree, those of youth being thicker than those of old age. Super-imposed on this normal variation, however, there are variations imposed by differences in climate from year to year. Just which climatic factor is reflected in the varied thicknesses of growth rings is a matter of judgement, but in areas where rainfall is the critical factor influencing growth, a ring thicker than usual indicates above-average rainfall, whereas in the higher latitude and altitude areas peripheral to the tundra regions such thick rings probably indicate high summer temperatures. Very long climatic records, extending several hundred years back in time, have been obtained from some of the large trees of the American west, but it has been shown that in many instances their evidence is suspect and the contribution of dendrochronology has perhaps been fairly summarised as a valuable but not infallible guide.

ARCHAEOLOGY

Another line of evidence derives from anthropology and archaeology. Briefly, there is now abundant evidence that considerable populations once lived where life is now insupportable. Furthermore these peoples kept, hunted or had intimate knowledge of animals which we now associate with much more humid environments. Perhaps the most spectacular example of such a population that has come to light has been described by Lhote from the central Sahara [6], where frescoes at Tassili and elsewhere afford clear evidence of the former presence of animal life characteristic of much more humid climes (see p. 136).

Studies of written records, and of the ancient wells, water works, and other evidences of the climatic regimes of historical periods, yield valuable data, and all the evidence from the various lines of investigation discussed above may be synthesised to give an overall picture, which is unfortunately far from precisely accurate, of the succession of immediate past climates. Methods of dating have been improved, particularly by the study of radio-carbon isotopes in carbon-bearing deposits.

Of considerable interest is the question of the effects of quite small fluctuations in climate, such as might affect existing settlements either adversely or favour-ably, and the probability that such fluctuations may be cyclic.

Effects of small climatic changes

In order to appreciate the full significance of even minor changes of climate in the arid zone, one must realise that the vital natural resources of surface water and pasture vary greatly from place to place. The Sahara is a good example to illustrate this.

Fig VII-2. Saharan rock painting at Sefar, Tassili n'Ajjer, Algeria. (*Photo J.-D. Lajoux*)

Several latitudinal zones mark the transition from humid through sub-humid and semi-arid to arid on both the northern and southern peripheries of this desert subcontinent. There are no very sharp lines of delimitation, but in a very general way some of the areas with 10–15 inches of rainfall per annum will still support a lush grassland vegetation and permit agriculture without irrigation. Drier areas yet, with only 5–10 inches of rainfall, will support no more than a sparse desert grassland, which may still, however, form a suitable environment for nomadic herders during all or much of the year. Even more arid regions with 2–5 inches of rainfall may provide some grazing when occasional rains serve to revive the plant cover. Finally the great dry interiors with one inch, a half inch or even less, support no plant life whatsoever.

During periods of climatic amelioration the marginal belts of this type will both expand and shift towards the interior of the desert, thus reducing the size, although never eliminating that core region. Conversely, periods of climatic deterioration will lead to a contraction and outward migration of the marginal zones, corresponding to an expansion of the barren interiors.

Climatic changes do not only produce latitudinal shifts of different ecological belts. Equally significant is the diverse pattern of minute but extremely important changes recorded in the highlands and mountains common to most of the world's dry regions. These areas of greater elevation enjoy a higher local rainfall even today, and during periods of moister climate, rainfall values there would increase appreciably, so creating large islands of more attractive country.

In the Sahara Desert, periods of greater rainfall noticeably favoured the various highland zones, and the resulting ecological effects were disproportionately large. Rains falling in the high country were channelled into wadis which frequently discharged on to the neighbouring lowland plains. But before such occasional waters finally evaporated into the subsoil, they provided comparatively bountiful conditions along the length of the valley. Sporadic floodwaters thus encouraged higher plant life, while the persisting soil moisture supported herbaceous vegetation for many months or even years. Furthermore, shallow wells dug into the valley bed frequently provided water for man and beast long after the last rains. Finally, percolating waters in the higher country provided springs at lower elevations.

Corroboration of this view comes also from the American south-west, where it has been calculated that the physiographically substantial changes in the Lake Bonneville and Lake Lahontan basins were a reflection of climatic changes which in absolute terms were of no great magnitude. In fact it is suggested that an increase in rainfall from the present 10 to about 18 inches per annum and a temperature decrease of about 5°C would suffice.

The pluvial periods of the Quaternary era

Pluvial periods have been defined as phases of widespread, long-term rainfall increase of sufficient duration and intensity to be of geological significance.

Evidence of such moist intervals has been particularly well studied in western Asia and on the African continent, although they are of course known elsewhere, but because the sequence of events is known with a measure of precision in these Old World areas this statement refers specifically to the areas mentioned. The pluvial periods recognised to date have affected not only the great desert areas and their margins, but also wide expanses of semi-arid grasslands stretching from the equator to mid-latitudes.

It is known, for example, that about five million years or so ago, shifting desert sands reached from the Cape Province to the Congo River, at a time when man's sub-human ancestors were evolving on the African continent. On later occasions, the present Kalahari dry country practically disappeared [7, 8]. In East Africa, moist periods, on occasions, supported lakes in regions where even drinking water is unavailable today. The famous Olduvai Gorge of Tanzania, which has produced the earliest associations of artefacts and cultural traces of early man, is one such site [9]. And in the Sahara there is pollen evidence of cypress, olive, pine, oak and other trees which thrive in the slightly wetter high country [10].

The pattern of pluvial periods is associated with the Pleistocene period, which began about one million years ago, when climates changed markedly in most latitudes and on most continents.[1] Higher latitudes were alternately subjected to continental glaciation and deglaciation, in a rhythm of glacial and interglacial periods. The lower evaporation during glacial episodes was reflected by improved moisture conditions in the higher latitude dry lands, as, for example, central Asia, Mongolia, or the Great Plains.

In lower latitudes, an even more complex succession of moister and drier pluvials and interpluvials succeeded one another. These changes apply particularly to the Sahara, Arabia, the Kalahari and Numib of southern Africa, and to the great Australian interior. Several problems have been raised by these lower latitude pluvials. Have they been synchronous with the higher latitude glacials or with the interglacials? Have such pluvials been contemporary on the different continents and, if so, have they been synchronous on both latitudinal peripheries of each desert?

Considerable progress towards the solution of these questions has been made by geological and geomorphological investigation, particularly in dry areas contiguous to an ocean coast. During various Pleistocene glaciations, millions of cubic kilometres of ocean water were locked in the continental ice masses so that world sea-level was lowered by about 100 metres or so. During the warmer interglacials this water was released and occasionally augmented somewhat by a reduction of the existing highland glaciers and possibly also of the ice caps of Greenland and the Antarctic continent. Thus, since the water content of the world's glaciers is ultimately derived from the oceans, periods of low sea-level

---

[1] An age of one million years, for some time accepted for the Pleistocene, may be an under-estimate and may also be varied according to stratigraphic reinterpretation. A period as short as 350,000 years has been suggested.

provide a good stratigraphic marker of glacier episodes, while periods of high (or higher) sea-level indicate interglacial periods. These glacio-eustatic fluctuations of sea-level, as they are called, have, however, no necessary relationship to the fluctuations of interior lakes.

Broad correlation between pluvial and glacial phases has been suggested on the basis of sea-level correlations, although in the case of the Mediterranean region it was possible to show that only the early parts of the low sea-level phases were moister, whereas the later parts were drier [11]. Again, in that area while some periods of interglacial time were quite dry, in the Sahara for example other periods were warm and moist.

The recent application of radio-carbon dating techniques to various samples from tropical Africa has shown that the last major pluvial episode of both East and South Africa began as long as 60,000–70,000 years ago, reached its maximum shortly thereafter, and waned after 38,000 years ago. Conditions were again moister from a little before 20,000 to about 12,000 years ago [12]. This second and last part of the last pluvial was less significant in the fully arid Sahara.

From the evidence available today, it can be suggested as highly probable that pluvial tendencies were contemporary in all drier latitudinal zones in the same way that glaciations were everywhere broadly synchronous. However, the correlation of pluvial and glacial is only a half-truth. There have indeed been moist cool intervals in the lower latitude dry areas. But there have also been dry cool intervals, moist warm intervals, and dry warm intervals [13]. For the time span of the last interglacial (ca 100,000-65,000 years ago) and last glacial (ca 65,000–10,000 years ago), the sequence recorded in the Sahara and on the Mediterranean borderlands seems to be broadly as follows:

Early last interglacial	Warm, comparatively dry
Late last interglacial	Warm, comparatively moist
Early last glacial	Cool, comparatively moist
Later last glacial	Cold, comparatively dry

The evidence in support of warm pluvials is mainly derived from fossil soils. Such soils are related to the red loams still forming in the tropical savannas today. Soils of this type are preserved as the *terra rossas* or red earths of the Mediterranean Basin, which no longer develop today, but which can be shown to be of interglacial age. They further include relict or fossil soils found on the terraces of the River Nile, in some of the Saharan highlands and on the coasts of Mauretania and Senegal, well beyond their present range of development. Each of these soils is testimony of former periods of warm, seasonally moist climate producing a rather intensive degree of chemical weathering. Similar fossil soils have been recognised in dry parts of East Africa.

Evidence for cool pluvials is of another kind. Such phases were not characterised by a close vegetative mat such as enabled the fossil red soils to develop, free from disturbances by erosion. Instead, intensive irregular rains provided great stream potantial to erode, remove and transport large quantities of rock, sand and soil [11, 14]. These were accumulated along the stream beds and are now preserved as widespread alluvial deposits, frequently dissected since to form river terraces. The cool pluvials were rather distinct phases of accelerated stream erosion, transport and deposition in areas with little or no stream activity today. Intensive rainwash erosion of unprotected soils and loose surface materials, together with longer and more effective stream discharge, seem to be characteristic of such phases.

There is only limited evidence available in support of interpluvial phases drier than today. Some of the rare instances are blown sands in many parts of southern and central Africa, suggesting an absence of vegetation on the Kalahari Sands during various occasions of the Pleistocene. In West Africa and in parts of the Sudan, fossil dunes, under considerable vegetation today, may be found ten or even hundreds of kilometres beyond the contemporary limits of blowing sands [10]. In the Mediterranean region, coastal dunes were able to migrate many kilometres into the interior during several cool interpluvials, into areas well forested under natural conditions today. The general scarcity of good, stratigraphically dated evidence of intensified aridity seems to reflect on the absence of good indicators. Geomorphic processes in the deserts today are comparatively slow, and barring sand dunes, only produce distinctive features in the landscape after long periods of time. Excepting the subordinate role of wind, most of these processes are at work in more humid areas today. They would consequently be difficult to recognise in an ancient context, unless perchance suitable material was available for wind activity.

The last 10,000 years

It was suggested above that the maximum of the last Pleistocene pluvial phase dates back to over 50 millennia ago, and that all pluvial characteristics were over well before the close of the Pleistocene period some 10 millennia ago. This implies that the climate of the arid zones had approached a mean not unlike that of the present. In the succeeding millennia climate continued to fluctuate a little and, although such fluctuations were seldom of geological significance, they may well have exerted considerable influence on human life.

There is no general scheme of Recent or Postglacial climatic fluctuations that would be acceptable in more than one region. This is largely a reflection of how little is known of the major climatic changes of the late prehistoric and the historical eras. One particular moist interval may, however, prove to be of rather general occurrence during the millennia of maximum Postglacial temperatures, approximately from the late sixth to the late third millennium [10]. Pronounced aridity in the second millennium B.C. may also be another such feature of more general validity.

An area in which Recent climatic changes have had somewhat spectacular ecological effects has been the Sahara Desert. It can be said that full aridity had set in by about 8000 B.C., so that there is no question of any so-called progressive desiccation since that time. Yet between about 5500 B.C. (or a little earlier) and 2350 B.C. the climate was moister than today. The evidence is of three kinds: faunal evidence, largely on the basis of prehistoric rock-drawings; botanical evidence, including a fair amount of fossil pollen; and some limited geological evidence. The details of this moist interval or subpluvial period may be provided in a summary way.

The two oldest groups of late prehistoric rock-drawings of the Sahara belong to a group of big game hunters, and to a culture based upon beef-cattle raising, the earlier known group of nomadic pastoralists. Both of these peoples left an invaluable record of themselves and their way of life, as well as of a manifold natural fauna in many of the now deserted Saharan highlands of the Hoggar, Tassili, Adrar, Air, Tibesti and Uweinat [6]. The animals depicted along the dry wadis of the Fezzan include not only gazelles, antelopes and ostrich, but also tropical savanna species such as the elephant, both the single and two-horned rhinoceros, hippopotamus, crocodile, an extinct water buffalo and the giraffe. As an example, the African elephant is a woodland or parkland form, requiring some 300–350 pounds of green fodder daily. Study of the natural distribution of the four most indicative species with modern rainfall distribution suggests that rhinoceros and hippopotamus do not occur in any areas with less than 6 inches rainfall, the elephant in areas with less than 4 inches, the giraffe with less than 2 inches. Such are obviously theoretical values, but even in view of the local micro-ecology they do provide rather conservative estimates. On the basis of this rich archaeologically recorded zoological evidence – which is supported by fossils in several areas – a tentative attempt to reconstruct the approximate rainfall distribution for the period ca 5000–2350 B.C. is made in Fig VII–3. The hypothetical lines of equal rainfall so obtained suggest an increased precipitation both on the northern and southern margins of the Sahara, where the marginal belts of vegetation shifted 100–250 km towards the core of the desert. The central highlands appeared as selected areas of water and pasture rising above the desert plains.

Floral evidence of this subpluvial phase is available from several areas. For example, three stumps of acacia, tamarisk and sycamore have been observed (and archaeologically dated) in the deserts east and west of the Nile in Egypt [7]. They indicate a savannah-like vegetation with small trees or thickets at edaphically suited localities such as are provided in wadi beds. In addition to literary sources there is historical documentation on several very ancient Egyptian tomb and temple reliefs of vegetative growth in the desert. This was, of course, limited to the more elevated country or to the marginal tracts, for the core of the Libyan Desert would have remained as lifeless as today. Similar evidence is available from the northern Sudan.

In the western Sahara several rock shelters, often with rock-drawings, have provided pollen grains of cypress, pine, evergreen oak, wild olive, hackberry, juniper, tamarisk and a number of Mediterranean-type shrubs, as well as the lotus and cereals [10, 4, 18]. Various radio-carbon dates of 3450, 3445, 3070 and

2730 B.C. have been obtained for some such beds. This evidence suggests an open subtropical woodland at edaphically favoured localities in the Saharan high-lands at the time.

Finally, the testimony of the rock-drawings and biological evidence is amply substantiated by local occurrences of geological sediments, including black organic soils found in wadi floors and formerly marshy depressions. These are

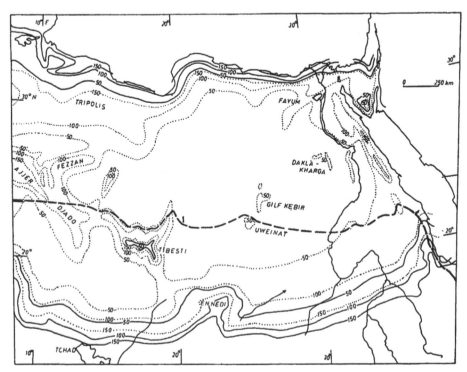

Fig VII-3. Modern and late prehistoric (5500–2350 B.C) distribution of precipitation in the eastern and central Sahara. Full lines indicate modern rainfall in millimetres; dotted lines indicate reconstruction of same for late prehistoric times. Heavy hatched line (1) indicates boundary between predominating winter and summer rains today. (*From 'Abhand-lungen der Akademie der Wissenschaften und der Literatur' (Mainz), Math. Nat. Kl., 1958, No. 1, Fig 2*)

particularly common in the Hoggar and Adrar areas, where they contain rich molluscan faunas as well as prehistoric fish-hooks and milling stones.

The subsequent period, ca 2350–800 B.C., was intensely dry. Geological and historical evidence in Egypt suggests that sand dunes invaded parts of the Nile Valley, that the Nile floods declined, that famines due to low floods were common, and that desert people were forced to immigrate into the Nile Valley. Conditions have remained more or less average during the last two millennia or so. As in other areas of the Near East, there is no sound evidence supporting a moister climate in Graeco-Roman times as has sometimes been suggested. This example of the Saharan area may give an impression of the kind of data available, and the degree of fluctuation verified.

Contemporary climatic trends

Meteorological records of a direct nature supplant geological and historical evidence in the late nineteenth and early twentieth centuries A.D. From all arid regions of the world, with the unique exception of the American south-west and northern Mexico, the records indicate a declining rainfall between about 1910 and 1940 (Fig VII-4). This was paralleled by a slight warm-up in higher latitudes. This warm–dry anomaly appears to have been interrupted or possibly reversed since 1940, however. The overall effects of this recent climatic fluctuation have

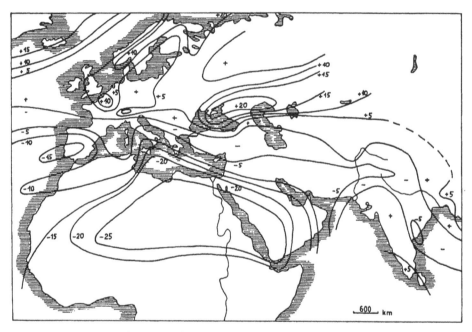

Fig VII-4. Precipitation anomalies 1881–1910 to 1911–1940 expressed in percentage deviations from the mean 1881–1910. Annual. (*From 'Geografiska Annaler', vol. 37, 1957, p. 107, Fig 3*)

had only limited impact on human existence in marginal lands. But the fluctuation has provided valuable actualistic experience for reconstructing interpluvial climates, and the observations provide at least some grounds for possible speculation on future development of precipitation trends in the arid zone.

A long-term weather prognosis advanced by one authority in 1951 [19] was a general amelioration of moisture conditions in lower latitudes during 1950–70. This forecast was made on the basis of expected sunspot developments, which have failed to materialise. Nevertheless the contemporary trend does at least not appear to be a negative one. Reliable explanations for the comparatively minor but ecologically important fluctuations within the general circulation of the atmosphere are still lacking. Until this is otherwise, no useful suggestions as to the future trend of the *dynamique* of arid zone climates can be offered.

Deserts of past geological eras

The deserts of the Quaternary era, in which we live, are perhaps more intensely arid than the dry lands of much of geological time, because of the intense mountain-building of Tertiary times, which created a more varied topography than obtained during long periods of relative quiescence between the orogenic revolutions. During such periods of stability the general circulation of the atmosphere was slack, with widespread regions of savannah type climate, relatively but not intensely arid.

It is probable, however, that in former geological eras with generally lower rainfall and higher temperatures, physiographic processes characteristic of very arid areas today could have been generated with less severe climatic conditions. Plants did not occupy the land until the Devonian, and even after that time the density of vegetative cover may reasonably be presumed to have been less than today especially as grasses did not appear until the Tertiary. Thus in the past, and especially in the more distant past, some areas which would not now be subject to physiographic processes characteristic of desert lands would have been affected by the sparseness of the vegetative cover; these are the biological deserts.

GEOLOGICAL CRITERIA OF ARIDITY

Many criteria have been advanced as evidence of past aridity, but few are conclusive and usually reliance must be placed on convergent lines of evidence. Dunes, including their external form, internal bedding structures, and size and shape of the component sand grains, are the most unequivocal evidence of the existence of deserts, coastal dunes excepted.

The outstanding example of an ancient desert which can be identified from dune forms preserved in the stratigraphic record comes from South America [20]. During Triassic times the Parana basin was an epeiric depression occupied by a vast sea of sands and dunes that covered an area of at least 1,300,000 sq. km. Preserved in the basin is a sequence of sedimentary and volcanic rocks resting uncomformably upon an ancient basement which includes Pre-Cambrian and Permian strata. The basal member of the sequence is of alluvial origin, the Piramboia sandstone. This is succeeded by the aeolian Botucatú sandstone which is intercalated with the alluvio-lacustrine Santana sandstone, and within which there is an extensive (1,200,000 sq. km) series of basaltic lava flows. There are aeolian sands above the lava in places. Reptilian and plant remains indicate an upper Triassic (Rhaetic) age.

The Botucatú sandstone is a fine-grained rock, some 70–90 per cent of its grains being in the diameter range 0·06–0·25 mm. It is predominantly quartzite, only about 5 per cent of its volume being felspathic. The large grains are well rounded, the small less well; they characteristically exhibit minute pitting and a veneer of red ferric oxide. These sedimentological features are typical of aeolian sediments.

A statistical study of the dune-bedding in the sandstone shows that the dunes formed and migrated under the influence of a northerly wind. In some few localities dune forms are perfectly preserved beneath the lava flows and conveniently exposed in railway cuttings. Individual dunes are up to 15 metres high. Arcuate slip faces can be traced in some places and are suggestive of barchans. The sandstones in many places display aeolian ripple marks.

The Botucatú Desert is but one of several postulated deserts of Permo-Triassic age, though it is the only one with dune forms both preserved and exposed. In other areas the evidence is derived from a number of other features, the chief of which are sedimentary petrology, dune structures, evaporites (salt deposits), wind-faceted pebbles, evidence of a desert varnish and of sandblasting, and a number of minor criteria.

The sedimentary characteristics of sandstones can be diagnostic of their origin and may be safely used. In particular, the grain size, range of size and shape of grain are indicative of the origin of the rock. There is a high coefficient of sorting and roundness. Aeolian sands range in size from 0·15 to 0·30 mm, and very commonly are well rounded to give a 'millet seed' texture. Aeolian sands very often possess a pellicule of ferric oxide which coats each grain and to some extent masks the minute pittings caused by collisions between individual grains.

Possibly more important and reliable are dune or aeolian bedding structures. In a vertical section the sequences of dune-bedding, separated by either planar or curved surfaces of erosion, display markedly contrasted angles of dip. From a study of such false bedding it is possible to deduce the wind direction at the time of deposition, and by tracing individual faces it is possible to reconstruct the morphology of a dune (Fig VII-5).

In the British Permo-Trias, sands that are of certain aeolian origin are fairly common. They are particularly abundant in the Lower Permian sand sea, where many dune structures are known: they were barchans driven by easterly winds. In the Triassic, however, only the Lossiemouth sandstone of Morayshire and a single lens of the Upper Bunter near Kidderminster are regarded as of certain wind-blown origin.

In the western USA dune-bedded sands are abundant. The Uinta (Lower Tertiary) sandstones of Utah and the Chuska sandstone of the same age, display dune structures to perfection in massive and precipitous bluffs. The Chuska sandstone was deposited by a southerly wind. Throughout Utah, Colorado and Wyoming there are aeolian sandstones ranging in age from the Pennsylanian (Upper Carboniferous) to the Upper Jurassic. The formations involved are the Entrada, Cow Springs, Navajo, Coconino, Tensleep, Weber and Casper sandstones [21]. Though some have doubted the aeolian origin of parts of these sequences, their structure and association in places with gypsiferous deposits make a desert origin probable. An analysis of their stratification shows that the dunes were moulded by a northerly wind.

The identification of some of the Permo-Triassic sediments of desert origin rested initially on the dominant red colour of the beds and on the presence of saline beds. A red colour *per se* has in the past been accepted as evidence of aridity, but brown, or yellow brown, is a more typical colour of desert sands, though some dune sands are red [22, 23]. The origin of the ferric oxide that imparts this hue is not known, but most workers are now agreed that red is more characteristic of weathering under humid tropical conditions. Some red-beds are accordingly interpreted as preserving their redness from the soils of a landscape

Fig vii-5. Bedding structures revealed in a wind-eroded wet sand dune, Kelso Dunes, Mojave Desert, California. Similar structures may be seen in fossil dunes. (*Photo R. P. Sharp*)

previously humid, as with the Tertiary laterites that exist in many parts of Australia, and which break down to give red soils under the present-day climate. The most that can be said is that arid conditions are non-reducing and thus favour the retention of the red colour.

Thick salt deposits, sometimes regarded as due to evaporation in continental desert basins, are now chiefly attributed to evaporation in marine environments, particularly in almost landlocked gulfs. They are, however, a clear indication of hot dry conditions on the nearby land mass, although not themselves of lacustrine origin.

Modern stony deserts both hot and cold are characterised by the presence of ventifacts or wind-faceted pebbles, the *Dreikanter* of the German literature. The presence of such faceted stones in ancient deposits is evidence of aridity, either hot or cold. In the Trias of the Nottingham district in the English midlands they occur quite commonly. They are virtually the only evidence of aridity in two Pre-Cambrian formations, the Torridon sandstone of the north-west highlands of Scotland, and the Jotnian sandstone of southern Finland, Sweden and the Kola Peninsula.

The Trias of the English midlands contains, besides ventifacts, liver-coloured pebbles which are probably stones with a veneer of desert varnish, well known from modern arid environments. Polishing, fretting and fluting by the wind armed with sand is another sign of arid conditions. The classic example of fossil sand-blasted rocks has been described from the Charnwood forest of Leicestershire, England, where the local Triassic strata lie unconformably upon a Pre-Cambrian sequence of crystalline rocks [24].

Breccias and conglomerates, although not of themselves evidence of desert conditions, are common in an arid environment where there is considerable relief. Ancient fanglomerates are known, for example the Brockrams of the Lower Permian of north-western England. Likewise possible relict calcrete has been recognised in the 'Cornstones' of the Lower Devonian of Wales and the Upper Devonian of Scotland.

KARL W. BUTZER

C. R. TWIDALE

References

[1] BROEKER, W. C. and P. C. ORR, 'Radiocarbon chronology of Lake Lahontan and Lake Bonneville', *Geol. Soc. Amer. Bull.*, Vol. 69, 1958, pp. 1009–32.

[2] RUSSELL, I. C., 'Geological history of Lake Lahontan', *U S Geol. Surv.*, Mon. 11, 1889.

[3] BERKEY, C. P. and F. K. MORRIS, *Geology of Mongolia*, New York, 1927.

[4] PONS, A. and P. QUÉZEL, 'Première étude palynologique de quelques palé-sols sahariennes', *Trav. Inst. Recher. Sahariennes*, Vol. 16, 1957, pp. 15–40.

[5] TRICART, J. and A. CAILLEUX, *Le modèle des régions sèches*, CDU, Paris, 1960–61.

[6] LHOTE, H., *The Search for the Tassili Frescoes*, London, 1959.

[7] CLARK, J. D., *The Prehistory of Southern Africa*, Harmondsworth, 1959.

[8] DU TOIT, 'The Mier Country', *S. Afr. Geogr. Journ.*, Vol. 19, 1926, pp. 21–6.

[9] COLE, S., *The Prehistory of Eastern Africa*, London, 1964.

[10] MONOD, T., 'The late Tertiary and Pleistocene in the Sahara and adjacent southerly regions, with implications for primate and human distribution', in F. C. Howell (Ed.), *African Ecology and Human Evolution*, Viking Fund Publ. Anthropol., Chicago, 1963, pp. 117–229.

[11] BUTZER, K. W., 'The last "pluvial" phase in the Eurafrican subtropics', *Proc. WMO–Unesco Symp. on Climatic Changes in the Arid Zones* (Rome, October 1961), 1962.

[12] CLARK, J. D. (personal communication).

[13] BUTZER, K. W., 'Palaeoclimatic implications of Pleistocene stratigraphy in the Mediterranean area', *Ann. New York Acad. Sci.*, Vol. 95, 1961, pp. 449–56.

[14] — 'Climatic-geomorphic interpretation of Pleistocene sediments in the Eurafrican subtropics', in F. C. Howell (Ed.), *African Ecology and Human Evolution*, Viking Fund Publ. Anthropol., Chicago, 1963, pp. 1–27.

[15] — 'Climatic change in arid regions since the Pliocene', in L. D. Stamp (Ed.), *History of Land Use in Arid Regions*, Paris, Unesco, 1961, pp. 31–56.

[16] — 'Das ökologische Problem der Neolitishen Felsbilder der östlichen Sahara', *Abh. Akad. Wiss. und der Liter., Math.-naturw. Kl.*, 1958, Nr. 1, pp. 20–49.

[17] — 'Die Naturlandschaft Agyptens wahrend der Vorgeschichte und des Dynastischen Zeitalters', ibid., 1959, Nr. 2, pp. 1–80.

[18] QUÉZEL, P., 'De l'application de techniques palynologiques à un territorie désertique. Paléo-climatologie du Quaternaire Récent au Sahara', *Proc. WMO–Unesco Symp. on Climatic Changes in the Arid Zones* (Rome, 1961), 1962, pp. 243–250.

[19] WILLETT, H. C., 'Extrapolation of sunspot–climate relationships', *J. Met.*, Vol. 8, 1951, pp. 1–17.

[20] ALMEIDA, F. F. M., 'Botucatú, a Triassic desert of South America', *XIX Int. Geol. Congr., Algiers, 1952*, Vol. VII, 1953, pp. 9–24.

[21] OPHYKE, N. D. and S. K. RUNCORN, 'Wind direction in the western United States in the late Palaeozoic', *Geol. Soc. Amer. Bull.*, Vol. 71, 1960, pp. 959–72.

[22] GERSTER, G., *Sahara*, London, 1960.

[23] DUNHAM, K. C., 'Red coloration in desert formations of Permian and Triassic age in Britain', *XIX Int. Geol. Congr., Algiers, 1952*, Vol. VII, 1953, pp. 25–32.

[24] WATTS, W. A., 'Charnwood Forest: a buried Triassic Landscape', *Geogr. Journ.*, Vol. 21, pp. 623–36.

ADDITIONAL REFERENCES

BARGHOORN, E. S., 'Evidence of climatic change in the geologic record of plant life', H. Shapley (Ed.), *Climatic Change*, Harvard U.P., 1953, Ch. 20, pp. 235–48.

BUTZER, K. W., *Environment and Archaeology: an Introduction to Pleistocene Geography*, London and Chicago, 1964.

GERSTER, G., *Sahara*, London, 1960.

F

GROVE, A. T., 'The ancient erg of Hausaland and similar formations on the south side of the Sahara', *Geog. Journ.*, Vol. 131, 1958, pp. 528–33.

HARE, F. K., 'The causation of the arid zone', L. D. Stamp (Ed.), *History of Land Use in Arid Regions*, Unesco, Paris, 1961, pp. 25–30.

HILLS, E. S., 'Die Landoberfläche Australiens', *Die Erde*, Bd 3–4, pp. 195–205.

McKEE, E. D., 'Problems on the recognition of arid and of hot climates of the past', A. E. M. Nairn (Ed.), *Problems in Palaeoclimatology*, London, Ch. 9, pp. 367–77.

REIFENBERG, A., 'The struggle between the desert and the sown', *Desert Research Proc. Symp.* (Jerusalem, May 1952), Jerusalem, Res. Council of Israel, 1953, Special Publication No. 2, pp. 378–89.

RUSSELL, R. J., 'Climatic changes through the ages', *Climate and Man*, USDA Yearbook of Agriculture, Washington, 1941, pp. 67–97.

SCHULMAN, E., 'Tree ring indices', *Compendium of Meteorology*, Boston, 1951, pp. 1024–9.

CHAPTER VIII

Plant Life in Deserts

The plant form

The desert environment impresses upon plant life conditions of extremely favourable warmth and light and extremely unfavourable paucity of moisture. Wherever or whenever water is available, plant life blooms in remarkable profusion. The desert plants exhibit special characters of structure (morphology), behaviour (physiology), and sociability (synecology).

Several systems have been proposed for the classification of desert plant forms. Adamson [1] examined the possibility of using Raunkiaer's classes of life form for several widely separated deserts. His data are reproduced in Table VIII-1. 'Except for the high therophytic (ephemeral) percentage in all of them, there is little agreement, so little in fact that to regard them as expressing in any way a direct relationship to climate seems quite out of the question.' Raunkiaer's theory is that the life form of a plant is an expression of its response to climate. Life forms are classified on a basis of the position and protection of the perennation buds [2, 3]. Adamson's analysis shows that this classification, based on a single criterion, does not provide a satisfactory expression of the diversity of form exhibited by desert plants. Zohary [4] notes that 'It is not the position and protection of the renovation buds that affect the water ecology of the plant, but the dimensions of the transpiring organs regularly lost by the plant in the critical season, that is most essential for the maintenance of desert summer vegetation'. He distinguishes ten morphological types within the desert vegetation of the Near East. Criticising Raunkiaer's terminology as 'inadequate for the classification of plant forms of desert zones', Cabrera [5] proposed a system which includes two divisions: the *Haloxiles* (woody plants with persistent stems and secondary growth) and the *Herbs* (without woody stems above the ground); and ten main subdivisions with numerous types. Shreve [6, 7] enumerates twenty-five types as 'the principal life forms of the North American desert'. These are quoted in Table VIII-2 to show the great morphological diversity in desert plants.

The remark that may be made on all the systems of life-form terminology is

TABLE VIII - I *Comparison of the life-form spectra of some of the deserts of the world, after Adamson (1939)*

Locality	Total number of species	Megaphanerophytes	Nanophanerophytes	Chamaephytes	Hemicryptophytes	Geophytes	Therophytes	Saprophytes	Epiphytes[1]	Helophytes and hydrophytes
Normal	400	23	20	9	27	3	13	1	3	1
ASIA Transcaspia	768	—	10	7	27	9	41	—	—	5
N. AFRICA Central Sahara	169	—	9	13	15	5	56	—	—	2
Timbuctu	138	11	12	36	9	3	25	—	(1)	3
USA Death Valley	294	2	21	7	18	2	42	3	—	5
AUSTRALIA Ooldea	188	19	23	14	4	1	35	—	(4)	—
SOUTH AFRICA Whitehill	428	1	8	42	2	18	23	5	(1)	—

[1] Epiphytic parasites (Lauranthaceae).

that the actual behaviour of the plant may vary according to local conditions of habitat. This leads to the distinction between life form and growth form. The former is the inherent morphological structure acquired by the plant under average conditions obtaining within its range; but the individual plant may acquire a different form in response to some local conditions. Species with this malleability are provided with an added asset that enables them to survive. This aspect of morphological (and physiological) elasticity is a marked feature of desert plants [8, 9]. A few examples may illustrate this point.

Panicum turgidum, a perennial desert grass which is widespread in the Middle and Near East deserts, acquires an evergreen growth form on deep surface deposits where some water is stored throughout the year. On shallower deposits it acquires a deciduous growth form: the aerial parts may remain dead-dry for the whole summer, or even for consecutive years with scarce or no rain; soon after the next shower new buds sprout with notable rapidity. *Zilla spinosa* also acquires an evergreen growth form in many of the main wadis, whereas on gravel plains it may acquire a distinctly deciduous form. Several other examples may be quoted. Such species are valuable indicators of the habitat conditions, especially the water regime.

Reference may be made to certain perennial species that succeed in a good

TABLE VIII - 2 *Life forms of the Sonoran Desert (after Shreve, 1951)*

Form	Examples
EPHEMERALS:	
Strictly seasonal:	
Winter ephemerals.	1 *Daucus pusillus, Plantago fastigiata*
Summer ephemerals	2 *Tidestromia lanuginosa, Pectis papposa*
Facultative perennials	3 *Verbesina encelioides, Baileya multiradiata*
PERENNIALS:	
Underground parts perennial:	
Perennial roots	4 *Pentstemon parryi, Anemone tuberosa*
Perennial bulbs.	5 *Hesperocallis undulata, Brodiaea capitata*
Shoot base and root crown perennial .	6 *Hilària mutica, Aristida ternipes*
Shoots perennial:	
Shoot reduced (a caudex):	
Caudex short, entirely leafy:	
Leaves succulent	7 *Agave palmeri, Dudleya arizonica*
Leaves non-succulent	8 *Nolina microcarpa, Dasylirion wheeleri*
Caudex long, leafy at top:	
Leaves entire, linear, semi-succulent	9 *Yucca baccata, Yucca brevifolia*
Leaves dissected, palmate, non-succulent	10 *Washingtonia filifera, Sabal uresana*
Shoot elongated:	
Plant succulent (soft):	
Leafless, stem succulent:	
Shoot unbranched	11 *Ferocactus wislizenii, Echinomastus erectocentrus*
Shoot branched:	
Shoot poorly branched:	
Plant erect and tall	12 *Carnegiea gigantea, Pachycereus pringlei*
Plant erect and low or semi-procumbent and low . . .	13 *Pedilanthus macrocarpus, Mammillaria microsperma*
Shoot richly branched:	
Stem segments cylindrical . .	14 *Opuntia spinosior, Opuntia arbuscula*
Stem segments flattened . .	15 *Opuntia engelmannii, Opuntia santa-rita*
Leafy, stem not succulent . .	16 *Talinum paniculatum, Sedum wootoni*
Plant non-succulent (woody):	
Shoots without leaves, stems green	17 *Holacantha emoryi, Canotia holacantha*
Shoots with leaves:	
Low bushes, wood soft . . .	18 *Encelia farinosa, Franseria dumosa*
Shrubs and trees, wood hard:	
Leaves perennial	19 *Simmondsia chinensis, Larrea tridentata*
Leaves deciduous:	
Leaves drought-deciduous:	
Stems specialised:	
Stems indurated on surface .	20 *Fouquieria splendens*
Stems enlarged at base . .	21 *Idria columnaris, Bursera microphylla*
Stems normal:	
Stems not green	22 *Jatropha cardiophylla, Plumeria acutifolia*
Stems green	23 *Cercidium microphyllum, Parkinsonia aculeata*
Leaves winter-deciduous:	
Leaves large	24 *Populus fremontii, Ipomoea arborescens*
Leaves small	25 *Olneya testosa, Acacia greggii*

season in 'flowering and setting good fruit in their first season, so that if they are killed by drought of the following summer they have completed a full life cycle as potential annuals, and this ability may be an important biological factor in the survival' [10]. As example, Haines mentions: *Diplotaxis harra, Caylusea hexàgyna, Euphorbia cornuta* (= *E. kahirensis*), *Farsetia aegyptia* and *Zilla spinosa*.

There are three main growth forms of desert plants: ephemeral annuals, succulent perennials and non-succulent perennials [11].

EPHEMERAL ANNUALS

The floras of desert areas comprise a considerable proportion (50-60 per cent) of short-lived herbs: the ephemerals. An ephemeral is capable of completing its full life cycle within an average period of 6-8 weeks. Their growth activity is restricted to the brief moist period, a process which saves the plant the adversity of water shortage which besets the perennial plants. It is often said that the ephemeral 'avoid rather than withstand' the dry season.

The desert ephemerals exhibit little or none of the xeromorphic habits of the perennials. Their main morphological attribute is their small size and their shallow roots; their principal physiological adaptation is their active growth: speed of germination, rapidity of growth, and early flowering and maturity.

The behaviour of seeds is of particular importance. The ephemerals live through the dry season in the form of seeds. The dry season may, in extreme deserts, be prolonged to a number of years. The viability of seeds seems to tide the plant over such prolonged episodes. The germination behaviour of seeds in response to certain amounts of rainfall is a phenomenon that has been the subject of field observations and laboratory experiments; see Went [12]. He states that

when there are rains of 10 mm or less in the desert, not a single seedling will appear. A 15 mm rain may bring forth a limited number of seedlings in small gullies where the water ran in a continuous stream, thus simulating a heavier rainfall. Extensive germination occurs only after precipitation of 25 mm or more. . . . Cloudbursts will loose as much as 75 mm of rain within an hour's time; after one of these, germination is very small, because large amounts of water have the same repressing effects as observed in the laboratory.

The figures may vary but the fact remains that the seeds seem capable of gauging the amount of rainfall. This is attributed to the washing of certain germination inhibitors that form at the surface of the seed coat. Excessive rainfall seems to remove certain diffusible growth-promoting substances in the seed, which are leached out at a slower rate than the more easily leached inhibitors, and hence the decrease of germination.

The exactness of the requirements of germination is apparently a valuable adaptation, for seedlings will certainly succumb unless germination is limited to conditions that will allow them to proceed with their growth. Desert ephemerals germinate with hesitation but their subsequent growth needs to be fast.

One of the observed features of the desert ephemerals is the regulation of size under the influence of habitat conditions. Referring to the ephemerals of the Sonoran Desert, Shreve [7, p. 121] states:

Height and bulk are chiefly determined by the amount of soil moisture available during the growing period, and, in the case of winter ephemerals, by the number of warm and sunny days following the rainy period. Individuals of the same species may come to maturity at a height of 1 cm with a single flower or may grow to a height of 2 or 3 dm, bearing hundreds of flowers. Such divergence of behaviour is due to the amount of precipitation.

Within the American deserts Shreve [6] distinguishes between winter annuals which appear in winter and early spring, and summer annuals which appear after the first heavy rain of the summer. He suggests that the restriction of two groups of ephemerals to their respective seasons is dependent on the optimum temperature for germination of their seeds [7]. For winter ephemerals the optimum is between 60° and 65°F, and for summer ephemerals between 80° and 90°F. He also shows the difference in the geographical distribution of the two sets of ephemerals in association with the seasonal pattern of rainfall. The distinction, in the north African deserts, between the ephemerals of the Mediterranean belt (winter rainfall) and those of the tropical deserts with summer rainfall (e.g. Sudanese deserts) is similar.

But within the strictly winter rainfall desert of the Mediterranean region, the flora comprises a number of summer annuals. Referring to these plants, Negbi and Evenari [13] write: 'Their rhythm of life seems to be out of gear with the seasonal cycle of climate and rainfall. . . . The problem therefore arises of how these plants can survive the dry rainless summer in a physiologically active state and how they manage to get the necessary water.' They conclude that germination is independent of conditions of temperature and dependent exclusively on rainfall, and that these summer annuals are segetal-ruderal plants where the natural vegetation is disturbed and competition with winter annuals and perennials is eliminated.

SUCCULENT PERENNIALS

Succulence is the morphological expression of the proliferation and enlargement of the parenchyma tissues of the plant organs: stems and/or leaves. This structural feature, together with the physiological feature of low transpiration rates, apparently enables the accumulation and storage of water during the rainy season. This is generally accepted as stored water that may be consumed during episodes of drought. Daubenmire [11, p. 154] writes: 'Every plant collector is familiar with the remarkable tenacity with which most succulent plants retain their moisture in a plant press. Sedums collected in flower commonly set seeds before drying. . . . A stem of *Ibervillea sonorae*, which was stored dry in a museum, formed new growth every summer for eight consecutive summers, decreasing in weight only from 7·5 to 3·5 kg.' See types 7 and 11–16 of Table VIII-2.

The cacti of the American deserts and the cactus-like *Euphorbia* spp. of the tropical African and Arabian (Yemen, etc.) deserts are a special group of desert succulents. They are marked by their form and their physiology. The stomata are closed during the day and open at night: their transpiration rates are low during the day when evaporation stress is high. The closure of stomata during the day is associated with peculiar aspects of photosynthesis. The carbon dioxide assimilated is partly derived from organic acids produced in anaerobic respiration [14].

NON-SUCCULENT PERENNIALS

The majority of desert perennial species are non-succulent. These are the hardy plants that endure the full stress of the desert environment. They comprise a variety of morphological forms: woody herbs, grasses, shrubs, trees, etc. (see Table VIII-2). The morphological, anatomical and physiological features that seem to enable these plants to tolerate the climatic and/or soil aridity are discussed in the section on xerophytism.

These perennials comprise three main categories of growth form, usually referred to as evergreen, drought-deciduous and cold-deciduous. They may better be described as biologically active throughout the year, biologically dormant during the dry season, and biologically dormant during the cold season. The shedding of leaves is only one aspect of life, but many desert plants maintain their life processes after shedding their foliage.

Many of these perennials have hard seeds that may not readily germinate. Went [15] writes:

The seeds of many shrubs that grow exclusively in washes (paloverde, ironwood, the smoke tree) have coats so hard that only a strong force can crack them. Seeds of paloverde can be left in water for a year without sign of germination, but the embryo grows out within a day if the seed coat is opened mechanically. In nature such seeds are opened by the grinding action of sand and gravel. A few days after a cloudburst has dragged mud and gravel over the bottom of a wash, the bottom is covered with seedlings. It is easy to show that this germination is due to the grinding action of the mud flow: for instance, seedlings of smoke tree spring up not under the parent shrub itself but about 150 to 300 feet downstream. That seems to be the critical distance: seeds deposited closer to the shrub have not been ground enough to open, and those further downstream have been pulverized.

The hard seeds of many desert plants (*Alhagi maurorum*, *Prosopis* spp., etc.) need to be soaked in concentrated sulphuric acid for one or two minutes before they may germinate. The passage of such seeds through the alimentary canal of animals may soften them, and seedlings are often noted sprouting from camel or goat droppings.

The post-germination growth of perennials is very different from that of ephemerals. Here the seedling produces the first few leaves, then the above-ground growth stops for several months while the root system proceeds penetrating deep into the ground. Eventually the shoot system resumes growth but the root system often maintains its more extensive volume.

Xerophytism

The complex of the adaptive attributes of plants subsisting with small amounts of moisture is termed xerophytism. Ephemerals evade rather than withstand drought and many authors do not consider them true xerophytes. The following is a summarised survey of the characters of desert perennials. In this summary the features are dealt with under morphological, anatomical and physiological characters. It is, however, obvious that such distinction is a matter of convenience and does not mean that the one set of characters is independent of the other. The morphological features are the outside expression of the internal physiological processes, and in turn may be instrumental in alleviating the stress of the arid environment on the physiological functions. The same may be said about the anatomical structure.

MORPHOLOGICAL CHARACTERS

1. Root growth

The development of an extensive root system as compared with a lesser shoot is one of the common features of the desert perennials. This is obviously a fortunate set-up, as the moisture revenue collected by an extensive root system is drawn upon for the consumption of a reduced shoot and a healthy water balance is more likely.

Studies on root growth are often handicapped by difficulties in handling the deeply penetrating roots and their branches and also by the labour involved. Our understanding of many aspects of plant life in the desert will remain incomplete till the accumulation of ample information on the root growth of various species under various habitat conditions.

The main extension of the roots may be vertical, horizontal or both [16, 17, 18]. The character of the root seems to depend on the habitat conditions. Deeply penetrating roots are common in deep alluvial deposits (e.g. wadi fill), and often reach a permanently wet soil or a groundwater store. Daubenmire [11, p. 157] quotes Meinzer, who records that roots of mesquite may extend to 'at least 65 feet, and alfalfa roots reaching 129 feet below the surface'. Polunin [19, p. 451] quotes Rubner as referring to the roots of tamarisks that 'could be followed during the building of the Suez Canal in places to a depth of 50 metres'. Roots of desert perennials penetrating to 10–15 metres are not unusual.

Horizontally extending roots are common in sandy habitats and in shallow soil overlying harder substrate. Rope-like roots may extend for several metres (5–20 metres) not far below the sand surface. Oppenheimer [20] quotes Morello in his studies on the poor steppes of Argentina where *Larrea divaricata* (usually a phreatophyte) forms horizontal roots 6 metres long but only 30–50 cm deep. He also quotes Sarup as describing extremely long horizontal roots reaching 88 feet developed by *Prosopis spicigera*, a psammophyte of the Great Indian Desert.

Of particular interest in this respect are the so-called rain roots. These are fine surface feeders that are produced by the woody part of the root just below the ground surface, in response to light showers. They may also be produced during periods of dew formation. These roots eventually wither and may reappear shortly after a new supply of water has become available. A comparable type of root is described by Zohary and Orshansky [21] in their study of *Zygophyllum dumosum* (a flat-rooted plant); the root system penetrates deeper with short-lived thread-like extension roots after exhausting the water reserve in the upper soil layer.

Attention has been drawn by Zohary and Orshansky [22] to certain lithophytic desert plants that may inhabit bare rocks: rhizophagolithophytes are plants with roots which dissolve the rock while entering it. These authors find that certain limestone rocks have high water-storage capacity, and quote *Podonosma syriacum*, *Ballota rugosa*, *Varthemia iphionoides* and *Stachys palaestina* as examples of these plants. To these may be added *Thymus capitatus* and several others. These are not to be confused with the chasmophytes which root into rock crevices.

2. Shoot growth

It is the general experience of desert students that plants develop root systems larger than their shoot. This low shoot-to-root ratio is established as early as the seedling stage. Oppenheimer [20] quotes Sveshnikova and Zalensky as finding that the weight of roots of several plants of the mountain deserts of Pamir surpasses that of the shoots 30 to 50 times. This is remarkable. Ratios of 1 : 3·5 and 1 : 6 are recorded by many workers.

Reduction of the transpiring surface is also affected by several other attributes: small leaves (microphylly), no leaves (aphylly), shedding of foliage, shedding parts (or the whole) of the shoot branches, rolling of leaves (grasses), etc. Orshansky [23] estimates that the reduction of the transpiring surface in the summer (rainless season) amounts to 87, 76 and 62·8 per cent in *Zygophyllum dumosum*, *Artemisia monosperma* and *Helianthemum ellipticum*, respectively. Comparable estimates are given by Oppenheimer [20] and Zohary [18]. Such seasonal reduction of the exposed surface of the plant will entail reduction of the water output during the dry season and is obviously of great importance in the water economy of perennial plants.

Attention may be drawn to two aspects of morphological plasticity of many desert plants. (1) Several spinescent species produce relatively broad leaves and reduced spines under less arid conditions, and small leaves (soon shed) and large pungent spines under normal desert conditions (*Alhagi maurorum, Nitraria retusa, Lycium arabicum, Zilla spinosa, Euphorbia cuneata*, etc.). (2) The roots of many desert plants produce adventitious buds sprouting into new branches when partly exposed by erosion: *Lycium arabicum, Zilla spinosa, Echinops spinosissimus, Pityranthus tortuosus, Iphiona mucronata, Heliotropium luteum*, etc. This ability has been observed in so many desert plants that it may be a general feature. It is not to be confused with the deeply seated root axes that bear adventitious buds normally (e.g. *Alhagi maurorum*).

Differences are noted between individuals of *Alhagi maurorum* in the coastal habitat (succulent leaves) and those growing inland; and between individuals of *Echinops spinosissimus* of the maritime habitats (trailing branches and succulent leaves) and those growing in the inland deserts. Other examples may be quoted. Without prejudice to the notion of ecotypes, these differences are further evidence for the morphological plasticity of the desert plants.

ANATOMICAL CHARACTERS

Several anatomical characters prevail in desert plants: heavy cuticularisation and cutinisation that may be supplemented by wax incrustations, resinous and varnish-like covering; profuse lignification and richness in sclerenchymatous elements; high ratio of volume to surface (compactness); abundant hairs; and special arrangement of stomata in recesses or grooves. Cuticularisation is the formation of a surface plaster-like layer of cutin, the cuticle. Cutinisation is the impregnation of the cell wall with the substance cutin, a process comparable to the impregnation of textiles with waterproofing material. Cutin is a mixture of fatty substances that may form a watertight layer. The thick cuticle, the cutinisation of the walls of the epidermal (and sometimes the hypodermal) cells, the waxy mantle and the thick hairy cover may minimise the loss of water through the epidermal cells and limit the moisture exodus to the controllable stomatal apertures. The sunken stomata or the grouping of stomata into special crypts (e.g. the leaf of *Nerium oleander*) or grooves (e.g. *Retama raetam*) provide protected situations not under the immediate influence of atmospheric aridity. These are supposedly adaptations that may help in reducing the rate of transpiration. The preponderance of lignified elements provides the organs with efficient mechanical support, saving its tissues from collapse under conditions of wilting or water deficit in the softer cells (for more details see Stover [24, Chap. 8]).

The anatomy of roots has received only modest attention, but several features have nevertheless been reported. The peripheric layers are often transformed into protective corky tissues. The inner cortex of many grasses (*Bromus rubens, Lygeum spartum*, etc.) may become spongy and form sleeves that are water absorbing and storing tissues. The appearance of dense long root hairs that agglutinate the soil particles and form compact tubules around the roots, provides added protection.

PHYSIOLOGICAL CHARACTERS

The hardiness of the desert plants is, in the final analysis, an attribute of the living cells and the behaviour of their protoplasm, which enables its survival under conditions obtaining in the desert. As Stocker [8] states, 'Unfortunately, our knowledge concerning the effect of drought conditions on the plasma is very incomplete.' Still less is our knowledge of the special features of the plasma which enable survival under prolonged periods of water deficit. Reference may here be made to desert lichens that seem to survive extreme conditions of dehydration.

It is often stated and supported by convincing measurements that the following features are universal among desert perennials. The osmotic pressure is high in drier habitats and drier seasons. The protoplasm comprises a high proportion of hydrophyllic constituents which seem capable of binding some moisture. The plant is capable of regulating, seasonally and diurnally, its transpiration rates especially under conditions of water shortage; this ability is enhanced by the anatomical and morphological attributes.[1]

Desert vegetation

LANDFORM AND.VEGETATIONAL PATTERN

Plant life in the desert shows particularly intimate relations between landforms and plant growth. The significance of landforms is primarily due to their controlling influence on the moisture regime and making the habitat accessible (or not) for the destructive agencies of grazing and human interference.

Physiographic features control the run-off and the processes involved in surface drainage, collecting and redistributing the available water. The levels lowest in relation to local topography receive water and soil collected from extensive areas.

Plant growth confined to the channels of the drainage systems distinguishes the *run-off deserts* from the *rain deserts* [18]. In the latter the rainfall is sufficient to maintain a sparse general vegetation, in the former the local rain is insufficient to support any perennial vegetation except in channels of the drainage or run-off systems.

Elevation, exposure and angle of slope may increase or reduce the effectiveness of the limited water resources. The contrast between the north-facing and the south-facing slopes is obvious. Even on rocky slopes where little flowering plant growth exists, the growth of lichens marks the directions with unfailing consistency. In other words, landform patterns are associated with microclimatic patterns that are effective environmental features. This may be a universal condition, but is especially notable in deserts, where plant life is in a precarious balance with the environment. It is natural, under such conditions, that small environmental differences should assume great importance.

The features of the ground surface have their perceptible effects on plant life.

[1] Detailed reports on the physiology of plant–water relationships in arid and semi-arid conditions are included in references [25, 26]. See also [14, 27, 28].

A compact rock surface affords little possibility of plant growth except for certain lithophytes including lichens. A fissured rock surface will permit the growth of chasmophytes, which send their roots into the crevices. A rocky surface veneered with residual rock fragments (*hamada*) provides a habitat with peculiar features different from those of the exposed rock surface. The rock fragments afford no room for seedling growth, but there may be small pockets of soft materials amongst the fragments where seedlings may grow. A rocky surface covered with a duricrust (rock detritus cemented by capillary rising salts) is usually very sterile. An erosion pavement is a rocky surface overlain by a layer of residual mixed rock waste with a surface mantle of lag scree. The plant cover is here confined to the shallow drainage runnels. The gravel desert is distinguished from the erosion pavement, the surface deposits being mainly transported material, not fragments produced *in situ*. Surface gravels are usually globose, not angular as are the surface screes of the erosion pavements. The mature forms of both geomorphological units are the result of the removal of soft material by deflation or washing and the accumulation of coarse lag material at the surface. The lag material becomes so closely strewn that a concrete-like layer (desert armour) is produced, which is impenetrable to plant roots. Geochemical processes form a sub-surface layer (or layers) of gypsum that is an added cause for the sterility of these deserts. A surface mantle of soft material will afford a favourable sub-stratum for plant growth. The material may be alluvium or aeolian sand sheets and dunes, or again a sedentary soil developed on rock. These are the desert soils.

The properties of the soil are of primary importance in determining the character of the vegetation. As Shreve [7, p. 21] states:

The physical texture of the soil, its depth, and the nature of its surface are equally important. . . . The profound influence of the soil upon desert vegetation is to be attributed to its strong control of the amount, availability, and continuity of the water supply. This fundamental requisite of plants is the most effective single factor in differentiation of desert communities.

The thickness of the soil layer has a great effect on the moisture regime. A shallow soil is moistened during the rainy season but is subject to complete desiccation in the long rainless season. This condition allows for the growth of ephemerals. A deep soil allows for the storage of some water in a permanently wet, deep-seated layer which provides the deep roots with a continuous supply of water. The soil depth necessary for the establishment of this permanently wet layer depends on the texture and stratification of the soil, and the rainfall and local drainage [29, 30].

Plant growth in aeolian sand is of a special character. Plants that inhabit the sand formations are of two categories, sand binders and sand dwellers. The former group is of primary importance as they build phytogenic mounds and hillocks and are instrumental in arresting the movement of the greater dunes and hence rendering them less mobile. These plants produce adventitious roots on the portions of their stems that are covered by sand, and new shoots replace the buried plant bodies. In this way the plant growth copes with the sand accumulation: plant growth keeps afloat on the sand inundation. The sand dwellers, on the other hand, are passive associates of blown sand.

The very mobile sand dunes of the barchan type are always sterile. The rate of their movement is too fast to allow for the settling down and storage of water in the sand, a phenomenon that controls plant growth on sand dunes. This process may be taken into consideration if the sand dunes are to be artificially

Fig VIII-I. A general view of a part of the mist-oasis of Erkwit (Sudan). Note the cactus-like *Euphorbia abyssinica* in the foreground, the *Acacia raddiana* in the valley bottom in the midground, and the hill in the background covered with an open growth of *Euphorbia* succulent-shrub form of vegetation

stabilised: the rate of dune migration needs to be checked mechanically before the appropriate binders are introduced.

The influence of mountains in causing orographic rains and other forms of water condensation is particularly noticeable in coastal deserts where chains of mountains fringe the coastal plain. The mist-oases of Erkwit, Elba and other parts of the Red Sea coastal mountains are examples of limited areas, associated with higher mountains, that support vegetation types much richer than the surrounding low country (see Figs VIII-I, VIII-2 and VIII-3). For details see [31, 32, 33].

Geological formations have a direct and an indirect effect on the plant growth. Each type of rock has a particular weathering form dependent on its structure, composition and lithology. The weathering form affects the features of the

Fig VIII-2. General view of the mist-oasis of Gebel Elba mountains. Note the plant growth on the slopes and in the valley. A tree of *Dracaena ombet* is seen in the foreground

Fig VIII-3. General view of the ground at the north-eastern feet of Gebel Elba (mountain range on the Sudano-Egyptian border). Note the *Acacia* scrubland within the drainage channel and the sterile coastal plain in the background. The mountain slopes in the foreground are covered with *Euphorbia cuneata* scrub

surface deposits and hence the water relationships and the plant growth. The rock formation influences the chemical and the physical composition of the soil, and regional geological structure often determines the pattern of the drainage systems and of the subterranean water reservoirs, all of which has a fundamental bearing on plant growth.

Water is evidently the most important ecological factor in the desert. We have already enumerated some of the aspects of the relationships between landform and the water regime. The second most important ecological factor is the destruction of vegetation due to human interference, including domestic grazing. This factor is also controlled, at least partly, by landforms that may make the habitat inaccessible to the destructive human-induced agencies. Relict patches of *Juniperus phoenicea* woodland are found in refugial sites in the mountains of Hallal and Yelleg of Sinai, otherwise it is absent in Egypt. There is evidence that it was previously much more widespread. Other examples may be quoted.

DYNAMICS OF DESERT VEGETATION

The common belief is that a desert is a virtually lifeless habitat with a few sparsely distributed depauperate plants. This is not true. The desert is a lively habitat with ever-changing aspects. The changes are seasonal, accidental and successional.

The seasonal aspects are the apparent expression of the phenological diversity of the plants and the seasonal rhythm of the climate. Let us consider the north African winter rainfall desert. The majority of ephemerals germinate in January, complete their life cycle by March and are dry in April. Certain succulent ephemerals (e.g. *Mesembryanthemum forskalei*) may persist till July; *Zygophyllum simplex* may persist for a complete year. The life span (and the size) of the ephemeral depends on the volume of the water resource, which varies from year to year and from place to place. Certain evergreen perennials may bear flower all the year (e.g. *Pityranthus tortuosus*, *Stachys aegyptiaca*, etc.), others have a restricted flowering season (e.g. *Zygophyllum coccineum*, summer; *Haloxylon salicornicum*, autumn). The winter deciduous perennials produce their new sprouts in the spring, flower by May and June, stay on throughout the autumn and dry in the winter (e.g. *Alhagi maurorum*). The summer deciduous perennials shed their foliage in July or earlier and go into their dormant phase. Certain species of this group (e.g. *Gymnocarpos decandrum*, *Fagonia kahirina*, etc.) may produce their new sprouts in the autumn prior to the outset of the rainy season. It seems that the plants are released during the autumn from the stress of extreme atmospheric dryness of the summer. Under the autumn climate the plants seem capable of balancing their water resources with their water expenditure. Dew may help to maintain this balance.

Seasonal changes also involve several of the physiological activities of the plants, notably the transpiration rates [4], and seasonal changes in the activity of the soil micro-organisms (including the nitrogen-fixing biota) and of the desert fauna will also influence the plant life. Many desert animals hibernate during the winter and the vegetation is thus saved, at least partly, from wild grazing.

The annual fluctuations of the climatic factors, especially rainfall, are one of the pronounced features of the desert climate. This entails fluctuations in plant growth. In years of heavier rainfall rich growth of ephemerals covers extensive areas with a green carpet. In other years, the ephemeral growth is very much reduced or may be absent. Many of the drought-deciduous perennials may fail to appear in years of exceptionally low rainfall. Zohary [34] refers to the desert species of *Poa* which fail to sprout in rainless years, produce only basal leaves in years with limited rainfall, and develop flowering culms in rainy years only.

Successional phenomena have been described in humid regions, but there was a belief that they do not exist in arid regions. Shreve [6], for instance, states:

One of the principal reasons for the abeyance of successional phenomena is the almost total lack of reaction by the plant on its habitat. The existence of a plant in a given spot for many years does nothing to make that spot a better habitat for some other plant or some other species. Fallen leaves and twigs are blown away, or small accumulations of organic matter are washed away. . . . Also little happens to change the character of the soil or its water-holding capacity. Only in the rich stands of desert vegetation is there a local amelioration of conditions due to the presence of large and long established perennials.

It is true that this autogenic type of vegetational succession is, apart from the building of phytogenic sand mounds, of little significance in the desert. Another type, allogenic succession, is one of the keys for understanding the desert vegetation. By allogenic succession is implied that successive waves of plant growth occupy an area due to gradual cumulative changes in the habitat produced essentially by physical processes independent of the plant growth. Desert vegetation manifests successional progressive and retrogressive changes. The former are due to the accumulation of soft material and the gradual building up of the surface of the soil; this may be attributed to water transportation or to wind transportation, or to both. The vegetational change concomitant with the gradual building up of soil is mainly the change from chasmophytes (on rock surface), to ephemerals (on thin soil), to succulent plants (on shallow soil), to non-succulent perennials (on less shallow soil), to a desert grassland, to a climax scrubland (*Acacia, Tamarix*, etc.) [35]. The climax vegetation will be dealt with later on.

The retrogressive changes are caused by the destruction of the surface soils and the removal of the products by exceptional torrents or storms. The habitat is rendered more susceptible to this adversity as a result of the artificial attenuation of plant growth due to cutting or grazing. Another example of retrogressive change is the gradual maturation of the gravel desert. The original mixed fluviatile deposits provide a habitat favourable for the growth of a desert scrubland. The mature desert armour stage, which results from the gradual removal of the soft ingredients and accumulation of lag gravel at the surface, is almost sterile. In between these two ends there are transitional phases of gradual impoverishment of plant growth. The post-maturation development of a sub-surface gypsum horizon adds to the causes of sterility.

GENERAL FEATURES

The general features of desert vegetation may be summarised in words of Shreve [7] as follows.

The outstanding characteristics of most desert communities are the low but unequal stature of the plants, the openness of stand, and the mixture of dissimilar life forms. . . . The low stature is responsible for the scarcity of truly subordinate plants. . . . The openness gives ample opportunity for the establishment of new individuals of the dominant species or for invasion by other species, but it is an opportunity which is rarely embraced. The lowness and openness combine to cause a relative scarcity of plant litter. The little that is produced is to a great extent eaten by termites and only to a very small extent reduced by fungi and bacteria. By the added agency of wind and sheet floods the plant litter is prevented from permanently enriching the spot where it falls. . . . Consequently this most important of the reactions of plants upon their habitats is reduced to a minimum in the desert. . . . The frequency in the desert of extensive communities which are simple in composition is not due to the poverty of the perennial flora so much as to the severity of the physical conditions. . . . In such simple communities the presence or absence of a particular species, or a change in its relative abundance, becomes important in determining the physiognomy. There are virtually no uncommon plants. . . . The same few species occur over and over again, and the recurrence of a particular habitat brings a repetition of its characteristic community. . . . It is not possible to use the term *Climax* with reference to desert vegetation. Each habitat in each subdivision of a desert has its own climax. . . . It is merely the particular group of species which, in somewhat definite proportion and with a fairly definite communal arrangement, is able to occupy a particular location under its present environmental conditions.

The desert vegetation, though open in its structural organisation, shows definite layering. The shrub layer is well developed only in a few scrubland types inhabiting especially favoured habitats. The suffrutescent layer, comprising low woody perennials 1–3 feet in height, is the main part of the permanent framework of the vegetation. The ground layer comprises a few dwarf or trailing perennials, and is enriched by the ephemeral growths of the wet season. The perennial framework forms an apparently homogeneous plant growth. Within this framework the ephemerals usually form a mosaic pattern of small patches, each dominated by one species.

The layering is more obvious in the subterranean growth. The perennials have deeply penetrating or horizontally extending roots. The ephemerals produce shallow-seated roots. The root growth is perhaps the main plant biological factor operative in the organisation of the plant community: the apparently open plant growth may be biologically closed due to root competition.

The vegetation within any stretch of the desert comprises a number of recognisable community types. These types are discernible in the field and repeat themselves with reasonable similarity. The bases of definition are necessarily elastic but combine the following:

(a) Each community type is characterised by one (or more than one) dominant species which is the most abundant perennial, the growth of which gives the vegetation its apparent homogeneity.

(b) The floristic assemblage of each type comprises a number of associate species that are present in the majority of the stands (sample plots), but not necessarily confined to a particular community type. (This is probably the main phytosociological difference between the community concept in desert vegetation and that in humid climate vegetation.)

(c) The main difference between the plant growth of the various community types is due to differences in the relative abundance and density of the species and not necessarily due to differences of absence or presence of species.

(d) Each community type needs to be referred to a discrete habitat type as a prerequisite of its identity. Plant growth (phytocoenosis) alone, though of special importance, is not enough to provide terms for recognising entities.

TYPES OF VEGETATION

We shall quote a few examples of attempts at the classification of desert vegetation in different parts of the world before suggesting the general principles that may be adopted. In his study on the vegetation of the Sonoran Desert (North America), Shreve [7] recognises seven subdivisions. For each he gives three sets of equivalent designations: the character of the vegetation, the preponderant or most distinctive genera, and the geographical location or some outstanding geographical features. The following are the seven subdivisions.

1. Microphyllous desert
   Larrea–Franseria region
   Lower Colorado Valley
2. Crassicaulescent desert
   Cercidium–Opuntia region
   Arizona Upland
3. Arbosuffrutescent desert
   Olneya–Encelia region
   Plains of Sonora
4. Arborescent desert
   Acacia–Prosopis region
   Foothills of Sonora

5. Sarcocaulescent desert
   Bursera–Jatropha region
   Central Gulf Coast
6. Sarcophyllous desert
   Agave–Franseria region
   Vizcaino Region
7. Arbocrassicaulescent desert
   Lysiloma–Machaerocereus region
   Magdalena Region

Within each of these main subdivisions he describes several types related to the landform pattern.

In a survey of the vegetation of the arid and semi-arid regions of Latin America, Cabrera [5] suggests the following system.

1. Desert: vegetation absolutely or practically nil; climate extremely dry (e.g. Peruvian coast and northern Chile).
2. Scrub steppe: vegetation of bushy or tussocky shrubs sparsely distributed, and interspaced by bare soil during the dry season. Climate very dry (cf. the dominant type of vegetation in Puna and Patagonia).
3. Cactus scrub (cardonal): vegetation of cereiform or *candelabrum Cactaceae* and low shrubs. Climate very dry (cf. the cactus scrub of Peru and the Ante-Puna, Argentina).
4. *Monte:* vegetation of taller and thicker-growing shrubs. Climate dry (e.g. north Mexico and Monte province, Argentina).
5. Thorn forest (espinal): vegetation of tall bushes and low trees with leaves or folioles of reduced size. Climate dry (e.g. the mesquite woods of central Chile).
6. Sclerophyllous scrub: dominant type of vegetation shrubs and small trees with coriaceous leaves. Climate less dry (e.g. vegetation of central Chile).
7. Grassy steppe: dominant vegetation consists of clumps of xerophil grasses fairly widely spaced. Climate slightly more humid (e.g. steppes of western Patagonia).

Reviewing the studies on the Australian vegetation, Davies [36] quotes the definition of major formations and their main subdivisions. The classification depends on a synthesis of the plant growth, the soil type and the climate (see Table VIII-3).

The classification of desert vegetation needs to take into consideration the geographic limits of regions, the geomorphologic pattern within the region, and the structural organisation and the floristic composition of the plant growth. The geographic limits of a biologic region are set by the climatic (and probably the palaeoclimatic) features; see for instance Emberger [37], north-west Africa; [38], Australia. Within the frames of the region the plant life is intimately related with the microclimatic and soil conditions which depend on the landform pattern. Several ecosystems may be recognised within each climatic division. Each ecosystem is a geomorphological unit associated with a biological assemblage.

TABLE VIII-3 *Relationships of arid plant formations and subformations to various climatic data and to soils, Australia (from Davies, 1955)*

Meyer ratio	Formations	No. of months $P/E > 0.5$	Mean annual rainfall, inches	Soils
15–25	Desert formations:			
	Sclerophyllous grassland	0	8	Desert sandhills
	Desert steppe	0		Stony deserts
25–40	Arid formations:			
	Sclerophyllous grass steppe[1]	0–2		Desert sand plains. Stony hills and plateaux. Brown soils of light texture
	Mulga scrub[2]	0–2	5–8	Desert loams
	Shrub steppe	0–2	5–8	Desert loams
40–50	Grass scrub[1]	2–3	10–15	Brown soils of light texture
	Arid scrub[2]	2–3	8–10	Desert loams
50–100	Semi-arid formations:			
	Savannah[1]	0–5	15–30	Grey and brown soils of heavy texture. Tablelands and ranges
	Mallee scrub[2]	4–5	10–15	Solonised brown soils
100–150	Savannah woodland[1, 2]	4–8	15–30	Tablelands and ranges, rendzinas and black earths, podsols, red-brown earths, terra rossa
	Mallee heath[2]	6–7	15–20	Sand sheet with solonised subsoil Lateritic sand plain

[1] 70–85° T isotherms.   [2] 50–70° T isotherms.

Within each ecosystem several community types may be defined on the basis of a plant growth type (phytocoenosis) and a habitat type; see the sections above on landform patterns, and general features. Let us consider an example that may illustrate these general principles.

In their ecological classification of the vegetation of the Sudan, Harrison and Jackson [39] divide the northern dry part into desert and semi-desert. This is a climatic division. The desert is the region with an average annual rainfall of 75 mm. The semi-desert region has the annual rainfall ranging from 75 to 300 mm.

Within the desert region perennial plant growth is confined to the dry water courses. Widespread patches of ephemeral herbs and grasses spring up after rare rain showers and give rise to the valuable *gizzu* grazing (type A.12 Sudan of Rattray [40]).

The vegetation of the semi-arid region is subdivided into five major types.

(i) Acacia tortilis–Maerua crassifolia *desert scrub.*

(ii) Acacia mellifera–Commiphora *desert scrub.*

'These two subdivisions occur where there is a sharp division into hard-surfaced, off-flow soils more or less completely bare of vegetation, at one extreme, and on-flow soils that have scrub bushes and even some trees at the other extreme.' Type (i) occurs in the east and type (ii) in the west. A gradual transition from one to the other exists between longitude 30° and 32°.

(iii) *Semi-desert grassland on clay*

'The typical dark cracking clay has medium or small cracks with a loosely friable and crumbly, almost spongy, surface texture.'

(iv) *Semi-desert grassland on sand*

'Large areas of vast undulating *gozes* (stabilised sand dunes) have a uniform vegetation, but there are marked differences . . . caused probably by differences in grazing pressure.'

(v) Acacia glaucophylla–A. etbaica *scrub*

This represents the vegetation of the Red Sea Hills.

This is a broad subdivision related to the main systems of landform and soil. Within each subdivision there is, as the quoted authors state, 'a number of distinct types of vegetation of too small an extent to map separately'.

The area of the semi-desert is about 190,000 sq. miles (ca 485,000 sq. km). An area of about 1,700 sq. miles (ca 4,400 sq. km) is surveyed by Kassas [41]. This area lies within the limits of type (i). Recognisable vegetation assemblages characterise the various landform units: hills, plains, gravel pavements, sand drifts and Khors (dry water courses). Within each of these vegetation assemblages one may differentiate a mosaic of various community types on the different slopes and the different parts of the same slope of a single hill. Plant growth varies according to the depth of the sand drift and to the dimensions of the catchment area of the Khor.

Parts of the Red Sea Hills were also studied in some detail [31, 32]. Ten vegetation types organised in successive zones are described. These would be subdivisions of type (v) Acacia glaucophylla–A. etbaica *scrub*. The main ecological factors concerned with the differentiation of the ten vegetation types are the altitude and the distance from the sea. These factors control the volume of the orographic rain.

VEGETATION-FORM

1. *Classification*

Desert vegetation, as all other types of plant growth, may be classified into plant communities. Each community type is dominated by one or more species. It is true that the dominants of the vegetation of the humid climates have some controlling influence on the associate species, but in the desert vegetation the dominant may not have such an ecological influence; the dominant is the most abundant species whose growth gives character and apparent homogeneity to the community. The growth form of the dominant gives the community its general appearance (physiognomy), the associates add detail to the structural features of the community. Hence it becomes possible to recognise several vegetation-forms that are linked to the growth forms of the dominants.

The desert vegetation comprises a number of principal vegetation-forms, each form embracing innumerable community types that differ in their floristic composition. These differences are due first to differences in the regional floras, and secondly to differences in the details of the habitat conditions. Yet all the types of a vegetation-form category bear some similarity in their ecological relationships and land use potential.

The main categories of desert vegetation-forms are:

A. 1. Accidental vegetation-form
B. Ephemeral Vegetation
   2. Succulent-ephemeral form
   3. Ephemeral grassland form
   4. Herbaceous-ephemeral form
C. Suffrutescent Perennial Vegetation
   5. Succulent-halfshrub form
   6. Perennial grassland form
   7. Woody-perennial form
D. Frutescent Perennial Vegetation
   8. Succulent-shrub form
   9. Scrubland form

2. Features and indicator significance

A.1. Accidental vegetation-form

In extreme deserts rain is not an annually recurring phenomenon. In the Dakhla district of the Libyan Desert (Egypt), for instance, 10 mm rain may fall as a cloudburst once every 10–11 years. This will cause the growth of many plants in specially favoured localities where run-off water accumulates. These plants are mostly ephemerals or potential ephemerals; they may also include a few perennials that may subsist for more than a year on the stored water; but the plant growth will usually last for a short season, then dry back to a long dormancy that extends for several consecutive years. Travellers across these extreme deserts will notice the dry straw of these plants, and will wonder at the growth of which it is evidence.

B. Ephemeral Vegetation

Ephemeral vegetation characterises deserts with recurrent annual rainfall. The species population is mainly of ephemerals but may also include perennial species that acquire an ephemeral growth form. We here refer to the ephemeral nature of the vegetation and not to the habit of the species.

This vegetational category indicates conditions that allow for no storage of moisture: soil wetness is maintained during a season but not the whole year. This may be due to the scantiness of rainfall, or to the soil conditions not allowing for the storage of moisture (e.g. thin surface deposits; see Fig VIII-4).

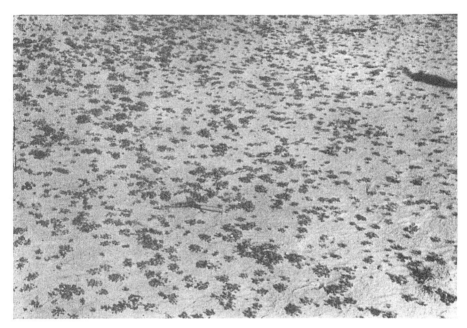

Fig VIII-4. The widespread growth of ephemerals on a thin sheet of sand overlying a gravel bed ˙

Ephemeral vegetation is universally the pioneer stage of secondary successions, arid and humid countries alike [42, p. 377]. In the less arid habitats it may soon become suppressed by the perennial vegetation, but in the arid habitats it may persist for a prolonged period.

2. Succulent-ephemeral form

Succulent ephemerals include such species as: *Zygophyllum simplex*, *Mesembryanthemum forskalei*, *M. crystallinum*, *M. nodiflorum*, *Aizoon canariense*, *A. hispanicum*, *Trianthema crystallina* (winter ephemerals), *Salsola kali*, *S. volkensii*, *S. inermis* (summer ephemerals), *Halopeplis amplexicaulis*, *Salicornia herbacea* (salt marsh ephemerals), etc. These are plants with a growing season longer than the usual 6-8 weeks of the non-succulent ephemerals. The prolonged life span is apparently due to the water stored in the plant tissues.

The desert succulent ephemerals (winter type) tolerate conditions of coarse surface sediments: erosion pavement, gravel desert, hamada, etc. The summer succulent ephemerals indicate, according to Negbi and Evenari [13], conditions where soil and natural vegetation have been disturbed.

3. Ephemeral grassland form

Ephemeral grasses are innumerable and in the Middle Eastern deserts include species of *Aristida, Eragrostis, Bromus, Poa, Schismus, Schoenfeldia, Stipa, Tragus, Cenchrus,* etc. Ephemeral grassland growth may cover extensive stretches of ground especially on shallow sand drifts. The importance of this form is notable in the warmer deserts. Reference has been made to the *gizzu* grazing of the western Sudan, mostly *Aristida* ephemeral grassland. According to Rattray [40, p. 50] herds and flocks will journey 300 to 400 miles (500-650 km) to take advantage of this ephemeral growth. Reference may also be made to the grass market of Omdurman (Sudan), to which caravans of camels and donkeys carry stacks of dried ephemeral grasses (see Fig VIII-14). Similar grassland growth appears after rain on many of the drier sandy parts of the Arabia, *Stipa tortilis* steppe of Vesey-Fitzgerald [43].

4. Herbaceous-ephemeral form

The growth of this form produces a mosaic pattern of patches which may be described as micro-oases amidst the otherwise sterile expanses of desert. Its

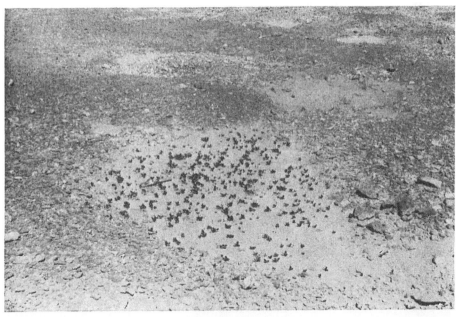

Fig VIII-5. Ephemeral growth restricted to small patches of depressed ground where some silt accumulated, otherwise the scree-covered erosion pavement is sterile

growth is associated with patches of soft deposits accumulated in especially favoured locales which provide a briefly sustained water supply (see Fig VIII-5). For more details see [7, pp. 119–34].

C. Suffrutescent Perennial Vegetation

This is the most widespread type of desert vegetation. It is distinguished by the permanent framework of perennial species. The plant growth comprises two main layers: a suffrutescent layer (30–120 cm high) and a ground layer. The former gives character to the vegetation and usually includes the dominant species; the ground layer comprises some of the dwarf and prostrate perennials and is enriched during the favourable season by the growth of ephemerals. There may be distantly spaced individual bushes or trees but their population is too thin to contribute to the characterisation of the vegetation.

5. Succulent-halfshrub form

Examples of this form are: the community types dominated by the various species of *Haloxylon, Anabasis, Zygophyllum*, etc., that are widespread in the Middle East deserts; and the salt-marsh communities dominated by species of *Salicornia, Arthrocnemon, Suaeda*, etc. This form of vegetation seems to indicate extreme conditions under which perennial vegetation may be established. The self-stored water seems to provide the plant with an internal water reserve when the soil fails to provide an ample supply of water, and thus tide the plant over the long dry season.

6. Perennial grassland form

Examples of this form include community types dominated by *Lasiurus hirsutus, Panicum turgidum, Pennisetum dichotomum, Poa sinaica, Hyparrhenia hirta*, etc. This vegetation-form is probably the most valuable for grazing potential. It usually indicates soil conditions that allow for the storage of some soil moisture in the sub-surface. The form is an effective soil binder and builder and may be successfully used for stabilising mobile dunes or sand sheets (see Figs VIII-6, VIII-7 and VIII-8).

Fig VIII-6. A general view of the *Panicum turgidum* growth (perennial grassland form) in the Cairo–Suez desert, Egypt

Fig VIII-7. A general view of the *Imperata cylindrica* growth (perennial grassland form) on a coastal sheet of sand, Gulf of Suez

Fig VIII-8. A general view of a grassland plain showing the effect of overgrazing. Note the single bush of *Acacia tortilis*

The different community types seem to indicate different soil types. For instance, *Lasiurus hirsutus* growth indicates shallow sandy soil, *Panicum turgidum* indicates deeper sandy soil, *Pennisetum dichotomum* growth indicates a calcareous soil, etc. The coverage of the plant growth may be taken as an index of grazing intensity (see Fig VIII-8).

7. Woody-perennial form

This form includes community types dominated by: *Zilla spinosa, Noaea mucronata, Artemisia* spp., *Astragalus* spp., etc. It indicates a transitional phase between the extremely adverse conditions of form 5 and the less adverse conditions of form 6.

This form also includes several types of desert saxatile communities. The plant growth is usually thin in the arid country and may be dense in the less arid habitats. Zohary [18] enumerates the following as dominant species of the lithophytic and chasmophytic plant communities: *Arenaria graveolens, Dianthus pendulus, Micromeria serpyllifolia, Ballota rugosa, Hyoscyamus aureus, Michauxia campanuloides, Varthemia iphionoides* and *Centaurea speciosa*; see also [44].

D. Frutescent Perennial Vegetation

This category includes the scrublands of the desert and semi-desert habitats. The plant cover comprises three main layers: a frutescent layer (120–300 cm high), a suffrutescent layer, and a ground layer. The first layer is dense enough, and it gives the vegetation a special character that distinguishes it from the previously described forms.

8. Succulent-shrub form

This form comprises the cactus communities of the American deserts, and the communities of the succulent giant euphorbias of the tropical African and Arabian deserts. It may also include the *Haloxylon persicum* scrub of the Middle Eastern deserts (saxual woods [18]) (see Fig VIII-1).

9. Scrubland form

This is the most highly organised form of the desert vegetation, and is present in especially favoured habitats. It indicates the highest levels of water revenue in the desert and also the most effective regime for water storage in the soil. The surface deposits are usually deep enough for the establishment of a sub-surface reserve of moisture; or the topographic features are such (high coastal mountains) that an ample supply of orographic rain is available [31, 32, 33].

This form includes community types dominated by species of *Acacia, Tamarix, Ziziphus, Pistacia, Retama, Larrea, Prosopis*, etc. (see Fig VIII-3).

The desert ecosystem

In the previous pages we have discussed the main features of the desert vegetation and its relation to the physical aspects of the environment. It is appropriate at this stage to consider the biotic aspects of the environment: the ecological phenomena may better be conceived in their complex union between the organisms (the biome) and the environment (the habitat). The biome is the whole complex of plants and animals (including man) living together in a social mass. The ecosystem comprises the biome and its habitat.

ANTHROPOGENIC FACTORS

By this we mean the direct and indirect influences of man's action on the ecological complex, be it by deliberate action or otherwise. Anthropogenic influence is universal but its outcome is most dramatic in the desert, where plant life subsists under adverse conditions of aridity and maintains a precarious equilibrium with its environment. Under such circumstances the balance may be easily upset and a chain of change initiated. The present vegetation bears evidence

Fig VIII-9. Camels feeding on the pungently spinescent *Acacia raddiana*

of the anthropogenic influence through the ages. The following are the principal aspects of this influence (see Figs VIII-9, VIII-10, VIII-11, VIII-12, VIII-13 and VIII-14).

(*a*) Complete destruction of the natural vegetation is carried out for cultivation. This is more pronounced in the semi-arid fringes of the desert where an annual crop (e.g. barley, millet) or an orchard crop (e.g. olive) may be cultivated. In more arid country the ground of specially favoured parts of the drainage systems (wadis and their flood plains) may be cleared for cultivation that is

Fig VIII-10. A general view of *Calotropis procera* on the sandy plain to the west of Port Sudan. The goats have finished almost every other plant except this poisonous shrub

Fig VIII-11. A caravan of camels carrying *Acacia* wood to the market of Port Sudan

Fig VIII-12. Relicts of *Tamarix aphylla* scrub on remnants of silt terraces of Wadi El Ghweibba (Gulf of Suez district)

Fig VIII-13. Remnants of a destroyed *Tamarix* thicket, Wadi Sheikh, Sinai. Note the Pleistocene silt terraces in the background

practised in years of good rainfall. The farming practices may comprise a variety of water-spreading methods, e.g. gravel mounds and strips, embankments, etc. In the case of annual-crop cultivation, the soil is left bare for a part of the year or for several consecutive years. Soil erosion will then be unhampered.

(b) Partiál destruction of the natural vegetation is the result of cutting, grazing,

Fig VIII-14. A general view of the desert plain west of Omdurman, Sudan. Note remnants of the ephemeral grassland growth and the caravan of donkeys carrying stacks of dried grass to the market of Omdurman

etc. These are usually selective processes and will cause one or all of the following: (1) reduction of the total plant cover, (2) change of composition, (3) change of dominance.

(c) The influence of grazing on the total plant cover is accentuated in exceptionally dry years. Many of the species that are avoided in the normal years are grazed in drier years (see Fig VIII-9). Man is no exception in this respect. Many wild grasses and other plants provide cereal foods under famine conditions, e.g. *Dactyloctenium aegyptium, Hyparrhenia* spp., *Oryza* sp., *Amaranthus* sp. and *Nymphaea lotus* [45].

(d) Whether the plant as a whole, its seeds or its subterranean part is sought is an important aspect of the selective action of destruction. Consider, for instance, three desert plants that are collected for their medicinal value: *Hyoscyamus muticus, Urginia maritima* and *Colocynthis vulgaris*. The shoots of the first species are cut, the bulbs of the second are dug out, and the fruits of the third are collected. The net outcome of these practices is extreme reduction of the population of the first two; *Colocynthis vulgaris* is, however, not rare.

Let us compare the growth of two bushes in the Red Sea Hills of Egypt: *Moringa aptera* and *Acacia raddiana*. The seeds of the former are collected for their valuable ben-oil and its trees are guarded. *Acacia* wood is cut for fuel and charcoal manufacture and the trees are destroyed.

(*e*) Certain species are not grazed in their fresh stage but may be eaten when dry, e.g. *Anabasis setifera* and many other desert succulents. These may be cut, dried and stored for feed.

(*f*) Moderate grazing or partial thinning may have some beneficial effects. The former may mean the artificial reduction of the transpiring surface and may thus be a cause of survival of the individual. Partial thinning may release some of the associates of the full competition, and may even allow for the self-regeneration of the thinned species. For instance, the roots of the creosote (*Larrea tridentata*) seem to excrete toxic substances which prevent its own seedlings from establishing themselves; but 'we commonly find young creosote bushes along roads in the desert, where road builders have torn up the old bushes' [15; 7, p. 159].

(*g*) Soil erosion: removal of surface deposits of soft material is a universal sequel of the destruction or reduction of the plant cover. This is often an irreparable process, and it may become impossible to regenerate the natural vegetation.

(*h*) The final outcome of the uncontrolled or ill-advised human interference may be the expansion of the desert ecosystem over regions that are not naturally deserts, hence the so-called man-made deserts. This is usually followed by the migration of the human population towards less arid or less destroyed regions and the story will be repeated. We may quote the instance of the Dongotona Hills (Sudan) described by Jackson [46]. He shows that the widespread erosion and the drying up of perennial streams have taken place within comparatively recent times, due to the invasion by the Dongotona tribes into the higher hills by pressure of successive waves of migration from the south-east. He adds, 'The rapidity of erosion and its results on the water supply show how swiftly the country can sustain immense damage.' Similar examples may be quoted from northern Nigeria, the Middle East, etc.

OTHER BIOTIC FACTORS

As plants live together they exercise two aspects of social relationship, dependent unions and commensal unions [3].

The dependence of the one-sided parasite on its host is an intimate form of communal relation. The destruction of the host will likewise mean the eradication of the parasite. *Loranthus* spp. parasitise on species of *Acacia*; *Cistanche* spp. on species of the succulent desert chenopods; *Cuscuta* spp. on a variety of desert plants. The mutual parasitism is well represented in the growth of lichens, which are the hardiest members of the desert plant life, and in the various forms of mycorrhiza and the bacterial association with plant roots. The nodule bacteria, universally associated with the Leguminosae, are also recorded in members of the Zygophyllaceae of the deserts [47]. The nitrifying power of the desert soils, due to the activities of micro-organisms, was the subject of many investigations.

The association between the climber and its support is not so specific except for the growth form. It is, however, of interest to note that some of the climbing species may persist after the destruction of the support. In such instances they may change their habit, but may still be taken as an indication of previous existence of their supporting shrubs. The climbing *Cocculus pendulus* acquires a trailing habit, *Ochradenus baccatus* a bushy form, *Pergularia tomentosa* a self-twining form, etc.

Some of the delicate desert plants (e.g. *Launaea nudicaulis*) survive only within the growth of woody bushes.

Commensalism involves the struggle for space, nutrition, water, light and other requirements, that is to say, competition. This is a universal phenomenon, perhaps much more obvious in the less arid countries where vegetation is dense. Bonner [48] draws attention to what he calls the chemical warfare between plants: some plants secrete chemical compounds which are toxic to other species. The leaves of *Encelia farinosa*, a species of the low hot deserts of the United States south-east, contain a toxic compound (3-acetyl-6-methoxyl benzaldehyde) which is toxic to many of the associate species, especially ephemerals. The roots of *Parthenium argentatum*, a rubber-bearing shrub of the American deserts, give off a substance (cinnamic acid) that is toxic to its seedlings. This is comparable to the instance of the creosote bush to which reference has been made.

CLIMAX COMMUNITIES

The climax, by definition, is the stable type of vegetation marking the close of a series of developmental stages (seral types of vegetation). It is in dynamic equilibrium with the prevailing climate and with the mature stage of soil development conditioned by this climate. But within the deserts the microclimates which are conditioned by landform are important factors; the soils are often too mobile to allow for the development of what may be considered mature soils; and anthropogenic factors have been actively changing the whole ecosystem to types that have become relatively stable. For these reasons one may better adopt, in desert ecology, the notion of a climax that is the potential type of plant growth that may be established if the habitat is protected against anthropogenic agencies for a period of time, say 100 years.

The theoretical visualisation of such a climax type will always be in way of approximation, but is an important corollary of the estimate of land-use potential. Its reconstruction will depend on the study of the relict patches of what seems to be the original vegetation, the study of the habitat conditions associated with the various types of communities and the possible developmental relation between these habitats, studies of the ecological requirements of the various species, and the most valuable information obtained from exclosure experiments (see Fig VIII-15). As an example reference may be made to Kassas and Imam, who

Fig VIII-15. A general view of an exclosure established to the west of Omdurman, Sudan, six years before the photograph was taken. Note the contrasting difference between the plant life on the two sides of the fence

Fig VIII-16. A general view of the growth of *Haloxylon salicornicum*. Note the sand hillocks built around this succulent chenopod. The ground between the hillocks is covered with gravel and is sterile.

studied the community types of the desert wadis, reconstructed the types of the climax communities, and suggested the applicability of their conclusions in land reclamation of desert wadis.

M. KASSAS

References

[1] ADAMSON, R. S., 'The classification of life-forms of plants', *Bot. Rev.*, Vol. 5, 1939, pp. 546–61.

[2] RAUNKIAER, C., *The Life Forms of Plants*, Oxford, 1934.

[3] BRAUN-BLANQUET, J., *Plant Sociology*, English ed., New York, 1932.

[4] ZOHARY, M., 'Hydro-economical types in the vegetation of Near East Deserts', *Biology of Deserts*, Inst. Biol., London, 1954, pp. 56–67.

[5] CABRERA, A. L., 'Latin America' (review of research), *Plant Ecology*, Unesco, Arid Zone Research, Vol. 6, 1955, pp. 77–113.

[6] SHREVE, F., 'The desert vegetation of North America', *Bot. Rev.*, Vol. 8, 1942, pp. 195–246.

[7] — 'The vegetation of the Sonoran Desert', *Carnegie Inst. Wash.*, Pub. No. 591, 1951.

[8] STOCKER, O., 'Physiological and morphological changes in plants due to water deficiency', *Plant Water Relationships in Arid and Semi-arid Conditions*, Rev. Res. Unesco, 1960, pp. 63–104.

[9] EVENARI, M., 'Plant physiology and arid zone research', *The Problems of the Arid Zone, Proc. Paris Symp.*, Unesco, Arid Zone Research, Vol. 18, 1962, pp. 175–195.

[10] HAINES, R. W., 'Potential annuals of the Egyptian desert', *Bull. Inst. Desert*, Vol. 1, 1951, pp. 103–18.

[11] DAUBENMIRE, R. F., *Plant and Environment*, New York, 1947.

[12] WENT, F. W., 'The effect of rain and temperature on plant distribution in the desert', *Desert Research* (Jerusalem Symposium), 1953, pp. 230–40.

[13] NEGBI, M. and M. EVENARI, 'The means of survival of some desert summer annuals', *Plant Water Relationships in Arid and Semi-arid Conditions, Proc. Madrid Symp.*, Unesco, Arid Zone Research, Vol. 16, 1962, pp. 249–59.

[14] MAXIMOV, N. A., *The Plant in Relation to Water*, London, 1929.

[15] WENT, F. W., 'The ecology of desert plants', *Scientific American*, April, 1955.

[16] CANNON, W. A., 'General and physiological features of the vegetation of the more arid portions of southern Africa, with notes on the climatic environment', *Carnegie Inst. Wash.*, Pub. No. 354, 1924.

[17] KACHKAROV, D. N. and E. P. KOROVINE, *La Vie dans les Déserts*, French ed., 1942.

[18] ZOHARY, M., 'On hydro-ecological relations of the Near East desert vegetation', *Plant Water Relationships, Proc. Madrid Symposium*, Unesco, Arid Zone Research, Vol. 16, 1962, pp. 199–212.

[19] POLUNIN, N., *Introduction to Plant Geography*, London, 1960.

[20] OPPENHEIMER, H. R., 'Adaptation to drought: Xerophytism', *Plant Water Relationships in Arid and Semi-arid Conditions*, Rev. Res. Unesco, Arid Zone Research, Vol. 15, 1960, pp. 105–38.

[21] ZOHARY, M. and G. ORSHANSKY, 'The *Zygophylletum dumosi* and its hydro-ecology in the Negev of Israel', *Vegetatio*, Vol. 5/6, 1954, pp. 340–50.

[22] — — 'Ecological studies on lithophytes', *Pal. Journ. Bot.*, Jer. Ser., Vol. 15, 1951, pp. 119–28.

[23] ORSHANSKY, G., 'Surface reduction and its significance as a hydroecological factor', *Journ. Ecol.*, Vol. 42, 1954, pp. 442–4.

[24] STOVER, E. L., *Anatomy of Seed Plants*, Boston, 1951, Ch. 8.

[25] *Proc. Madrid Symp.*, Unesco, 1962.

[26] *Reviews of Research*, Unesco, 1960.

[27] KRAMER, P. J., *Plant and Soil Water Relationship*, New York, 1949.

[28] CRAFTS, A. S., H. B. CURRIER and C. R. STOCKING, 'Water in the physiology of plants', *Chronica Botanica*, Waltham, Mass., 1949.

[29] KASSAS, M., 'Habitat and plant communities in the Egyptian desert', I. Introduction, *Journ. Ecol.*, Vol. 40, 1952, pp. 342–51.

[30] — 'Landforms and plant cover in the Egyptian desert', *Bull. Soc. Géog. d'Égypte*, Vol. 24, 1953, pp. 193–205.

[31] — 'The mist oasis of Erkwit, Sudan', *Journ. Ecol.*, Vol. 44, 1956, pp. 180–94.

[32] — 'Certain aspects of landform effects on plant water resources', *Bull. Soc. Géog. d'Égypte*, Vol. 33, 1960, pp. 45–52.

[33] TROLL, C., 'Wüstensteppen und Nebeloasen im Sudnubischen Küstengebirge, Studien zur Vegetations und Landschaftskunde der Tropen', *Zeitschrift der Gesellschaft für Erdkunde zu Berlin*, Nr. 7/8, 1935, pp. 241–81.

[34] ZOHARY, M., *Plant Life of Palestine*, New York, 1962.

[35] KASSAS, M. and M. IMAM, 'The wadi bed ecosystem', *Journ. Ecol.*, Vol. 42, 1954, pp. 424–41.

[36] DAVIES, J. G., 'Australia (review of research)', *Plant Ecology*, Unesco, Arid Zone Research, Vol. 6, 1955, pp. 114–34.

[37] EMBERGER, L., 'Afrique du Nord-Ouest (review of research)', *Plant Ecology*, Unesco, Arid Zone Research, Vol. 6, 1955, pp. 219–49.

[38] — 'La place de l'Australie Méditerranéenne dans l'ensemble des pays Méditerranéens du vieux monde', *Biogeography and Ecology in Australia*, W. Junk, Den Haag, pp. 259–73.

[39] HARRISON, M. N. and J. K. JACKSON, 'Ecological classification of the vegetation of the Sudan', *Forest Bulletin*, No. 2, New Series, Khartoum, 1958.

[40] RATTRAY, J. M., *The Grass Cover of Africa*, FAO Agricultural Studies, No. 49, 1960.

[41] KASSAS, M., 'Landforms and plant cover in the Omdurman desert, Sudan', *Bull. Soc. Géog. d'Égypte*, Vol. 29, 1956, pp. 43–58.

[42] CLEMENTS, F. E., *Plant Succession and Indicators*, New York, 1928.

[43] VESEY-FITZGERALD, D. F., 'The vegetation of central and eastern Arabia', *Journ. Ecol.*, Vol. 45, 1957, pp. 779–98.

[44] DAVIS, P. H., 'Cliff vegetation in the eastern Mediterranean', *Journ. Ecol.*, Vol. 39, 1951, pp. 63–93.

[45] FERGUSON, H., 'The food crops of the Sudan and their relation to environ-ment', *Proc. Phil. Soc. Sudan*, 1953 (1954) Conference.

[46] JACKSON, J. K., 'The Dongotona Hills, Sudan', *Emp. For. Rev.*, Vol. 29, 1950, pp. 139–42.

[47] SABET, Y. S., 'Bacterial root nodules in the Zygophyllaceae', *Nature*, Vol. 157, 1946, p. 656.

[48] BONNER, J., 'Chemical sociology among plants', *Scientific American*, March 1949.

ADDITIONAL REFERENCE

ECKARDT, F. E. (Ed)., *Methodology of Plant Eco-physiology*, Proc. Montpellier. Symp., Unesco, Paris, 1965 (Arid Zone Research xxv).

CHAPTER IX

Animals of the Desert

Deserts by and large do not support so large a number of animal species as other areas do, but the animals that live there are often highly adapted, sometimes bizarrely so, to their environment. To attempt a systematic treatment of such animals would be far beyond the scope of the present chapter, so that our approach will be rather to consider the desert environment in so far as this presents certain peculiar problems for animal life, to derive therefrom certain general principles so far as this is possible, and then to illustrate and amplify these by examples. A good deal of information about the physiological ecology of desert animals is now available, and the reader who requires a fuller treatment of these matters than is possible here may be referred to Schmidt-Nielsen's admirable *Desert Animals* [1] and to reviews by various authors in *Adaptations to the Environment*, being Section 4 of the *Handbook of Physiology* [2, 3, 4, 5, 6, 7].

Some of the most important factors which characterise terrestrial (as opposed to aquatic) habitats are also those which, in an extreme form, pertain to deserts. It might be expected, therefore, that those groups of animals which are already well adapted for terrestrial life in a general way, will be better represented in deserts than those which are not. On these grounds we can see at once why mammals, reptiles and birds among the vertebrates, and insects and arachnids among the invertebrates, include nearly all the inhabitants of deserts, for these same groups are also well adapted to general land life while others are not.

There are basically three problems confronting animals which evolve from aquatic to terrestrial habitats: the necessity to breathe air, to conserve water and to avoid, tolerate or control extremes of temperature. These, of course, are also the problems, often in an acute form, that face animals which live in deserts. Accordingly it will be useful now to consider the means adopted by each of the main groups of terrestrial animals in solving these problems, for this may be expected to throw light on many of the extreme adaptations found in desert animals.

Water relationships

Three major phyla of animals have, during the course of their evolutionary history, invaded the land. These are the vertebrates, the arthropods, and the molluscs. The vertebrate answer to the respiration problem was to replace gills by internal lungs. This occurred in two stages: in the amphibian stage the lung was not good enough as an organ for total respiratory exchange and a proportion of the oxygen uptake took place through the skin, which therefore had to remain moist. In reptiles, however, a more efficient lung was developed, which was able to take care of all oxygen requirements, and consequently the skin could be made relatively impermeable to water, with consequent advantages in water conservation. Birds and mammals inherited their lungs from reptilian ancestors.

The mere fact that a lung is internal is by itself of no importance in water conservation. What is important is that the lung need be only partly ventilated when the animal is at rest – and this of course does conserve water because only a small part of the moist respiratory membrane is exposed to the drying effect of each inhalation. When the need arises, ventilation becomes deep and the whole surface is available for oxygen uptake.

Land arthropods use an entirely different system for solving the same problem. Insects have developed an internal system of ramifying tubules, the tracheae, which carry (largely by diffusion) oxygen to all parts of the body. This appears to be a very satisfactory system so far as respiration is concerned, but again, as a means of water conservation it works only because the rest of the insect's surface is relatively waterproof – sometimes surprisingly so – and because the external openings of the tracheae, the spiracles, are capable of being closed when the insect is at rest, so that water vapour is not lost from the respiratory surfaces all the time. In the arachnids (spiders and scorpions), which seem to have evolved as terrestrial arthropods separately from the insects, respiratory exchange occurs in 'lung books'. These are sets of leaf-like respiratory surfaces set in internal cavities whose narrow openings are also occlusible. The general integument of arachnids is rendered impermeable in a similar way to that of insects, by the presence and maintenance of a thin wax-like layer at or near the surface.

The crustacea are arthropods which have but few representatives on land, probably because even the terrestrial members breathe by what are in effect gills. The whole integument is also used for oxygen uptake, and is much more permeable to water than that of insects, so that the twin problems of respiration and water conservation have not been efficiently solved.

Water may also be lost in the excretion of waste nitrogen. In aquatic forms nitrogen is excreted as ammonia, which is economic because no carbon is lost, though a moist integument is necessary for its elimination. In terrestrial animals, ammonia is replaced by urea or uric acid as the end product of nitrogen metabolism. Both these materials are relatively non-toxic, and although urea requires more water for its elimination, uric acid is very nearly insoluble and therefore requires hardly any free water.

Birds and reptiles excrete waste nitrogen in the form of uric acid. Mammals, however, use urea, probably for the following reason. Uric acid, being very insoluble, crystallises out very readily when water is withdrawn from the urine to concentrate it. This happens in the terminal portion of the kidney tubules, and consequently, if uric acid were present, blockage would occur in these ducts, in the fine ureters and in the urethral duct from the bladder to the exterior. In birds and reptiles, the concentrated urine does form a solid gel, but this can be squeezed by peristalsis from the collecting ducts of the kidney directly to the cloaca, which, of course, opens widely to the exterior.

Insects excrete waste nitrogen as uric acid, and in circumstances where water conservation is important they lose virtually no water at all in this way. Arachnids use guanin, which is just as effective.

The view that the environment, particularly the embryonic environment, of each group of animals determines the nature of its nitrogen end product, is no longer tenable, for there are too many exceptions, but we can at least agree that uricotelic animals (those whose end product is uric acid) are well adapted in this respect to land life.

Temperature relationships

Questions of size have great importance here. Evaporation of water necessarily involves the absorption of heat (internal or external), so that if evaporation occurs at the surface of an animal's body, some of this heat at least will be taken from the body and consequently the animal will be cooler. It is clear, then, that loss of water has its compensations, and whether or not it is mainly beneficial or harmful depends upon the balance of a number of conditions, chief of which are the water balance and the temperature stress of the animal at the particular time concerned.

Now in general it is easier for a large animal than for a small one to make use of water evaporation for cooling, because it has a larger volume and hence a greater water reserve, per unit area of surface. It can, therefore, suffer a given rate of water loss per unit area for a longer time before its water content is reduced to a lethal level. Thus a flea could tolerate a loss of 5 mg/cm^2/hr for only 15 minutes before losing 10 per cent of its body weight of water, while a man could tolerate 4,500 times this rate (or an equivalent rate for 4,500 times as long) and suffer the same proportional loss of water (see Fig IX–1).

Since heat is absorbed at the site of evaporation, it is important that this shall be at the surface of the body, and not separated from it by an insulating layer. If an animal has a thick fur coat whose spaces become flooded with liquid when the animal sweats, then evaporation will occur at the outer surface of the fur, and (since fur is a good heat insulator) much of the heat absorbed will come from the surrounding air rather than from the animal's body. The same argument reveals the disadvantage of a subcutaneous layer of fat in an animal which needs to cool itself by sweating from the skin. On the other hand, a hair layer which prevents direct solar radiation of the skin, while still permitting movement of water vapour, is clearly advantageous both in reducing the heat load and in ensuring that heat is withdrawn from the skin and not from the surrounding air.

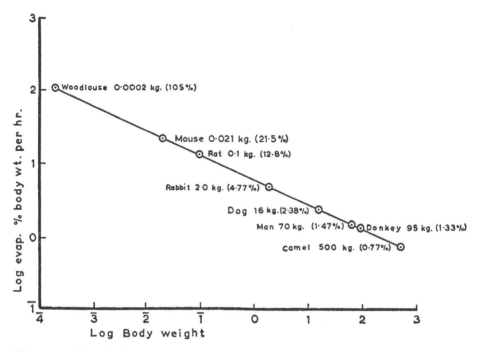

Fig IX-I. The relation between body weight and the amount of water which must evaporate per hour from the body to preserve a constant temperature in desert conditions. (*Data in part from Schmidt-Nielsen* [18])

Water and temperature relationships in arid areas

So far what has been said refers to life on land generally. We may now summarise the argument and see particularly how it applies to what are in fact extreme terrestrial conditions – desert areas.

Loss of water associated with nitrogen excretion is the only form of water loss that has no possible compensation in a hot environment, for the heat loss associated with evaporation occurs at a distance from the body. Water loss from the respiratory surfaces is inevitable although it may be reduced by minimal lung ventilation, as in birds, reptiles and mammals, or by occlusion devices, as in insects and arachnids. When respiratory membranes are external, as in the skin of amphibians or the gills of woodlice, this source of water loss is often large and may prevent the colonisation of dry areas on land.

However, loss of water from the integument of an animal is probably inevitable. In small animals, such surface transpiration must be reduced to a minimum if they are to exist in a dry habitat. If it is not so reduced, then surface cooling may occur, and may be taken advantage of but only for a short time, and such animals will be confined to cryptozoic niches. Larger animals can afford surface transpiration, and indeed often use this as a temperature-controlling mechanism.

It seems then that of the three variables: size, nature of the integument and nature of the respiratory surface, only certain combinations are feasible in a well-adapted land animal, and, *a fortiori*, in a desert living form. Small size combined with an occlusible respiratory system and an impermeable integument is a satisfactory combination for land life, because a high metabolic rate is unnecessary and because such animals can make use of small shelters in crevices and so forth. Large size, a ventilated respiratory membrane and a permeable integument is also satisfactory, but other combinations are less so.

To these interdependent variables must be added a number of independent ones – that is variables whose optimum value always lies at one end of the possible range. Thus it is always an advantage in arid areas to lose as little water with nitrogen excretion as possible, and to have a high upper lethal temperature. Such optimum values are not always attainable, perhaps because of an animal's evolutionary history. Thus mammals owe their success to maintenance of a constant temperature, and to permit the body temperature to rise as high as it may in some insects or reptiles would adversely affect the whole organisation. Again, amphibia and mammals excrete soluble urea – and improvement here lies only in the extent to which the urine may be concentrated. Nitrogen cannot be excreted entirely without water in the form of uric acid by these animals, because the amphibian kidney is incapable of concentrating urine and for reasons discussed above in respect of mammals.

A further, but less general consideration concerns the shape of animals, and we may take this here because it is involved with temperature tolerance. One way in which an animal can acquire a larger surface area without changing its volume is, of course, to change its shape. A mammal, for example, with long thin extremities – legs, ears, and neck – will present a greater surface for convective cooling by the air (when the air temperature is lower than that of its body). In fact a number of desert mammals do conform to this plan, which has come to be known as 'Allen's rule'.

There are several other more or less general problems which desert animals face, and to which they often respond in characteristic ways – locomotion over shifting sand is one of these – but it will be more convenient if we take these as they arise during the remainder of this chapter.

Large mammals

One of the best-known desert animals, both popularly and scientifically, is the camel – about whose physiology and performance we now have considerable information as a result of the work of the Schmidt-Nielsens [8] and others. We shall consider this animal as an example of a large mammalian desert animal, and see how far our general principles are borne out.

The physical conditions of desert environments are described elsewhere in this book, so that they need not be repeated here, save to recall two important points. Firstly, that there is a wide fluctuation in temperature (and consequently of relative humidity) each day, and secondly, that temperatures in sheltered places such as the bases of vegetation, rock crevices and burrows below the soil, are much more equable than in the open.

Camels conform to our theoretical picture of a desert animal fairly well. Being large they cannot lead a cryptozoic life, but they can afford to reduce the heat load by sweating. Their main energy store, fat, is concentrated in the hump and not deployed in a subcutaneous layer all over the body. They are 'spindly' in shape, and they have a covering of hair which is not so dense as to become saturated with sweat.

But this is by no means the whole story, for this remarkable animal possesses a number of physiological adaptations which in combination enable it to survive in the open desert for longer periods than perhaps any other comparable animal. It is capable of drinking, when the opportunity occurs, very fast and very deep. A camel may tolerate a loss of water equal to 25 per cent or more of its body weight and replace this by drinking. One of Schmidt-Nielsen's camels which weighed 325 kg drank 103 litres of water in 10 minutes and this was evenly distributed through the tissues in two days. Donkeys are almost as good as camels in this respect, but most mammals, man included, die after losing about 12–15 per cent of their body weight in water.

When man and other animals lose water, this occurs both from the tissues and from the blood, so that the blood becomes more viscous until, at a critical point, the heart is unable to circulate it fast enough to transfer central body heat to the surface for cooling. At this point the animal rapidly dies in what Adolph and Dill [9] call 'explosive heat death'. Camels, on the other hand, are able to tolerate a much greater water loss because, owing to some physiological process whose mechanism is not fully understood, water is withdrawn mainly from the tissues and the gut, so that the blood suffers hardly any concentration. In a camel which lost 50 litres of water, the blood volume was reduced by only 1 litre.

MacFarlane [10] and his co-workers have recently found that in conditions leading to severe dehydration, camels lost 20·5 per cent of their body weight in 9 days while merino sheep lost 23·5 per cent in 4 days. Of this, the cells and gut contents contributed water to the extent of 81·2 per cent of the total weight loss in camels, while only 46·2 per cent was derived from these sources in the sheep. There is evidence too, that by far the largest proportion of the water was derived from gut contents in the camel.

The precise mechanism whereby this occurs certainly deserves further study. Since camels can also absorb huge quantities of water very rapidly and tolerate the osmotic stress which must result, it would seem that there is some form of physiological 'water space' in the tissues. Camels certainly do not store water in bulk either in the hump or elsewhere. The hump is fat – and although it is perfectly valid chemistry to point out that the complete oxidation of 100 g fat yields 107 g water, whereas oxidation of 100 g carbohydrate yields only about half this amount, yet there is no real advantage in storing fat, so far as water conservation is concerned, because 100 g fat needs more than twice as much oxygen as an equal weight of carbohydrate does to oxidise it. Extra oxygen means extra breathing, and this means extra loss of water from the moist respiratory membranes. Actually, for the production of equal amounts of *energy*, the combustion of carbohydrates yields rather more water than fat does.

'Metabolic water' is indeed an extremely important source of water for desert animals. It is possible, as we shall see later, for some animals to live on an entirely waterless diet, and obtain sufficient water for their needs from the oxidation of their food. But the oxidation of all forms of food (fat, carbohydrate or protein) produces water in all animals – there is nothing peculiar to desert animals about this. What is distinctive is their capacity to reduce the loss of water in other ways and so to bring their water output down to no more than their water income from metabolism.

Another adaptation shown by camels is that unlike most mammals, they permit their body temperature to vary over a wide range. During the cool night a camel's temperature may fall to 34°C and it rises slowly during the heat of the day to as much as 41°C before sweating begins. The large volume of a camel thus acts as a heat buffer, for more than 3,000 kilocalories of heat must be absorbed (or lost) to change its temperature through 7°C.

Small mammals

Before we turn to more particular aspects of desert adaptation it will be useful to compare the water and temperature relationships of small desert mammals with those of the camel.

Since they are small, these animals are not able to regulate their body temperature in a hot environment by sweating, and consequently they avoid true desert conditions by digging burrows in which they remain during the heat of the day, and by foraging for food at night. As we might expect on theoretical grounds, they do not have sweat glands, although there is a safety device whereby evaporative cooling can come into play in an emergency. Thus the normal temperature of the American kangaroo rat, *Dipodomys merriami* (a widespread inhabitant of extreme deserts), is 36–37°C, and this (as in the camel) is permitted to rise in warm air. But when the animal's temperature approaches about 42°C (which is lethal) a copious foaming at the mouth occurs, which wets the fur, evaporates, and so reduces the body temperature. Such rats have been known to survive 43°C for 20 minutes, while the common white rat dies at 39°C. During such emergency cooling kangaroo rats lose as much as 15 per cent of their total body water. Clearly they could not continue in this way for long, and in fact the kangaroo rat is not exposed in real life to these extreme conditions.

Kirmiz [11] has found that jerboas (*Dipus*) fall into a state of deep sleep at extreme temperatures about 35°C (which cuts metabolic heat production to a minimum) and salivate copiously at 40°C.

Carpenter [12] has worked on the heat and water relationships of two species of kangaroo rat, *Dipodomys merriami* (mentioned above) and *D. agilis* (which is limited to the relatively mesic coastal ranges of southern California). By measuring temperatures in their burrows at different times of the year, Carpenter has been led to believe that the 'difficult' aspect of the environment for both these kangaroo rats is not heat at all, but cold. This underlines a most important point about desert climates which is apt to be overlooked – dry deserts, particularly at night in the winter time, are often extremely cold. Carpenter believes that this is reflected in the physiology of kangaroo rats in so far as both these species have much better thermal insulation between body and surrounding air than any other mammal of equivalent size. This is very useful at low temperatures, because the rate of metabolism need not be increased to produce more heat until the outside temperature is below 25°C, and even below this the increase is small compared with that in other mammals. But in hot environments, of course, good thermal insulation is in some circumstances a disadvantage. Carpenter's measurements suggest, however, that the animals, by remaining in their burrows during the heat of the day, can stay within their 'thermal neutral zone', i.e. they do not have to waste energy and water in keeping cool. Their powers of temperature regulation at high temperatures, however, are not nearly so good as those of the antelope ground squirrel, *Citellus leucurus*, studied by Hudson [13].

However, it is essentially in water conservation that kangaroo rats and jerboas excel. Although they will drink if water is available, they can live indefinitely on a dry diet. They achieve this by several mechanisms, including the production of a very highly concentrated urine. Comparative figures in Table IX–I (taken from

TABLE IX - I *Maximum urine concentration*[1]

	Electrolytes	Urea
Man	2·2%	6%
White rat	3·5%	15%
Kangaroo rat	7·0%	23%

[1] From B. Schmidt-Nielsen [14].

Schmidt-Nielsen [14]) show that their maximum urea concentration is nearly four times the value in man, and the total electrolyte concentration is twice as great as that of sea water. Sperber [15] and others after him have pointed out that these desert rodents have remarkably long kidney papillae, which extend down into the ureter. It is tempting to see in this (as Chew [16] points out) an association with Wirz's [17] theory about concentration of the urine (see p. 196). If concentration proceeds along the terminal portion of the kidney duct, it seems not unlikely that the longer these ducts are the greater the concentration that can be achieved.

Water is also conserved by producing almost dry faecal pellets. The kangaroo rat loses only 0·76 g water in faeces in conditions where a white rat loses 3·4 g. The lack of sweat glands has already been mentioned, and the amount of passive transpiration of water vapour from the skin is also very low. Finally, loss of water from the respiratory tract is relatively low, as the following considerations show. In man the amount of water lost in this way, if the air is perfectly dry, is 0·84 ml water for each ml oxygen absorbed (assuming that the expired air is saturated with water vapour). In the kangaroo rat, measurements by Schmidt-Nielsen [18] show that the total water loss (including that from the skin) is only 0·54 mg/ml O_2. The reason for this is illuminating: in kangaroo rats the temperature of expired air as it leaves the nasal passages is well below that of air in the lungs so that, although fully saturated, it contains less water than it otherwise would; less, in fact, than the expired air of man, which is at a higher temperature owing to his relatively broad nasal passages.

In addition to these powers of water conservation, the rat stores a quantity of food in its burrow, and this comes into equilibrium with the higher humidity there and consequently yields more water when it is eaten.

Kirmiz [11] has obtained a great deal of information about another small desert rodent, the jerboa (*Dipus aegyptius* = *Jaculus orientalis*), showing that it resembles the kangaroo rat in many physiological and behavioural adaptations to desert life.

It must not be thought that all desert-living rodents are as good at water conservation as kangaroo rats and jerboas. Some are incapable of living for long on dry seed, and they normally get the water they need from cactuses and other succulent plants on which they feed. Examples of such animals are the pack-rat, *Neotoma*, of the south-west United States, the sand rat of Africa, *Psammomys*, which feeds on succulent *Salsolla* leaves, and the subterranean mole lemming, *Ellobius*, of central Asia, which lives mainly on bulbs.

Further mammalian adaptations

STRUCTURE

We may now turn to consider adaptations of a somewhat different nature, shown by one or more desert mammals. As regards the question of shape and related topics, several so-called 'rules' have been suggested as generalised statements. There is Bergson's rule – that similar or related animals are smaller in warm regions than in cold ones; Allen's rule – that peripheral parts of animals in hot regions are extended; Wilson's rule – that coats are hairy rather than woolly in hot regions; and Gloger's rule – that the colour of animals from such parts is predominantly yellow to reddish brown. We need not, I think, be worried about the validity of these rules – they can be taken as useful talking points.

Certainly desert rodents show a number of characters in common, although some of them are but distantly related. The chief examples are set out in Table IX-2, which is derived partly from Buxton [19].

TABLE IX - 2 *Desert rodents and marsupials*[1]

Common name	Genus	Family	Desert habitat
Kangaroo rat	*Dipodomys*	Heteromyidae	N. America
Pocket mouse	*Perognathus*	Heteromyidae	N. America
Gerbil	*Gerbillus*	Muridae	Palaearctic
Gerbil	*Dipodillus*	Muridae	Palaearctic
Jird	*Meriones*	Muridae	Palaearctic
East African gerbil	*Taterona*	Muridae	East Africa
Jerboa	*Jaculus*	Dipodidae	Palaearctic
Jerboa	*Dipus*	Dipodidae	Palaearctic
Jerboa	*Allactaga*	Dipodidae	Russia
Flat-tailed jerboa	*Salpingotus*	Dipodidae	Palaearctic
Mongolian jerboa	*Euchoreutes*	Dipodidae	Mongolia
Spring hare	*Pedetes*	Peditidae	S. Africa
Kangaroo mouse	*Notomys*	Muridae	Australia
Marsupial rat	*Phascogale*	Dasyuridae	Australia
Marsupial jerboa	*Antechinomys*	Dasyuridae	Australia

[1] From P. A. Buxton [19] and other sources.

All the animals in Table IX–2 have a generally similar form – they are 'jerboa-like' (Figs IX–2 3, 4). They are rat sized (except for the larger spring-hares), with short fore legs and long spindly hind legs. They have a reduced number of

Fig IX-2. Jerboa, *Dipus aegyptius*, in the laboratory eating lettuce. (*Photo Dr J. P. Kirmiz*)

toes and long narrow feet. The tail is long and in the typical form bears a terminal tuft of hairs. The gait is characteristically biped.

This striking array of convergent characters can hardly be fortuitous, and indeed it is not difficult to suggest adaptive explanations. The short fore limbs are used for burrowing and for drinking, while the long hind limbs are obviously associated with a leaping gait, and serve to keep the animal's body well away from

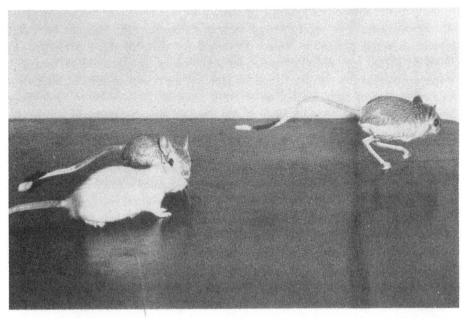

Fig IX-3. The jerboa, *Dipus aegyptius*. Notice the long tail with a tufted tip, used for balance and support, the long hind legs and short front legs associated with a bi-pedal gait, and the long ears. (*Photo Dr J. P. Kirmiz*)

Fig IX-4. The kangaroo rat, *Dipodomys phillipsi*. (*Courtesy of the American Museum of Natural History*)

the hot desert surface. The long tail assumes great importance as a balancing organ. These animals certainly conform with Allen's rule – but whether or not the peripheral extension serves the function of heat dissipation is in my view doubtful. It is also necessary to remember that there are other species of rodents, which do not live in deserts, but which also have many of the above characters. Examples are the jumping mice, *Zapus*, from North America and *Sicista* from Europe.

Many desert animals including such widely different forms as gerbils, kangaroo rats, desert hedgehogs, sand foxes, addax, gazelles, and even bats, are known to possess enlarged tympanic bullae. Evidence for this has been presented by Heim de Balzac [20] and others, but the significance of this common feature is obscure. So too is the significance of the fact that the colour of desert mammals is very often that of light sand. The obvious explanation in terms of camouflage cannot be sustained because most of the animals are nocturnal, and some are entirely subterranean.

Although most of this chapter is written with hot deserts in mind, it must not be forgotten that deserts may also be very cold. If we include such areas, then examples of animals conforming to Bergman's and Allen's rules are not hard to come by. Wright [21] has drawn attention to the differences between certain domestic animals from hot and cold deserts – thus the Kurdi sheep from the northern Iraq desert are rather large, with a good thick coat – but the legs are longer and the neck and ears less compact than in sheep from colder areas. At the other extreme are the desert sheep of the tropics, e.g. from the Sudan, where the body is far less compact, the legs, neck and ears exceedingly large, and the woolly coat is replaced by fine short hairs. A very similar series may be seen in goats, from the Angora goat of Turkey, through the Persian and Syrian desert forms to the tropical desert goat with its very large ears and elongated form. The fat-tailed sheep is another example of shape (in this case distribution of fat reserves) being associated with climate, and this is probably the reason why temperate beef cattle, *Bos taurus*, lay down subcutaneous fat, while the tropical *Bos indicus* stores it intermuscularly.

Finally we may point to the differences between the two-humped Bactrian camel (of the cold deserts of Turkestan) and the Arabian and Saharan form. The first is larger, more compact, with shorter legs and, significantly, a heavier coat of longer hair.

NUTRITION

A problem which confronts most animals is a seasonal shortage of food. So far as this is a general shortage, animals adapted to range more widely will of course possess an advantage, so that all the other adaptations of camels and donkeys, for example, which permit them to remain active and hungry even after long waterless periods, are useful in this respect.

There is a particular nutritional problem which confronts animals in arid areas, namely acute protein shortage, and this is overcome in camels by a further physiological trick. They retain a large proportion of urea which would otherwise be excreted, and return it to the stomach where the rich microflora and fauna are capable of converting urea to amino-acids which may then, presumably, be absorbed into the pathways of protein synthesis. A camel fed on dry dates and hay excreted only 1 g urea per day, which is equivalent to an intake of only 2·5 g protein a day – a figure very much lower, of course, than its true protein intake. When large amounts of urea were injected into a camel, less than one-tenth was recovered in the urine, while the stomach showed increased urea and ammonia concentrations.

It is well known that domestic cattle may be fed some of their nitrogen requirements in the form of urea, and it would be interesting to know whether some breeds of cattle are able in addition to retain urea from the urine for the same purpose. It seems likely that the ability to utilise urea in this way may turn out to be rather widespread, but its distribution has not yet been defined. Retention of urea also aids in water conservation so that this is a doubly valuable mechanism.

REPRODUCTION

An aspect of desert biology not so far referred to is reproduction. The effect of annual cycles of climates upon breeding activity is very well known. In some mammals, sheep for example, the rhythm is largely exogenous (affected by environmental factors) so that these animals when transported from the northern hemisphere to a climatically similar region in the southern hemisphere soon adapt their breeding cycle to the reversed conditions. In some desert animals, however, the rhythm seems to be more rigidly fixed by internal mechanisms. Thus Bodenheimer [22, 23] points out that camels from a winter rain area when transported to the summer rain regions of the Sudan cease almost entirely to reproduce. Volcani [24] has studied the reproductive physiology of camels in the Negev (southern Israel) and finds that both the rutting season (January to March) and the time of birth after a twelve months' gestation period, coincide with the short flush of green vegetation. Lactation lasts for a further 3–4 months, so that reproduction occurs only once in two years – a remarkable adaptation to desert life.

Certain north African gazelles behave similarly. Several species when kept in the Cairo Zoo on fresh green food all the year round nevertheless retained the reproductive cycle characteristic of their native areas, where birth time corresponds with maximum rainfall. *Gazella dorcas* from Bersheba, where most rain falls in February, gave birth in March, while *G. ruficollis* from Khartoum, where rain falls in August, gave birth in September.

The reproductive cycle of small desert rodents also seems to be confined to the short spring – even though some gerbils (*Meriones, Acomys*) are known to be capable of continuous breeding through the year if conditions are good. It seems that abundant fresh food is the trigger that sets the reproductive process going in these animals. The small mouse-like marsupial *Sminthopsis* from Australia, whose females have ten nipples, produces ten young at a birth in good years, but five or less in bad ones. However, information in this important field is still scrappy, and further research is needed before a satisfactory picture can be obtained.

Another field in which our knowledge is very inadequate, and in which research would be of great value, is the water economy of any large desert mammal save camels and donkeys. It is often suspected that several species of antelopes can go without water for a long time. African antelopes are believed to be able to survive on a water ration far lower than that necessary for cattle, and some species, like the gerenuk, appear never to drink at all. Buxton [19] quotes Chapman as saying that some desert antelopes seem not to need water, e.g. the oryx (*Oryx leucoryx*) and the addax (*Addax nasomaculatus*) (inhabitants of the Sudan) and others. If this is so we should like to know how they manage it, for the answer may (as Huxley [25] points out) have very important implications indeed, particularly in the field of conservation and utilisation of wild life for food in arid areas. The truth, as Bourlière [26] and others suggest, is probably that the addax, the oryx and the dorcas gazelle cannot go without water indefinitely, any more than camels can; but their needs are very limited. Certainly these animals feed on succulent plants such as water melons when they are available. Certain antelopes, e.g. the duikers, are known to lick the copious morning dew from the leaves of plants, and in this way to survive in areas where there is no standing water. This sort of thing may account for some at least of the stories about larger mammals doing entirely without water.

Birds

Birds at least have the advantage (for desert animals) of excreting uric acid. However, they do have a rather high evaporative water loss occasioned by their non-fossorial habits, and their high constant temperature which results in a need to ventilate the lungs. During heat stress much water is lost by panting, but the extent of this is reduced by permitting a degree of hyperthermia (as camels do).

It is now known that many marine birds excrete, via the nasal glands, extraordinarily highly concentrated salt solutions, and this would appear to be a useful achievement for desert birds too, but so far none have been found to do this. The subject is reviewed by Schmidt-Nielsen [27]. A few birds seem to be independent of free water. Examples quoted by Buxton [19] are the Houbara bustard, *Chlamydotis undulata*, the thrush, *Podoces*, and the larks, *Alaemon* and *Ammomanes*, besides others. These birds must obtain water from their food, and perhaps from dew.

But many desert birds are less independent, and visit the true desert away from streams only for short periods. Thus the Californian desert quail (*Lophortyx gambeli*) wanders out to forage in the desert, but requires to drink, and therefore returns to the scrub along river margins. Similarly, sandgrouse (*Pterocles*) of deserts fly in from many miles to water at a pool each day, and the desert-inhabiting mourning dove (*Zenaidura macrocera*) behaves similarly. During the breeding season these sandgrouse soak the breast feathers with water, and carry this to the nest where it is used for evaporative cooling and for providing the young with water.

The eggs of birds, particularly the small ones, are vulnerable to overheating by exposure to direct sunlight. Consequently desert birds adopt various devices to shelter their eggs. Some place them as far as possible in sheltered places: larks under bushes, for example, and wheatears in holes or small caves. Those that breed in the open such as the pratincole, *Glareola pratincola*, and a small tern, *Sterna saundersi*, either shade the first eggs by standing over them or, as in the sandgrouse, begin to incubate as soon as the first egg is laid.

However, these are rather special adaptations, and the overall picture, so far as it goes, suggests that in many desert birds adaptation takes the form of a great pliability in breeding season so that this corresponds with beneficial conditions no matter when or for how short a time these occur. Whereas in temperate climates, breeding, moulting, and migration are controlled by day length, Keast [28] has found that in the Australian desert nesting is stimulated, often very rapidly indeed, by the onset of rains. During a good (prolonged) rainy season, breeding may even be continuous and thus offset the effects of bad years when breeding is very restricted. The usual time of breeding also changes, and birds which will breed in the spring in Adelaide will breed in late summer, autumn or winter in the desert, because the rain there falls in the summer months.

Information is beginning to accumulate that some birds may possess special physiological adaptations to desert life. Poulson and Bartholomew [29], for example, find that savannah sparrows can tolerate a greater degree of dehydration and drink more strongly saline water than other birds. They manage this, it seems, mainly by tolerating a high salinity in the body fluids rather than by excreting a concentrated urine, for the bird kidney can produce urine with an electrolyte concentration no more than twice that of the blood. The mammalian kidney, it will be recalled, can do very much better than this. Further information is reviewed by Schmidt-Nielsen [1] and by Dawson and Schmidt-Nielsen [4].

Reptiles

Reptiles, like mammals and arthropods, form a major part of the desert fauna. They are basically well adapted for such a life because they excrete uric acid (although several are known to excrete greater or smaller proportions of urea), they have cleidoic eggs and are on the whole 'small' and can therefore readily find shelter if necessary. Recent work on the physiology of desert reptiles has been reviewed by Schmidt-Nielsen and Dawson [3].

EXCRETION AND WATER BALANCE

There is no evidence that the kidneys of reptiles can produce hypertonic urine (i.e. concentrate the urine to a point when its osmotic pressure is greater than that of the blood). Nevertheless the total urine flow is very small compared with that of mammals, and it is much lower when the animal is dehydrated than when it is normal. The Australian skink, *Trachysaurus*, when short of water produces 0·024 ml urine per 100 g of animal per hour and this is only about 5 per cent of the rate for white rats in similar conditions. The generally low rate of urine formation may be associated with the relatively small size and poor blood supply of the glomeruli in the kidneys, and the virtual anuria in dehydrated lizards is associated with a high salt concentration in the blood. Very high salt concentrations up to one and a half times the limit for mammals is readily tolerated by the Australian skink, *Trachysaurus*, and this is a valuable adaptation to arid areas.

In addition, the fact that uric acid is excreted instead of urea means that nitrogenous waste can be eliminated with the minimum of water loss even though the kidneys cannot concentrate the urine, for urine water is reabsorbed in the cloaca. Since uric acid, being so poorly soluble, begins to crystallise out (and thus to exert no further osmotic pressure) at very low concentrations, much water can be so absorbed before isotonicity between the cloacal fluid and the blood is reached. Evaporation from the skin is also low, and lower in desert forms than in others.

A further adaptation to desert life is shown by the desert monitor, *Varanus griseus*, a widespread species ranging from north Pakistan all the way across the desert lands to the west Sahara. This lizard, when injected with the equivalent of 22 per cent of its weight of water, excreted only 25 per cent of this in the following 7 days and stored the rest in the tissues; while a non-desert garter-snake, *Thamnophis*, in like circumstances eliminated half the extra water in 10 hours.

This subject has recently been reviewed by Chew [16], and it seems that so far as water balance is concerned, excretion of uric acid, tolerance of high blood salt concentrations, and the water-storage capacity of the tissues, are properties of great adaptive value to some reptiles. In addition, Norris and Dawson [56] have recently found that the chuckwalla, *Sauromalus hispidus*, which often feeds on halophytic plants, possesses nasal glands which secrete a salty fluid strongly hypertonic to the blood: an adaptation which calls to mind the nasal glands of marine birds and which is very appropriate for a desert reptile.

TEMPERATURE RELATIONSHIPS

Another kind of general adaptation to desert life is the ability of reptiles to control body temperatures by behavioural mechanisms – although this does not mean that all reptiles function best at a particular temperature.

The temperature limits and optima for many species have been established by Cowles and Bogert [30], and more recently by Mayhew [31], and these show a surprising range. The lowest temperature which the rattlesnake, *Crotalus cerastes*, voluntarily tolerates is 16°C, and the highest for the chuckwalla, *Sauromalus obesus*, is 42°C. Mayhew recorded even 47°C in the desert iguana, *Dipsosaurus*. On the whole the mean optimum temperature (in the sense of that most frequently observed in active animals) is lower than the normal mammalian temperature (which is about 38–40°C), so that reptiles are not consistently thermophilic. Nocturnal reptiles prefer a lower temperature than diurnal ones (29°C for the rattlesnake, *Crotalus atrox*, and about 40°C for the fringe-toed lizards, *Uma* spp.) and snakes in general prefer cooler temperatures than lizards do.

Brain [32] measured the lowest temperature at which panting starts, as well as the critical maximum temperature, in several species of nocturnal and diurnal geckos and lizards in the Namib Desert in South West Africa, and found that while panting occurs at much lower temperatures in nocturnal than in diurnal species (a result perhaps to be expected), nevertheless the critical maximum temperatures were similar for all species, including the apparently delicate nocturnal *Palmatogecko*. The range of temperatures within which the nocturnal species are active is remarkably low. Comparative figures obtained by Brain are shown in Table IX–3.

TABLE IX - 3 *Some temperature relationships of desert reptiles.[1] All temperatures in °C. It should be understood that there is no experimental evidence that 'panting' causes a lowering of body temperature in lizards.*

Species	Habit		'Panting' starts	Critical maximum	Activity temperatures
Ptenopus garrulus	Nocturnal gecko		37·4	44·2	11·6–20·0
Ptenopus carpi	,,	,,	34·8	42·7	10·2–19·0
Palmatogecko rangei	,,	,,	36·2	43·5	
Rhotropus afer	Diurnal gecko		42·0	43·9	28·5–36·5
Aporosaura anchietae	Diurnal lizard		42·2	45·1	
Meroles suborbitalis	,,	,,	40·8	44·0	24·2–39·1
Meroles cuneirostris	,,	,,	40·7	45·4	

[1] After Brain [32].

Under captive conditions, given an environment including a wide temperature range, the rattlesnake maintains its body temperature accurately between 31° and 32°C even though this snake can tolerate a wide temperature range. Such temperature control is effected by a variety of mechanisms. In part it is done by selecting, so far as possible, the right position in or on the soil, in rock crevices and so forth. Reptiles will also bask with all or part of the body exposed to the sun as necessary. If the temperature rises then they retreat to cooler depths, and in extreme conditions they possibly effect respiratory cooling by active panting (see Table IX-3). It must be emphasised, however, that there is no evidence that panting by lizards does effect cooling. In New World lizards at least there is evidence to the contrary, for Templeton [33] found but little increase in the rate of evaporative water loss in the desert iguana when that animal began to 'pant' at temperatures above 40°C.

During the cold desert night, diurnal reptiles shelter in the warmest place available, and often will not emerge until the morning sun has warmed them to the temperature necessary for full activity. The chuckwalla does not emerge unless this condition is fulfilled, and the fringe-toed lizard, *Uma*, remains buried in sand until the sun has warmed it.

Owing to their thin, non-fatty skin, and to pigmentation, the body of lizards absorbs and loses heat rapidly by conduction and radiation. A large-bodied lizard can therefore remain active for longer than a small one as the environmental temperature falls. In some diurnal lizards there seems to be no doubt that pigment movement in the skin alters the reflectivity of the surface advantageously – thus Atsatt [34] found that several lizards are darker in the cool morning than in the heat of the day. Cole [35], Norris [36] and others have confirmed and extended these results.

Diurnally active lizards are compelled to move over very hot soil, but the dangerous overheating which might be expected to result is offset certainly by their gait, where soil-to-skin contact is reduced by running on the heels with the toes reflexed upwards, and the body held high above the substrate; possibly, also, by the very light-coloured, highly reflective under belly, which is characteristic of desert lizards such as the flat-tailed horned 'toad', *Phrynosoma m'calli*, and the desert iguana, *Dipsosaurus dorsalis*, but does not occur to the same extent in non-desert forms such as the coast horned 'toad', *Phrynosoma coronatum*, and many others.

There is little doubt, however, that colour in these animals is often protective (cryptic). A good example is provided by the side-blotched lizard or Stansbury's swift, *Uta stansburiana*, which is black and lives on black larva ridges in the Pisgah crater in California, while *Uma scoparia*, the Mojave Desert fringe-toed lizard, which lives very near, but on white sand between the ridges, is very light-coloured.

OTHER ADAPTATIONS

Animals which live in deserts, particularly sandy deserts, often have special adaptations concerned with locomotion. Fringe-toed lizards, *Uma* spp., for example (see Fig IX-5), which live in completely plantless wind-blown sand, although they do not go far away from plants, can disappear into the loose sand extremely rapidly by a few active wriggling movements. These lizards possess well-developed scaly fringes along the toes, a character possessed by other sand-living reptiles such as the shovel-snouted lizard, *Aporosaura* (see

Fig IX-5. The fringe-toed lizard, *Uma inornata*. This is a sand swimmer, restricted in distribution to wind-blown sand in parts of Southern California's desert. (*Photo Dr W. W. Mayhew*)

Fig IX-6. The desert gecko, *Palmatogecko rangei*, during a moult. (*Photo Dr C. K. Brain*)

Fig IX-7), and the sand lizard, *Scaptira*, from the Namib Desert in South West Africa. A small gecko from this desert, *Palmatogecko*, even has completely webbed fingers and toes (see Fig IX-6). The habit of burrowing into sand is associated with a reduction of limbs, as in the limbless skink. *Typhlosaurus*, from the Namib. Some snakes, the desert side-winder *Bitis perengueyi*, for example, are particularly adept at sinking into loose sand (see Fig. IX-8).

Rapid burrowing is facilitated by the shovel-like snouts of both *Aporosaura* and *Scaptira*. When in danger these animals dive into the loose sand head first, and move through it by an action akin to swimming. The nostrils of these desert

Fig IX-7. The shovel-snouted sand lizard, *Aporosaura*, which dives head first into the sand. Notice the fringe of scales on the toes, and the flattened muzzle. (*Photo Dr C. K. Brain*)

sand lizards are protected by opening backwards or sideways from the raised swelling on which they are mounted. The ear openings too are either reduced or protected by a grill of scales.

The Colorado Desert fringe-toed lizard, *Uma notata*, has been shown by Norris [36] to possess a number of interesting special adaptations which enable it to live for long periods under the surface of loose sand. Before submerging this lizard inhales, and after submerging exhales, so that a small space is provided between its body and the sand beneath it for future respiratory movements. Furthermore the first exhalation through the nostrils blows away the finest sand particles (which would otherwise block the air channels) and allows only large grains, through which breathing is possible, to lodge near the nostril openings.

Many lizards and snakes hibernate during the winter, and upon emergence their preference temperature is lower than it is during the summer. Because of the dependence of activity on temperature, the times of day at which various species are most active is different. Furthermore, as Mayhew [37] has found, there may be in any one species two peaks of activity each day at the height of summer (when it is too hot in the middle hours), but only one in the spring and autumn months.

So far as breeding is concerned, information is only now becoming available. Mayhew's results [38, 39, 40] seem to show that, as in desert birds and other

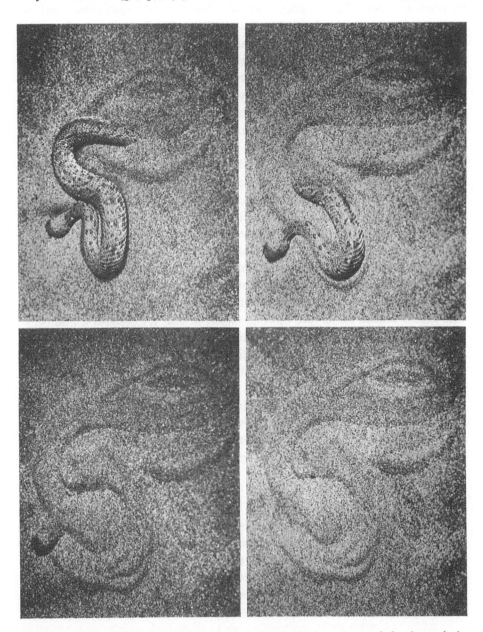

Fig IX-8. The side-winder, *Bitis perengueyi*, disappears into the sand, leaving only its short, black tail as a bait for prey. (*Photos Dr C. K. Brain*)

animals, reproduction is rather closely associated with food supply, which itself is determined in part by recent climatic effects.

Thus Mayhew finds that reproductive activity (as shown by testicular enlargement, motile sperms, eggs in the oviduct, etc.) of the two fringe-toed lizards, *Uma inornata* (from the Coachella Valley, California) and *Uma scoparia* (from the Mojave Desert), varies greatly from year to year according to the rainfall of the previous winter, for this determines the abundance of annual plants and the associated insects upon which these lizards feed. On the other hand, a third species, *Uma notata*, from the Algodones dunes, shows about the same reproductive activity each year, although the peak of activity may be delayed after dry winters. The reason for this seems to be that, although this species normally feeds on the insects of annual plants, when these are scarce it will eventually supplement its diet with insects from the perennial plants which survive in the dune area on moisture deep in the sand.

Finally, the long-tailed brush lizard, *Orosaurus graciosus*, which lives in the same locality as *Uma notata*, exhibits the same reproductive pattern each year because it feeds only on insects produced by the perennials.

So far as water supplies are concerned, many desert reptiles can live indefinitely without drinking, because of their good water-conservation mechanisms which have already been referred to. But further work may show that these animals are capable of using dew and other small collections of water without actually drinking, as Bently and Blumer [41] recently found in the Australian spiny agamid, *Moloch horridus*. This animal, which was previously believed to absorb water through the skin after rain, actually employs a very different mechanism. The skin is indeed rather like blotting paper, in that it possesses a number of very fine capillary channels. But the skin is not permeable to water: the capillaries lead to the lips, and it seems likely that the lizard secretes from the mouth a hygroscopic mucus which absorbs the capillary water and is then swallowed.

Apparent absorption through the skin has been described in *Uromastix*, the spiny-tailed agamid from Africa. Perhaps the mechanism here is the same as in *Moloch*. Clearly it is a very useful adaptation, since the skin can remain impermeable, while yet providing a collecting area for dew or rain.

Amphibia

The association of amphibia with deserts may seem somewhat incongruous – as indeed it is, for on the whole such animals are very dependent on water. It must not be forgotten, however, that deserts are by no means always and everywhere without water; and what concerns us at present is to take note of those adaptations which permit some amphibia to live in surprisingly arid areas. No amphibian so far reported is capable of living entirely without free water.

In arid areas, what water exists often has a rather high salinity owing to rapid evaporation. One might, therefore, expect to find that amphibians living in such areas would be able to tolerate higher salinities than normal. Such, however, does not seem to be the case, for most amphibians cannot tolerate salinities higher than between 1·0 and 1·5 per cent NaCl. The frog *Rana pipiens*, a widespread inhabitant of fresh water in temperate regions, also occurs in saline streams in the Californian desert, but it does so by utilising only those parts of the stream where fresh-water seepage reduces the salinity to a tolerable level.

A facility possessed by several, perhaps all, frogs and toads is the ability to store large amounts of water in the bladder. This, as Ewer [42] and others have shown, is a real advantage if the ponds dry up because the stored bladder water may be used to supplement water lost from the tissues by evaporation into dry air. The Argentinian salt-marsh frog, *Pleurodema nebulosa*, seems to have adapted this general facility in an interesting way to its own needs. It lives in the dry areas of north-western Argentina and spends most of its time in burrows. Only occasionally do pools form, and these are saline. The frogs enter these pools and absorb water like any other frog, but by secreting a very dilute urine into the bladder they maintain a high salt concentration in the body water – and this other frogs do not do. Consequently they are able to absorb water from higher salinities than other frogs, and this may well stand them in good stead when the salt marshes dry up. Ruibal [43], who made the observations on this frog, interprets the meaning somewhat differently. He believes that the adaptive feature lies in the fact that the urine is dilute.

Another way in which anurans differ among themselves in adaptation to arid environments concerns the rate at which they can absorb water through the skin after dehydration. Ewer [42] has shown that the terrestrial toad, *Bufo regularis*, rehydrates more rapidly than the aquatic clawed toad, *Xenopus laevis*, and Bentley, Lee and Main [44] have much evidence of this from Australian anurans. These workers used several species of each of the genera *Heleioporus* and *Neobatrachus* which occur in habitats representing a range of aridity (measured in terms of the number of rainless days in a year). They found that all species died at approximately the same degree of dehydration (a water loss of 40–45 per cent of body weight), so there is no relation between this and habitat. All species of *Heleioporus* also regained water at the same rate after desiccation; but the species of *Neobatrachus* showed significant differences in this respect, from 33·3 mg/cm²/ hr in *N. pelobatoides* to 99·4 mg/cm²/hr for *N. wilesmorei*. These rates correlate well with environmental aridity; for *pelobatoides* occurs in areas with 60–120 rainy days a year, while *wilesmorei* has less than 40 such days. *Heleioporus* spp. are burrowers in sandy friable soil, and the burrows, which are sealed as the frog proceeds, reach down to moist layers in summer – so that they are not likely to lose so much water as *Neobatrachus* spp. which live in clay soils where burrows are not so beneficial because they do not provide moist micro-environments.

The spadefoot toad, *Scaphiopus hammondi*, shows an interesting adaptation when it lives in arid areas. This animal lays, in small temporary pools, a large number of small eggs which develop into numerous small tadpoles, and ten or so large eggs which hatch later and feed upon the tails of the small tadpoles which, by then, have fed upon the microflora of the pool. The large tadpoles develop very rapidly and metamorphose before the pool dries.

By reason of these adaptations, and of others such as rapid larval development, aestivation and non-aquatic larvae, amphibians are by no means as scarce in deserts as might be expected. No amphibian has been shown to be able to produce a hypertonic urine, however, and consequently with a few exceptions they are absent from water whose salinity is greater than that of their body fluids. A striking exception is the remarkable crab-eating frog, *Rana cancrivora*, which can live in sea water. Gordon and others [45] showed that it does so not by producing a hypertonic urine but by retaining a very high concentration of urea and inorganic salts in the blood, and thus side-steps the problem in much the same way as dogfish and sharks do.

Chew [16], in a recent review, points to four physiological processes which are adaptive to terrestrial conditions: retardation of evaporation from the skin, enhancement of water uptake through the skin, antidiuresis and reabsorption of water from the bladder. The last three of these processes are probably hormone-controlled. But there is no clear relation between the degree of development of any or all of them and the degree of 'terrestrialness' of the species concerned.

Arthropods

Some arthropods, particularly insects and arachnids, as we have already seen, have a basic morphological and physiological organisation which confers efficiency for life on land; and since they are small, further adaptation to desert life need not be profound because they can avoid the extreme heat and dryness by becoming partially cryptozoic or nocturnal. Nevertheless, modifications of arthropod anatomy and physiology in adaptation to desert environments certainly do exist, though the field has not been well explored and future research will undoubtedly lead to many further discoveries. Information which is already available has been reviewed by Pradhan [46] and by Cloudsley-Thompson [2].

We may take locusts (Fig IX-9) as a typical, but by no means extreme, example of a desert arthropod. More is known about locusts than about any other desert arthropod, largely because they can do so much damage.

Locusts are in fact ordinary grasshoppers which have developed the capability of aggregation and migration to an extreme degree. Coupled with this is a pronounced polymorphism, in that the shape, colour and behaviour of these insects varies between the two extreme types or 'phases' – gregarious and solitary.

As Uvarov [47] has stressed, an essential feature of locust ecology is that different stages of the life history are spent in different environments. Eggs are laid in the ground, usually in bare sandy patches, the young hoppers require abundant vegetation for food and shelter, while the adults can survive (but not reproduce) with a minimum of food in an arid environment. This sort of condition – a patchy mosaic of vegetation cover – often occurs along the border between two major vegetational zones, such as forest and grassland, or savannah and desert.

At first sight the physiological requirements of locusts do not seem particularly suited to deserts. The female lays eggs only in moist sand (but she may oviposit below a layer of dry sand if she can reach down to moisture with the extended

Fig IX-9. Desert locust (*Schistocerca*) hoppers on the Red Sea coast of Eritrea (Ethiopia). (*Photo Anti-locust Research Centre – C. Ashall*)

abdomen). The developing eggs of the desert locust need to absorb from the soil more than their own weight of water, the hoppers need a plentiful supply of fresh green vegetation, and the adults cannot mature except in high humidity or with green food.

As in so many other cases, the answer to the question how such an apparently water-dependent animal survives in deserts lies in the great seasonal variation of deserts, which flush green for a brief period; and since the time of flush differs in different places, the migratory habit is clearly of enormous advantage. A good season, followed by a drought, will bring about a concentration of locusts into the few habitable areas. Such concentration leads to the development of the swarming gregarious phase.

Now as Rainey [48] has shown, the direction of flight of locust swarms is determined largely by the prevailing wind, and because these winds usually blow into convergence areas of air masses where rain falls, locusts by and large are also conveyed to these places, where conditions are admirably suited for egg-laying and hopper development. This theory also accounts for the observed annual movements of locusts from areas with winter or spring rainfall to those with the summer monsoon season.

Locusts, therefore, depend for their continued existence upon finding areas with plentiful green vegetation, and the fact that the adults go into diapause (arrested development) until such conditions occur is highly adaptive. But unfortunately the activities of man often assist the spread of locusts, when by burning or other means we produce a mosaic of vegetational cover, and also when we move into arid areas and, by irrigation, grow green crops, where locusts will congregate. There are numerous examples of this process: the Land Development Scheme of Abyan, Aden, is one quoted by Uvarov, and the intensive cultivation areas of the Tokar Delta on the Red Sea coast of the Sudan is another. In America, the Rocky Mountain grasshopper, *Melanoplus*, normally has but one generation a year in Arizona, but with the intensive cultivation of alfalfa by irrigation, the insect can get through several generations a year.

High temperature tolerance (not control), diapause, high productive ability and mobility are the characters that make the locust a very successful desert arthropod. But other arthropods without this combination of characters also exist in deserts, and we may now refer briefly to a few of them and their adaptations.

So far as water balance is concerned, it is clearly advantageous for a desert arthropod to possess a cuticle relatively impermeable to water, and on the whole this expectation is borne out. The rate of transpiration from most desert beetles and scorpions is low. Cloudsley-Thompson [49, 50, 51] has found that the fat-tailed scorpion, *Androctonus australis*, common in the north African desert, loses water from its surface at the low rate of $0 \cdot 014$ mg/cm²/hr, which is about one-hundredth of the rate in the wet tropical form *Pandinus imperator*. The large aggressive camel spider, *Galeodes arabs*, found in the desert around Khartoum loses water more rapidly than the desert scorpion, but much less rapidly than most spiders. There are exceptions, however, which remind us that the situation is not capable of being summed up by a facile generalisation. Thus the well-known temperate beetle, *Tenebrio molitor*, has one of the least permeable cuticles known, and there are woodlice which live in deserts.

The last statement underlines the very great importance of the existence of a range of microclimates, particularly for desert arthropods. C. B. Williams [52] showed many years ago how an animal, by moving through a very short distance, can avoid extremes of temperature and dryness. In one of his examples, the temperature of the surface rose to 56°C while 10 cm below it was only 34°C. At night the surface temperature fell very steeply, and in fact by moving up and down through a distance of only 30 cm an animal could live in a constant temperature throughout the daily cycle.

The behavioural adaptations of many such animals are more important than any physiological specialisations. Indeed it would be hard to point to any important physiological adaptation to desert life in arthropods which is not found in at least some non-desert forms.

Even so, it is rather remarkable that a woodlouse, *Hemilepistus reaumuri*,

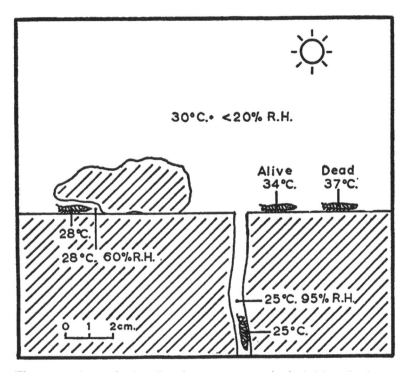

Fig IX-10. A set of microclimatic measurements in the habitat of a desert woodlouse, *Hemilepistus*, in Algeria. These animals avoid extreme conditions by digging holes – but if they are caught in the open by strong sunlight, evaporation of water cools them significantly

should be able to exist in a desert. Edney [53] found that in fact these animals avoid true desert conditions by digging holes, and emerging only when the temperature is low. Unlike other woodlice, they walk with the body raised well off the ground. They feed upon desert scrub and presumably obtain sufficient water therefrom. Certainly their rather permeable integument may be of value in a crisis – when they are caught outside their holes in strong sunlight, for example, for then their body temperature is significantly lower than it would be if they were dry. Measurements made in the field in Algeria near Bardia established this point (see Fig IX-10).

So far as morphological features are concerned, most diurnal desert arthropods

H

are cryptically coloured – this is particularly true of the many species of grass-hoppers. Those that are not, such as the tenebrionid beetles, are usually black. This strange fact (which applies not only to arthropods) has often been remarked upon, by Buxton [19] and others, but never satisfactorily explained. It is possible of course that black surface pigmentation prevents radiation energy from penetrating deep into the tissues. Heat would then be generated at the surface and,

Fig IX-11. A large white sparassid dune-spider, *Leuchorchestris*, which is nocturnal and probably feeds on geckos. (*Photo Dr C. K. Brain*)

if the ambient air is cooler than the surface of the animal, removed by convection. But the evidence for this is lacking. Bodenheimer [22] measured the amount of various wavelengths of radiation which passed through a variety of elytra and wings of insects. Black elytra markedly reduced ultra-violet transmittance, as did the white wings of *Pieris* butterflies, but the results as a whole are not readily explained. One difficulty in interpreting such information is that the structures vary in thickness as well as in opacity and colour. Nocturnal arthropods, however, are very often white or light-coloured.

In the Namib Desert of South West Africa, Lawrence [54] has reported finding several species belonging to three genera of tenebrionid beetles that have all developed a flat plate-like shape (see Fig. IX-12), which certainly assists in rapid burrowing into the loose sand by alternate sideways movements. Other tenebrionids there have compact rounded bodies with long spidery legs, which assist in running rapidly over the sand surface (see Figs IX-13, 14). A common feature in many of the spiders and insects found in the Namib is the development of large brushes of flattened hairs on the tarsi – they obviously function as 'snow shoes' for moving over the sand.

Curiously enough, one of the attractive features of extreme sandy deserts

Fig IX-12. Two flat, light-coloured tenebrionid beetles, *Lepidochora eberlanzi*, and *L. porti* (lower left), from the Namib Desert, South West Africa. (*Photo Transvaal Museum – Dr L. Schultze*)

such as the Namib is the relative paucity of the fauna and flora, for this permits a rather full knowledge to be obtained of food chains and interspecific population relations. Lawrence reports a preliminary study along these lines in the Namib. Here the bottom of the food chain seems to be the tough tussock grass *Aristida amabilis*. This is fed upon by a number of tenebrionid beetles, a silver-fish (*Lepisma*), of which there are very large numbers, and other arthropods. These intermediate links are fed upon by the limbless and blind sand skink, *Typhlosaurus*, the delicate web-footed gecko, *Palmatogecko* (see Fig IX-6), and other lizards. Ants (an aggressive *Camponotus*) and some smaller tenebrionids are fed upon by the tougher diurnal lizards. Of the several spiders, one, a large white sparassid dune spider, *Leuchorchestris*, probably feeds upon the slow-moving gecko by night (see Fig IX-11).

Fig IX-13. The long-legged *Onymacris unguicularis*, a tenebrionid beetle that runs swiftly over the hot sand in the Namib Desert. (*Photo Transvaal Museum – Dr L. Schultze*)

Fig IX-14. *Onymacris candidipennis*, a white tenebrionid beetle from the Namib Desert, South West Africa. (*Photo Dr C. K. Brain*)

Molluscs

Although the phylum Mollusca has a large number of terrestrial representatives in the form of land snails and slugs, most of these are confined to moist habitats. Nevertheless, there are a few pulmonate snails that exist in arid areas although little is known of the manner in which they are adapted to this way of life.

McMichael and Iredale [55] refer to certain species which occur in central Australia, and these are known to resist desiccation. A general adaptation is the capacity to aestivate, in which state they may survive for years. As soon as rain falls they emerge, breed quickly, and then return to their retreats in rocks, cracks and roots of trees. During aestivation these snails form a thick calcareous epiphragm, which may reduce evaporation from the foot and mantle.

The occurrence of large numbers of snail shells in deserts together with the rarity of the snails themselves, is testimony to the length of time occupied in aestivation. One species, *Chloritisanax banneri*, was found to be alive after nearly six years in a museum box!

Unfortunately little is known of the physiology of these species, although in others the presence of uric acid as an excretory product is well known. Studies on their water and temperature relationships would be well worth while.

Animal populations and phenology

In this final section we may attempt to bring together some of the main points already made, in terms of a rather more general picture of the animal life in deserts. By and large both the number of species and the number of individuals of desert animals are relatively low. Heim de Balzac [20] records only twenty-four true desert mammals in the north African Sahara, namely two each of hedgehogs, carnivores and antelopes, one hare and seventeen rodents. One of the rodents, Lichtenstein's jird (*Meriones libicus*), although considered to be fairly abundant, was present only at a density of thirteen per square mile according to an estimate at Beni-Abbas.

The larger ungulates, like the addax, dorcas gazelle, oryx and others, in the southern Sahara, occur in small troops, up to twenty head or so, in areas with little vegetation, but the same species in less extreme environments may occur in much larger herds of up to several hundred head. Troops of wild asses (*Equus hemionus*) in the Gobi Desert also vary greatly in size according to available food. In plentiful times, small family groups of a male and three or four females associate together to form large herds.

It is today a truism to say that an animal population is in dynamic equilibrium with other animals, plants and the inanimate environment. The study of these interrelationships lies broadly in the field of population dynamics – a field into which it is not our present purpose to venture, except to point to one important consideration. This is that both the numbers and the behaviour of animals are strongly affected by climatic factors both directly and indirectly. It is probably true to say also, that the greater the fluctuation of climatic factors, the greater the swing in animal numbers. In deserts, temperature, humidity and groundwater (among other factors) fluctuate more rapidly and over a wider range than they do elsewhere, and consequently we should expect to find those oscillations in animal numbers, which occur almost everywhere, to be exaggerated in desert areas.

First we may consider the diurnal changes. In the daytime most deserts give the appearance of being almost completely lifeless. The only animals to be seen are a few sun-tolerant beetles and lizards, and even these are fast runners, dashing from one patch of shade to the next. By night, however, the same area comes to life – small rodents, jerboas, kangaroo rats and the like – emerge from their holes, reptiles – snakes, lizards and geckos – are abroad, and a whole range of arthropods including insects, scorpions, spiders, phalangids and centipedes are plentiful. Observers report finding the sand dunes of the Namib Desert 'virtually covered' with the slow-moving off-white flat beetle *Lepidochora* – a species which during the day burrows deep in the sand to avoid the hot upper layers.

Many arthropods seem to emerge for rather short periods in each daily cycle. Thus Lawrence [54] reports that in the Namib one species of scavenger beetle appears just before sunset and is absent later, while others feed only in the early morning. No doubt these habits have an advantage in avoiding mutual interference.

Such daily cycles reflect, of course, changes in the position of animals rather than changes in numbers. So, too, do some of the annual cyclic changes which may be as striking, if not more so, than the daily ones. The annual appearance of very large numbers of animals is a common phenomenon in all arid areas and deserts. In many of these 'appearances', the animals are present in the area all the time, but are evident in the open only at one time of the year, usually during the short rainy season. Among the arthropods, the appearance of the beautiful scarlet giant velvet mites (*Dinothrombium*) just after the first rains affords a striking example. These animals, which feed on winged termites and whose appearance above ground coincides precisely with that of their prey, are virtually unobtainable at other times of the year. The larvae are parasitic on grasshoppers for a short time, but what happens during the remainder of the life cycle is not known. The giant millipedes, *Alloporus* and others, so common in arid areas during the rains, are also virtually unobtainable in all but two months of the year, and the same is true of the desert woodlouse, *Hemilepistus*, of North Africa and the Middle East.

Bodenheimer [22] has studied this phenomenon quantitatively in respect of Orthoptera in the desert and in the Mediterranean areas of Israel. The peak of activity (in the sense of the number of species present) for both groups is from March to June, but 72 per cent of all the desert species are active at this time, against only 42 per cent of the Mediterranean ones. The latter are distributed more or less evenly and widely throughout the rest of the year.

However, these annual flushes are the rule rather than exceptional, and examples from all terrestrial groups of animals abound. Reference has already been made to this sort of behaviour in amphibians, reptiles, molluscs and mammals in the course of this chapter.

An annual event of a somewhat different nature, also contributing to great changes in the desert population, is the migration of many species inwards from more mesic areas. Migrations of this kind are shown most strikingly by fast-moving animals: birds, mammals and winged insects, which move into the desert during the brief rains when both insect and plant food is abundant, and where breeding may take place. Examples of these movements have also been considered above.

A third category of events that results in violent population changes in desert areas comprises phenomena of reproduction and development. Because deserts are such difficult environments and because they are suitable for active life only for short periods each year (and sometimes not at all for several years), reproduction has to be rapid when the occasion offers, and development has to be rapid if the young stages are less hardy than the mature animal. This is why, for example, the tadpole stage of the desert frog *Pleurodema nebulosa* is abbreviated; why in the spadefoot toad, the majority of small tadpoles are sacrificed as food for the few rapidly developing large ones; and why in desert birds, breeding habits are very pliable and are rapidly stimulated by the onset of rain.

Violent annual climatic fluctuations may have very important effects upon the population structure of individual species. In humid tropical areas reproduction may be continuous so that individuals of all ages and stages of development occur together, whereas in deserts, adverse conditions for most of the time prevent reproduction except for a short burst, and thus lead to a population much more homogeneous as regards age and size.

Bodenheimer [22] has shown this to be true of the Levant vole, *Microtus guentheri*. This animal has an extremely high reproductive potential, but is subject to near extinction when it migrates to the borders of the desert. The reason for this is interesting. The vole becomes virtually sterile unless its food contains some proportion of fresh green plant material, but during eight or ten months of the year on the fringes of the desert, no green food at all is available, although the staple food, grains and roots, may be sufficient. Consequently reproduction ceases, older animals die off, and the population is reduced to very low numbers (only one or two pairs per square kilometre) and to a narrow age range.

Many desert insects, too, have only one generation annually. Bodenheimer quotes the case of certain large praying mantids, which occur in all stages of development during the summer and autumn in palaearctic deserts, but those that have not reached an advanced stage of development when the winter sets in are killed, so that the population in late winter consists only of last-stage nymphs, which become adult in spring and lay eggs all at the same time.

It is largely as a result of this almost explosive reproductive ability in good years, that so many desert or near desert species readily become pests. It cannot be said too often that one of the main dangers inherent in the exploitation of arid lands is that of providing, by irrigation and plantation, ideal environments for the rapid multiplication of species, particularly of birds and insects, that cause damage to human food. The classic example of such a pest is the desert locust, *Schistocerca*, which has already been referred to.

The non-gregarious north-east African grasshopper, *Phymateus viridipes*, is another example. The eggs are laid in difficult, non-arable land, but the adult females migrate towards cultivated areas, and do much damage to the leaves of vines, figs and cereals. Several scarabaeid beetles such as *Rhizotragus*, *Phylopertha*, *Polyphylla* and others, whose larvae occur permanently in semi-arid areas peripheral to deserts, sporadically cause very serious damage to cultivated areas at a distance from their breeding places. *Polyphylla*, for example, destroys the roots of cedar trees in Morocco.

Not only insects, but birds and mammals too can become pests in the desert environment. Crows (*Corvus* spp), perhaps because of the otherwise difficult environment, are forced into association with man and thereby congregated in large numbers, causing damage to cereals and dates. Quelea finches too, not as a result of association with man, but by virtue of their enormous reproductive capacity and gregarious behaviour, have become pests of first importance in semi-arid areas, feeding as they do on maize and other grains. Of the typically desert mammals, the gerbills, *Gerbillus* and *Meriones*, in some parts of North Africa, notably south-west Morocco, have become commensals with man and made serious inroads into his food supply. Even wild sheep and goats feed upon the cladodes of the spineless cactus *Opuntia inermis* for the sake of the water content, and thus destroy the young plants whose fruits form part of man's food supply in North Africa.

In summary, then, it seems that deserts and arid lands, when cultivated, are vulnerable to pests because of a combination of the following three factors. Firstly, animals normally inhabiting arid areas are particularly well endowed as a result of natural selection to resist drought and other extreme conditions and to take advantage of the slightest relaxation of the hard desert conditions by rapid reproduction and migration; secondly, because of the mosaic nature of the vegetation and microclimates of these regions, animals are forced to congregate into the few tolerable areas where life is possible; and thirdly, man, by his irrigation and plantation works, provides small areas where conditions must be nearly perfect in the midst of a vast inhospitable aridity.

There is, however, a different and much more hopeful aspect of desert biology. It would certainly be wrong to think that, simply because a species of animal is well adapted to desert life, it must inevitably become a pest. On the contrary, those physiological, morphological and behavioural features of animals which permit them to cope with the desert environment, provide part of the answer at least to several of the problems involved in developing arid lands for human habitation. Camels, for example, are such admirable animals for desert life in almost every way that it seems surprising that they have not been developed and utilised to a greater extent than they have.

In semi-arid areas again, particularly in Africa, there are many species of mammals, antelopes, equids and bovids, which can thrive in areas unfit for intensive domestic cattle production. They fit into the environment because of their water and heat relationships, perhaps because of their ability to convert urea into protein, certainly because they each feed upon different constituents of the plant cover, and because they can live with their parasites instead of succumbing to them.

One thing is certain – we can learn a great deal from desert animals, and we may look forward with confidence to results of both practical and fundamental importance from research in this field.

E. B. EDNEY

References

[1] SCHMIDT-NIELSEN, K., *Desert Animals*, Oxford, 1964, 277 pp.

[2] CLOUDSLEY-THOMPSON, J. L., 'Terrestrial animals in dry heat: arthropods', *Handbook of Physiology*, Section 4, 1964, pp. 451–66.

[3] SCHMIDT-NIELSEN, K. and W. R. DAWSON, 'Terrestrial animals in dry heat: desert reptiles', *Handbook of Physiology*, Section 4, 1964, pp. 467–80.

[4] DAWSON, W. R. and K. SCHMIDT-NIELSEN, 'Terrestrial animals in dry heat: desert birds', *Handbook of Physiology*, Section 4, 1964, pp. 481–92.

[5] SCHMIDT-NIELSEN, K., 'Terrestrial animals in dry heat: desert rodents', *Handbook of Physiology* Section 4, 1964, pp. 493–508.

[6] MACFARLANE, W. V., 'Terrestrial animals in dry heat: ungulates', *Handbook of Physiology*, Section 4, 1964, pp. 509–40.

[7] LEE, D. H. K., 'Terrestrial animals in dry heat: Man in the desert', *Handbook of Physiology*, Section 4, 1964, pp. 551–82.

[8] SCHMIDT-NIELSEN, B. and K., T. R. HOUPT and S. A. JARNUM, 'Water balance of the camel', *Amer. J. physiol.*, Vol. 185, 1956, pp. 185–94.

[9] ADOLPH, E. F. and D. B. DILL, 'Observations on water metabolism in the desert', *Amer. J. physiol.*, Vol. 123, 1938, pp. 369–78.

[10] MACFARLANE, W. V., R. J. H. MORRIS and B. HOWARD, 'Turn-over and distribution of water in desert camels, sheep, cattle and kangaroos', *Nature*, Vol. 197, 1963, pp. 270–1.

[11] KIRMIZ, J. P., *Adaptation to Desert Environment*, Butterworth, London, 1962, pp. 168.

[12] CARPENTER, R. E., *Copeia* (in press).

[13] HUDSON, J. W., 'The role of water in the biology of the antelope ground squirrel, *Citellus leucurus*', *Univ. Calif. Publ. Zool.*, Vol. 64, 1962, pp. 1–56.

[14] SCHMIDT-NIELSEN, B., 'Water conservation in small desert rodents', in Cloudsley-Thompson (Ed.), *Biology of Deserts*, 1954.

[15] SPERBER, I. L., 'Studies on the mammalian kidney', *Zoologiska bidrag fran Uppsala*, Vol. 22, 1944, pp. 249–431.

[16] CHEW, R. M., 'Water metabolism of desert-inhabiting vertebrates', *Biol. Rev.*, Vol. 36, pp. 1–31.

[17] WIRZ, H., 'The location of anti-diuretic action in the mammalian kidney', in *The Neurophypophysis* (Ed. Heller), Butterworth, London, 1957, pp. 157–170.

[18] SCHMIDT-NIELSEN, K. and B., 'Water metabolism in desert mammals', *Physiol. Rev.*, Vol. 32, 1952, pp. 135–66.

[19] BUXTON, P. A., *Animal Life in Deserts: a Study of the Fauna in Relation to the Environment*, Edward Arnold, London, 1923, pp. 196.

[20] BALZAC, HEIM DE, 'Biogéographie des mammiferes et des oiseaux de l'Afrique du Nord', *Bull. Biol. France et Belge*, Supp. 21, 1936, 477 pp.

[21] WRIGHT, N. C., 'Domesticated animals inhabiting desert areas', in Cloudsley-Thompson (Ed.), *Biology of Deserts*, 1944.

[22] BODENHEIMER, F. S., 'Problems of animal ecology and physiology in deserts', in *Desert Research*, Special publ. No. 2, Research Council of Israel and Unesco, Jerusalem, 1953, pp. 205–9.

[23] — 'Problems of the physiology and ecology of desert animals', in *Biology of Deserts*, Ed. Cloudsley-Thompson, 1954.

[24] VOLCANI, R., 'Seasonal activity of gonads and thyroids in camels, cattle, sheep and goats', *Thesis*, Jerusalem, 1952.

[25] HUXLEY, J. S., 'Eastern Africa: the ecological base', *Endeavour*, Vol. 21, 1962, pp. 98–107.

[26] BOURLIÈRE, F., *Mammals of the World. Their Life and Habits*, Harrap, 1955, 223 pp.

[27] SCHMIDT-NIELSEN, K., et al., 'Extrarenal salt excretion in birds', *Amer. J. Physiol.*, Vol. 193, 1958, pp. 101–7.

[28] KEAST, A., R. L. CROCKER and C. S. CHRISTIAN (Ed.), 'Biogeography and ecology in Australia', *Monographiae Biologicae*, Vol. 8, 1959, pp. 1–640.

[29] POULSON, T. L. and G. A. BARTHOLOMEW, 'Salt balance in the savannah sparrow', *Physiol. Zool.*, Vol. 35, 1962, pp. 109–19.

[30] COWLES, R. B. and C. M. BOGERT, 'A preliminary study of the thermal requirements of desert reptiles', *Bull. Amer. Mus. Nat. Hist.*, Vol. 83, 1944, pp. 265–96.

[31] MAYHEW, W. W., 'Taxonomic status of California populations of the lizard genus *Uma*', *Herpetologica*, Vol. 20, 1964, pp. 170–83.

[32] BRAIN, C. K., 'Observations on the temperature tolerance of lizards in the Central Namib Desert, South West Africa', *Cimbelasia*, No. 4, 1962, pp. 1–5.

[33] TEMPLETON, J. R., 'Respiration and water loss at the higher temperatures in the desert iguana, *Dipsosaurus dorsalis*', *Physiol. Zool.*, Vol. 33, 1960, pp. 136–45.

[34] ATSATT, S. R., 'Colour changes as controlled by temperature and light in the lizards of the desert regions of S. California', *Publ. Univ. of Calif.*, *Biol. Sci.*, Vol. 1, 1939, pp. 237–76.

[35] COLE, L. C., 'Experiments on toleration of high temperature in lizards with reference to adaptive coloration', *Ecology*, Vol. 24, 1943, pp. 94–108.

[36] NORRIS, K. S., 'The evolution and systematics of the iguanid genus *Uma* and its relation to the evolution of other N. American desert reptiles', *Bull. Amer. Mus. Nat. Hist.*, Vol. 114, 1958, pp. 251–326.

[37] MAYHEW, W. W., 'Photoperiodic responses in three species of the lizard genus *Uma*', *Herpetologica*, Vol. 20, 1964, pp. 95–113.

[38] — 'Reproduction in the sand-dwelling lizard *Uma inornata*', *Herpetologica*, Vol. 21, 1965, pp. 39–55.

[39] — 'Reproduction in the arenicolous lizard *Uma notata*', *Ecology*, Vol. 47, 1966, pp. 9–18.

[40] — 'Reproduction in the psammophilous lizard *Uma scoparia*', *Copeia*, 1966, pp. 114–25.

[41] BENTLEY, P. J. and W. F. C. BLUMER. 'Uptake of water by the lizard *Moloch horridus*', *Nature*, Vol. 194, 1962, pp. 699–700.

[42] EWER, R. F., 'The effects of posterior pituitary extracts in *Bufo carens* and *Xenopus laevis*, together with some general considerations of anuran water economy', *J. exp. Biol.*, Vol. 29, 1952, pp. 429–39.

[43] RUIBAL, R., 'Osmoregulation in amphibians from heterosaline habitats', *Physiol. Zool.*, Vol. 25, 1962, pp. 133–47.

[44] BENTLEY, P. J., A. K. LEE and A. R. MAIN, 'Comparison of dehydration and hydration in two genera of frogs (*Helioporus* and *Neobatrachus*) that live in areas of varying aridity', *J. exp. Biol.*, Vol. 35, 1958, pp. 677–84.

[45] GORDON, M. S., K. SCHMIDT-NIELSEN and H. M. KELLY, 'Osmotic regulation in the crab-eating frog *Rana cancrivora*', *J. exp. Biol.*, Vol. 38, 1961, pp. 659–78.

[46] PRADHAN, S., 'The ecology of arid zone insects (excluding locusts and grasshoppers)' in *Arid Zone Research*, Vol. 8, Unesco, 1957, pp. 199–240.

[47] UVAROV, B. P., 'The locust and grasshopper problem in relation to the development of arid lands', in White (Ed.), *The Future of Arid Lands*, 1956.

[48] RAINEY, R. C., 'Flying locusts and convection currents', *Bull. Anti-locust Res. Centre, London*, Vol. 9, 1951, pp. 51–70.

[49] CLOUDSLEY-THOMPSON, J. L., 'Some aspects of the physiology and behaviour of *Galeodes arabs*', *Ent. exp. & appl.*, Vol. 4, 1961, pp. 257–63.

[50] — 'Bioclimatic observations in the Red Sea hills and coastal plain, a major habitat of the desert locust', *Proc. Roy. ent. Soc.*, Vol. 37, 1962, pp. 1–36.

[51] — and M. J. CHADWICK, *Life in Deserts*, Philadelphia, 1964, 218 pp.

[52] WILLIAMS, C. B., 'Some bioclimatic observations in the Egyptian desert', in Cloudsley-Thompson (Ed.), *Biology of Deserts*, 1954.

[53] EDNEY, E. B., 'The survival of animals in hot deserts', *Smithsonian Report for 1959*, 1960, pp. 407–25.

[54] LAWRENCE, R. F., 'The sand dune fauna of the Namib desert', *S. Afr. J. Sci.*, Vol. 55, 1959, pp. 233–9.

[55] MCMICHAEL, D. F. and D. IREDALE, 'The land and freshwater Mollusca of Australia', in Keast *et al.* (Ed.), *Biogeography and Ecology in Australia*, 1959, pp. 224–44.

[56] NORRIS, K. S. and W. R. DAWSON. 'Observations on the water economy and electrolyte secretion of chuckwallas (Lacertilia, *Sauromalus*)', *Copeia*, No. 4, 1964, pp. 638–46.

ADDITIONAL REFERENCES

BARTHOLOMEW, G. A., 'The role of physiology in the distribution of terrestrial vertebrates', *Zoogeography*, Amer. Assoc. Adv. Sci., 1958, pp. 81–95.

BENTLEY, P. J., 'Adaptations of amphibia to arid environments', *Science*, Vol. 152, pp. 619–23.

CLOUDSLEY-THOMPSON, J. L. (Ed.), *Biology of Deserts*, Institute of Biology, London, 1954, 224 pp.

UNESCO, *Human and Animal Ecology*, Arid Zone Research Series No. 8, 1956.

WHITE, G. F. (Ed.), *The Future of Arid Lands*, Amer. Assoc. Adv. Sci. Publ. No. 43, Washington, D C, 1956, 453 pp.

ADDENDUM

It has recently been found that a desert cockroach, *Arenivaga* sp., which lives in the sand dunes of southern California, absorbs water vapour from unsaturated air in relative humidities of 82 per cent and above (Edney, 1966). The adaptive value of this process has yet to be determined, but it appears highly probable that these insects, when short of water, may readily restore the loss by moving downwards through the sand until they reach a region where the humidity is above 82 per cent. In addition, the cockroaches are able to regulate the osmotic concentration of their body fluid against the effects of considerable water loss.

REFERENCE

EDNEY, E. B., 'Absorption of water vapour from unsaturated air by *Arenivaga* sp. (Polyphagidae, Dictyoptera)', *Comp. Biochem. Physiol.*, 1966 (in press).

CHAPTER X

Man in Arid Lands:
I – Endemic Cultures

Considering the uses to which land can be put the term *arid* involves a negative value judgement.

Water, the vital commodity, is lacking or scarce, and even in semi-arid areas the excess of water loss by evapotranspiration over rainfall severely limits agricultural settlement. The striking discrepancy between the area of the arid zone (36 per cent of the world's land surface) [1] and the number of people living therein (13 per cent of the world's population) stresses the slight use made of arid lands (see Table x-1).

TABLE X - I *Distribution of population in arid zones*

	Popula-lation, millions	Per cent of population		Per cent of area		Average density of arid zone population
		Arid zone	World	Arid zone	Total land	
Arid Zone	384	100	12·8	100	36·3	7·9
Extreme arid	5	1·3	0·2	11·9	4·3	—
Arid	103	26·9	3·4	44·6	16·2	4·7
Semi-arid	276	71·7	9·2	43·5	15·8	13·0

(Compiled from 1960 figures for an approximate world population of three billions.)

The physical basis

Arid lands are distinguished by the special features of a number of natural factors basic to human occupance, of which water is the most important, although other factors, chiefly climatic, are also significant. These physical factors have been discussed in Chapter v and it will suffice here to note that, among topographic influences, mountains have a marked effect not only on local climate and weather, but also on human occupance.

Mountains introduce an element of lesser aridity into regions of all grades of dryness. The higher they rise, the more pronounced becomes their mitigating influence, which is especially pronounced on the windward side exposed to the moisture-bearing air. Consequently, the land adjoining these mountains, although climatically of various shades of aridity, benefits from the availability of groundwater on fans or in groundwater horizons fed in the moister mountain area. This creates a universal type of mountain border occupance and cultivation in arid lands, for which a most appropriate term has been coined in the dry *Nordeste* of Brazil: *pé da serra* cultivation.

WIDE SPACES

The second characteristic of most arid areas is wide open spaces. As noted before, more than one-third of the world's land area is arid. For obvious reasons the distribution of people in the arid zone is extremely unequal: 72 per cent of the arid zone population is found in the semi-arid area, whereas only 28 per cent is located in the arid and extremely arid areas. Moreover, these latter occupy 56·5 per cent of the arid zone, so that the share of population of the arid zone proper is only half its share of its area (see Table x-1). Again, most of these people are concentrated in very limited areas, oases or *pé de serra*-type locations, and the net result is that in all arid regions one finds extraordinarily large empty areas between the few inhabited spots. These extensive empty spaces and their sparse populations present a critical problem in arid zone occupance. The development of an area and the provision of the services considered necessary by a given human population depends on the economic level this population attains. The sparser the population, the lower this level will be. The provision of a decent level of services is therefore, *inter alia*, a function of the size of the area concerned. Arid areas are in this respect always at a disadvantage, their development being extraordinarily expensive in relation to units of population. This disadvantage is shared by oases, including mountain oases. Although these can often grow agricultural produce very reasonably, the long distances to market and high transport costs place on them a considerable economic burden. Distance, therefore, is a major obstacle to development and most arid zone populations cannot attain the standards of living found in related occupations in the humid zone. This creates a challenge which man attempts to meet in various ways.

THE CHALLENGE OF THE ARID ZONE

At the lowest level man fits into the framework of primitive conditions and adapts himself to them as well as he can. Like other primitive people these inhabitants of the arid zone are very conservative, which retards development even more. But even here, under the pressure of the environment, extraordinary achievements in adaptation and technology are made, which permit man to live in the desert's stark conditions. We find, therefore, at different times and in different regions both highly advanced civilisations and underdeveloped people in the truest sense of the word.

For high technology it is sufficient to recall the ancient and modern irrigation cultures of Mesopotamia, the Nile Valley and northern China. These are among man's most noble achievements – irrigation cultures based on rivers each one of which presents difficult technical problems. Such irrigation cultures require high standards of technique and administration, which must be maintained unceasingly. Any slackening of man's effort will bring about decay of dams, drainage systems or installations for lifting the water, or disorganise the highly complex and detailed organisation of water distribution. Deleterious effects on the land may follow, e.g. salinisation. To cure these is sometimes a task for generations, hardly ever to be achieved except by another high civilisation.[1] As the perseverance of man and his political organisation undergo changes for a variety of reasons, some of them directly related to the arid environment, we find extreme fluctuations of cultural achievement and standards in the areas referred to. The outstanding example is Mesopotamia, where the high levels of the Sumerian and Babylonian irrigation cultures, already achieved in the third millennium B.C., have not yet been reached again. Even in the river oases, where conditions of aridity have been locally removed, we find the same trend to extremes characteristic of the arid zone in general: they alternate between prosperity and a state of decay. Unsteadiness is the keynote.

The great river oases represent optimum conditions of 'arid' environment for man, as they provide two commodities lacking in the desert – river water and soil alluviated by the river's sediment load. Other physical factors are amply available here as elsewhere in arid lands: high temperatures and the world's longest hours of actual sunshine per year. Small wonder, therefore, that flourishing civilisations of this type achieve some of the world's highest densities of agricultural population: data for Egypt referring to the cultivated area only show densities between 330 and 910 per square kilometre (Census of 1957); figures for provinces of the Hoang-Ho valley of China (Census of 1953) show densities of 80 to 400 for the province as a whole; the cultivated area, therefore, should have densities well over 300. Such figures are matched only in very exceptional cases elsewhere, mainly in the western parts of the Netherlands where overall densities are between 280 and 920 (Census of 1956) or the Po Valley of Italy.

The extremely uneven distribution of arid zone population is fittingly demonstrated by Egypt. If we relate its population to the cultivated area only, the country has an overall population density of 660 per square kilometre. If, however, we relate this same population to the total area of Egypt – including its empty desert areas, which are nearly 29 times as great as the cultivated land – the overall population density dwindles to 23.

---

[1] It may be, in fact, that many of the ancient civilisations based on irrigation were destroyed by salinisation: see Nace and Tison (International Cooperation in Scientific Hydrology, *I.C.S.U. Review*, 5 (1963)).

Man's occupance of arid areas

Today, as usually in the past, the deserts proper and their borderlands hold but little attraction for man unless he has some special motive. The arid area proper, with the exception of oases, is a habitat not desired by most people. Much of it, therefore, is an area of retreat, inhabited by marginal populations. Some of these withdrew into the arid area before the advance of more powerful and more advanced populations in adjoining regions. From the Middle East cases are even known where sedentary people have taken to the nomadic way of life, temporarily or for good, in such circumstances. On the whole, people living permanently in the arid zone proper are either highly advanced or rather underdeveloped.

A special incentive is required to bring about advanced and intensive development. This might be the occurrence of important mineral resources, whose exploitation makes it worth while to undergo the hardship involved. Another common reason is provided by the vast extent of arid lands located between densely settled non-arid areas. Routes have to pass through arid areas. Their alignment and the volume of travel on them depend on a variety of factors: demand for traffic in both terminal areas, value of the goods to be transported, technology of desert travel as against that of alternative routes often of much longer mileage, and security conditions along the way.

This last fact has had a varied influence. The wide and uncontrolled areas of desert naturally attract anyone wishing to avoid the eye of authority. Slave-caravans crossing the Sahara until the nineteenth century, and drug traffickers in the Near East at the present day, preferred and still prefer desert routes. The isolation and the lack of effective control created a time-honoured extortion racket in the deserts of Asia and North Africa. Until the introduction of motor travel, any caravan, large or small, wishing to travel there had to buy protection from the nomadic tribes roaming the desert, otherwise these same tribes would raid the caravan. No purpose, however holy, would save the traveller from these raids. Until the nineteenth century the annual pilgrim caravans to the holy cities of Mecca and Medina were often raided by Bedouins, of the same Moslem faith as the pilgrims, who looted their goods and often sold the pilgrims themselves into slavery. Some nomads indeed considered raids and robbery as part of their normal way of life, a perfectly proper and fitting activity, and protection money was considered a regular and legal source of income.

Despite the great open spaces, desert routes are often no less clearly defined than a modern highway. Of this anyone who ever flew over a desert has seen evidence – the routes converging on oases and water-holes.

Both mineral exploitation and trade routes are unstable foundations for permanent occupance even in humid regions, for a mineral resource can be exhausted or become obsolete for technological or economic reasons, and trade routes may shift. Settlements based on either of these foundations are, therefore, likely to be unstable in the long term. But in the arid zone, because of the lack of other means of support, the difference is not infrequently between a high level of development, quite beyond the standard of arid areas in general, and total abandonment. Gold-rush towns and ghost towns mark these two extremes, which sometimes are but a few dozen years apart. Another reason for the limited local impact of some mineral economies is the relatively small number of people required for all but very big mining ventures. An outstanding example is the small number of men engaged in the oil industry of the Middle East – a number quite out of proportion to the economic value of the industry.

NORTHERN CHILE – A MINING AREA IN DECLINE

The desert of northern Chile is a fitting example of an arid mining area, its difficulties, its investments and installations, its rise, prosperity and part-decay (Fig x-1). Its flourishing dates back to the second half of the nineteenth century, when nitrate from this desert found a ready market as fertiliser around. the world. To mine the nitrate, which is available practically on the surface, the desert lacked everything: labour, transportation, supplies and services to maintain the labour force and the mining process. The major towns, mining centres or ports get their water by pipeline from the borders of the Cordillera de los Andes. The pipeline to Antofagasta, the largest of these towns (population ca 100,000 in 1960), is almost 350 km long. In earlier days water was supplied in barrels by ship and railway and delivered to the mines by mule carts. Fresh vegetables and fruit were scarcely less precious than water.

Bowman [2] quotes a fitting description by Gilliss of a typical though rare market scene at Cobija in 1851:

It was a matter of no little interest to witness the avidity of the population on landing the garden-stuff brought from Arica. Probably within ten minutes after the first boat-load of bags had been landed, all over town Indians, including soldiers, might have been seen stripping the rind from green sugar-cane . . . housekeepers bearing away piles of ears of maize, sweet potatoes . . . an hour later the beach—which had served as the impromptu market-place—was again bare.

Not only water but also soil is lacking locally to grow agricultural supplies. To meet this latter deficiency a special type of truck gardening has been developed in the coastal towns. The cargo boats which delivered Chilean nitrate to Australia had to return empty across the South Pacific. Instead of returning with water in ballast, many ships, therefore, brought a cargo of soil, which was spread on to well-fenced truck gardens. So both in Iquique and Antofagasta in Chile vegetables are grown on Australian soil! At Iquique there are about thirty truck gardens of this type, known as *quintas*, up to a quarter hectare in size. More recently four larger *quintas* have been installed, of a half-hectare area each. They are located on the former site of a municipal garbage dump. Over this a layer of sand was spread to which cow manure was applied. After six months' treatment a biologically sufficient soil stratum has been created to start cultivation. The half-hectare quinta is irrigated with up to 18 cubic metres of water per day. The

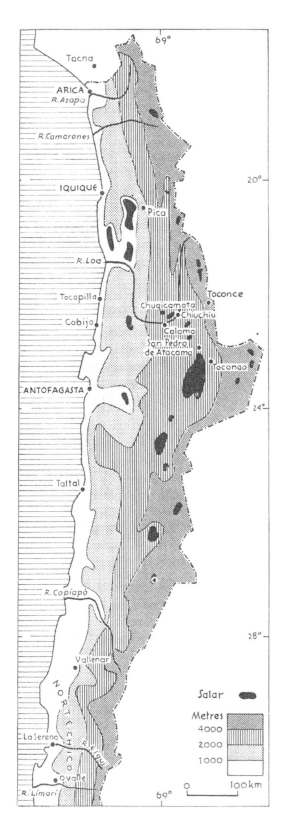

Fig X-1.
The northern desert of Chile.
(*N. Z. Baer, Dept. of Geography,*
The Hebrew University)

water is applied by hose to each plant, six men being employed per half-hectare quinta. A variety of vegetables is grown, including spinach, lettuce, cauliflowers and carrots. Even with heavy morning cloud in winter, characteristic for the coastal area, the noon and afternoon sun is fierce enough to have a crop ripening within 28–38 days.

The collapse of the nitrate industry about 1920 brought economic disaster to the desert of north Chile, a disaster from which it has not fully recovered after forty years. Fortunately, the development of other mineral resources, copper at first and iron later, has revitalised the regional economy, but evidence of regression is clear throughout the region. Iquique, the second largest city, shows an extremely slow rate of development, its population having increased by only 25 per cent between 1907 and 1960, from 40,000 to 49,500. Antofagasta, the largest and booming town of the region with a population of close to 100,000, grew by 51 per cent during the same period. Even this is a slow rate of growth for cities today. Santiago de Chile, the capital, increased its population by 235 per cent from 1920 to 1959. The main square of Iquique boasts a great opera house in which famous artists appeared in the heyday of the nitrate period: today it serves as a cinema.

Much more lamentable is the case of Taltal, once another busy port. Today many houses are in wretched disrepair, the population numbers barely a few thousand, and no train runs on the rails of the feeder line still leading into Taltal. The few surviving shops in this sleepy town do not open before 10 a.m., for want of customers. And ironically, here, as in many a gold-rush town, the Teatro Alhambra was built in 1921 when the nitrate industry was already in decline. All these ghost or semi-ghost towns have one thing in common: outsize cemeteries.

THE CENTRAL NEGEV –
RUINED TOWNS ALONG AN ABANDONED TRADE ROUTE

About the beginning of the Christian era there existed in what is today the southern part of the kingdom of Jordan, east of the Rift Valley, the important trader kingdom of the Nabataeans (Fig. x-2). Their capital was at Petra, famous today for the architecture cut in the rock around the ruined city. The Nabataeans conducted a considerable trade between Mediterranean and eastern lands, and from Petra their trade routes led to the Mediterranean ports. At the level of transportation technology of that time, a trade route of 170 to 300 km could not be operated without a staging and service area roughly halfway along. The highlands of the central Negev conveniently filled these requirements, being located about halfway to the coast and at a higher elevation than the coastal desert. Here, therefore, at altitudes between 300 and 650 metres, six towns were founded to supply the needs of caravans.

These towns were of no negligible size. Considered estimates of their population arrive at 3,000–8,000 each. Their inhabitants were wealthy. Not only did

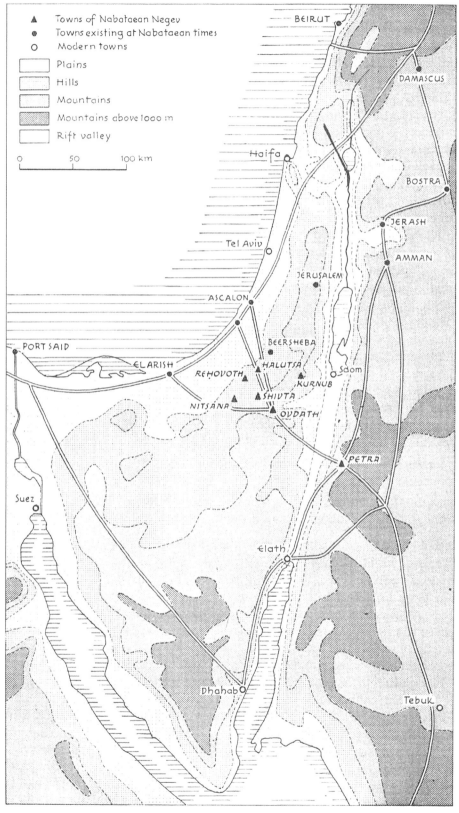

Fig x-2. Towns and roads of the Nabataean Negev and adjacent areas. (*N. Z. Baer*)

many of the houses have two and some of them three storeys, but the towns were graced by conspicuous public buildings. Shivtah, for example, had three churches, one of them attached to a monastery and decorated inside with marble; at least in one case the roof had been built of large logs of wood, obviously imported. The streets of the town were well paved and every house was fitted with a cistern.

Aside from their main task of servicing the caravans, some of the townspeople engaged in subsidiary agriculture nearby. By most ingenious methods [3, 4] they cultivated a sizeable area in a region which has today an annual rainfall ranging from 70 to 120 mm according to the location of the town. By all evidence, rainfall at that time was of the same order of magnitude as at present. This agriculture was based on an amazing knowledge of the mechanism of desert climate. All the fields were on the very bottoms of the valleys, but only on those of secondary ones, never in a major valley. A system of channels, sometimes only low walls, conducted the rainwater to the fields. To increase the run-off and at the same time apparently to speed up the sedimentation of soil on the originally rather barren valley bottoms, they raked the stony surface of the desert pavement into rows of low stone heaps; this permitted the rain to flush the powdery soil under-lying the hammada crust to the valley bottom, thus speeding up the accumulation of cultivable soil.

All this depended apparently on a clear appreciation of the nature of the desert, where rainfall comes in a very few localised downpours of high intensity. The desert pavement makes the surface rather impervious, and both factors combine to bring about a high ratio of run-off. It was this run-off, of possibly a maximum of half a dozen rainfalls per winter (and sometimes even less), on which the ancient Nabataean farmer based his agriculture.

Today all this is in ruins, as it has been for well over a thousand years. The wealthy towns of the central Negev and their fields swiftly decayed after the collapse of the Nabataean kingdom and the withdrawal of the subsequent Roman frontier settlement. The traveller crossing the Negev Desert today wonders what brought men to settle here in ancient times. He has another good reason to wonder. Israel, the land of the Bible, abounds in archaeological sites, but all are superficially destroyed and it takes an archaeological excavation to uncover the remains of the walls. Only here in the central Negev did the ruined towns survive as full-sized ruins, slowly deteriorating over more than a millennium. The two exceptions were the two northernmost of the ancient towns which were largely dismantled in the nineteenth and early twentieth centuries, when Bedouins made a good profit by selling the well-cut masonry to builders who used it in the construction of the cities of Gaza and Beersheba. Elsewhere in Israel an ancient site abandoned or destroyed for any reason was reoccupied a number of times; these towns in the central Negev were just abandoned, to remain ghost towns through the centuries.

THE INSTABILITY OF SETTLEMENT IN THE ARID ZONE

The Norte Grande of Chile and the central Negev of Israel have been treated here at some length as examples of mining settlement and settlement along trade routes in arid areas. Both were spectacular achievements at the time when they prospered. Both areas represent specialised developments responding to special challenges. In both cases much ingenuity and effort made settlement possible under adverse conditions, so adverse indeed that the moment the extraneous need for this development came to an end the whole project collapsed. Unfortunately, there is no lack of similar examples from many other desert areas all over the globe, because this type of arid zone occupance and development responds to extraneous needs, not to those of the arid zone itself. The arid zone in these cases is passive, not active. If a mining area or an important trade route undergoes severe deterioration in humid lands, the settlements concerned may suffer regression, but not large-scale abandonment. After a while the population will adapt itself to the change and develop another branch of economy. Arid zone people lack this alternative at the same scale. Although arid zone occupance often involves multiple use of the land's resources, these frequently are so limited as to make readjustments of towns created to serve a special need rather difficult. The cessation of this need frequently affects the whole occupance pattern of the area concerned. For this reason all comparable areas in the arid zone provide examples of abandoned settlement and not only of stages of prosperity and decline. The many 'lost cities' from Zerzura to Loulan in the Lop Desert[1] are mute evidence of this.

The two areas cited above – the Norte Grande and the Negev – have another feature in common, in distinct contrast to humid lands. The settlement pattern of humid lands forms a kind of pyramid with villages at the base, smaller towns higher up and large cities at the top. Both the instances cited from the arid zone and others likewise, lack this normal pyramidal structure. They have one type of settlement only, generally towns, responding to the purpose the settlement pattern is serving. Again, the absence of settlements of various grades performing different tasks deprives arid zone settlement of the possibility of adaptation to changing conditions.

Admitting then the essential instability of such special arid zone developments, it remains to survey the much less spectacular peoples of these areas. These people never live under the glamorous conditions the central Negev of Israel or the Norte Grande of Chile experienced at the time of their prosperity. They are few in number and live literally on the edge of the habitable world. Their life is most precarious, and 'the least accident may betray them' as Bowman illustrates [2, pp. 68–9],

by the history of a line of settlements in the Chacarilla valley. It was at one time a fertile and frequently visited district. But early in the seventies, as nearly as we

---

[1] This ghost town lies on an ancient route across a salt bed in the Tarim Basin in the Lop Desert of Chinese Turkestan. Once irrigated by the Kuruk Darya, the area is now one of abandoned cultivation and dead forests.

could determine, a great flood came down the gorge, broke down the irrigating ditches, cut up the terraces, or deposited infertile sand, gravel, and even boulders upon them, overwhelmed orchards, and so generally devastated the farms and discouraged the inhabitants that all but a remnant of them moved away. The shock which such an occurrence gives to a self-contained oasis is always terrific and sometimes fatal.

Nevertheless, the common desert dweller commands the interest of the student of arid zones no less than the one involved in special, spectacular technological projects. Here people often of no formal education have to sharpen their wits on a hard environment to make a living. Primitive as many of them are by our standards, they have achieved most ingenious feats of adaptation.

ADAPTATION OF PRIMITIVE PEOPLE TO DESERT LIFE

Until quite recently some very primitive people were living under arid conditions, and a few — a very few — are still there today, mostly in semi-arid rather than truly arid areas. The two most primitive contemporary groups are some tribes in central Australia and the Bushmen of the Kalahari of South Africa. Both people made their living originally as food collectors and hunters, and as such developed amazing skill in tracking and hunting their prey, some of it very elusive and some as large as elephant or giraffe. Both peoples live in most primitive dwellings, if indeed one may so call them. The most rudimentary type of dwelling of all arid zone people is that used by central Australian tribes. They just put up a wind-break or breakwind, a shelter made of two forked sticks set upright in the ground, supporting a horizontal branch against which bark or branches of bushes are leaned [5]. Thomson gives this description of the breakwinds used by the Bindibu near the saline Lake Mackay [6]:

On the windward side . . . a low wall of sparse brush was built, often in crescentic form, about 2 ft high. In the lee of each breakwind the earth was scooped out at right angles to the shelter in the form of one or more gravelike depressions, a few inches below ground level, each just wide enough to take the body of one member of the family. Fires were maintained on the ridges between each of these depressions, and in this way the Bindibu found shelter from the keen winds that swept the desert at ground level. In the morning, the family often moved out from the lee of the shelter to sit in the morning sun. Some distance from the breakwinds used for sleeping . . . rough platforms of brush had been made of branches arranged transversely to make a day-time shelter or canopy for a mother and children and as a rack for food or possessions.

PHYSIOLOGICAL ADAPTATIONS OF CENTRAL AUSTRALIAN TRIBES

This behaviour has, rightly, aroused considerable interest. For people living in the wet tropics and not yet subject to the social conventions of 'modern' dress, to live and sleep naked would seem to provide maximum comfort, considering the general high levels of temperature, humidity and cloudiness. Even so, most primitive men in the tropics live in houses and many kindle a fire at night to keep comfortable in sleep. By contrast, man in arid lands is exposed to intense solar radiation and has to endure the extreme diurnal variations of temperature occurring on earth (see Chapter III). Everyone who has camped out in arid areas has experienced the contrast between teeth-chattering pre-dawn cold and the scorching heat of mid-morning, sometimes as early as 8 a.m. The primitive people of central Australia live and sleep naked in an area where early-morning temperatures drop in winter as low as −10°C, whilst around noon the same day 30–34°C might be recorded. Summer shade maxima will be in the upper forties (plus or minus 110° on the Fahrenheit scale). In an investigation carried out by Stanton Hicks and others in the 1930's in the Musgrave Ranges to determine the physiological aspects of such a mode of life, the subjects were placed in tents, to avoid the difficulties of measuring the cooling power of the environment out of doors, in the presence of a number of open fires. For comparison, white men, stripped naked, were placed in the same position (prone on the ground) in exactly similar tents alongside, and were subjected to the same measurements at the same time. The experiments were carried on while the temperature was falling in the late afternoon.

The features of the adjustment by the native to meet these circumstances, as far as it was possible to determine them, are as follows:

(1) As the result of the use of small fires to which he lies very close, areas of intensely high skin temperature are found on the body of the sleeping natives. It is possible that in this way a large amount of radiant heat is absorbed so that the balance can be preserved despite the fact that the other parts of the body are so intensely cold. The magnitude of the factor was not determined, but it appears probable that the conception of a rapid loss of heat from the cold parts, notably the extremities, balanced by heat absorption at hot areas is a correct one.

(2) The native has an extremely active control of his skin circulation, and heat loss from a cold part is minimised by a diminution of the circulation through that part. Moreover, it appears that this control is more definitely localised than is the case in the white man. This combination of the maximum limitation of circulation in the cold areas and consequently of heat loss, with a considerable uptake of heat from the areas heated by fires, may constitute the means by which the native is enabled to maintain his body temperature.

(3) Despite the cold environment the metabolism of the native does not increase. Comparison of the responses of white and natives under identical conditions in this respect was not made, but there is no doubt that whites under conditions approaching those of the natives would show such a change.

The natives experienced the subjective sensations of cold just as keenly as did the whites. The explanation of the apparent anomaly of a native being able to sleep despite subjective sensations of cold may lie in the possibility that the wakefulness of a cold white subject is not due to the stimuli arising from the skin, but to the state of sympathetic tone that arises from cold. The increased metabolism is only one manifestation of this increased adrenal activity. In the native, however, no such general response occurs, and in consequence the vagotonic state necessary for sleep is not reversed. The native maintains his body temperature by a local reflex controlling the skin circulation and the general autonomic system is not involved. These results of Hicks [7] have been evaluated by Wulsin [8].

THE SIP-WELLS OF THE BUSHMEN

No less interesting than the unconscious adjustment to climate of the central Australian tribesman is the skill the Bushmen of South Africa display in securing and storing water in the dry environment of the Kalahari, where there is nothing but damp sand from which to procure it. This is how a recent observer describes the operation of these sip-wells [9]:

The procedure is as follows, and is usually carried out by an old woman. A certain plant serves as indicator of damp sand at the spot chosen and the woman proceeds to scrape a hole in the sand to arm's length. She then takes a grass stem or hollow reed, surrounds one end of it with fine roots or grass, and places that end in the bottom of the hole. The sand is then carefully replaced round the reed and firmly tramped down. After an hour or two the woman begins to suck through the reed, producing, of course, a semi-vacuum in the wad of grass buried below. The sucking has to be maintained for anything up to an hour before the pressure of the atmosphere has pressed the films of water round each sand-grain to the bottom of the reed. The water then comes up into the woman's mouth, and she inserts a second and shorter reed in the other side of her mouth leading down to a small hole at one end of an ostrich egg. She alternately sucks with one side of her mouth and expresses the water from the other side down the reed to fill the ostrich egg shell. A number of these are filled, carefully sealed with clay, and buried in different parts of their hunting territory for use when the dry season comes upon them. One wonders how the sip-well technique could have been devised. An old traveller of the Kalahari, an Australian who was left there after the Boer War, described to me how he and his companion with four donkeys were kept alive for a fortnight by the ministrations of a single old Bushman woman with her sip-well.

Water supply is the basic technical problem facing man in the arid zone. He has, therefore, devoted a great amount of ingenuity and effort to providing water, and many of his finest achievements in arid lands are related to water engineering. Whenever and wherever man was able to ensure its supply on a sustained basis he could achieve sedentary occupation. Where water was available only seasonally or in minimal amounts, occupation could be but temporary. There is, however, a marked difference between these achievements and the often mediocre level of civilisation in other respects in arid regions.

TYPES OF WATER USE AND DEVELOPMENT OF
PERSONAL AND DOMESTIC WATER SUPPLY

The sip-wells of the Bushmen serve to highlight the first and most important use of water, that is to sustain life in man. The Australian aborigine is aware of every rock-hole that retains water, some of which, the gnamma holes in granitic rocks, are in otherwise waterless areas. He also scrapes in the damp sand of rivers, digs out frogs engorged with water, taps the sap in certain plant roots or in callus growths on trees, and in all ways short of permanently modifying the natural scene is an expert in self-preservation in the desert.

A most ingenious installation is the so-called air well. These are piles of fairly large-sized smooth-surfaced stones, limestone being the favourite material. On the stone surfaces inside the pile very small amounts of water are formed by condensation, especially when the temperature difference between the inner stone and the air rises.

Such air wells were apparently known to the ancient Greeks and were reportedly used by them in the Crimea. Some Bedouin tribes in southern Tunisia customarily erected them. Recently, Chaptal experimented with a modern design of air well in the vineyard region of Montpellier, France. He built pyramids of broken calcareous stone over a base of 9 square metres enclosed in a concrete cube. Air entered through holes in the outer cube and the water was collected at the bottom. 'The dry stone produced water every day, but in much greater quantity during the hot, dry summer days than during the humid days of winter. In fact, in the winter the amount generated was almost negligible; in the summer the stone yielded nearly 2½ litres of pure water in 24 hours' [10].

Wells dug in rock or in the sands and gravels of dry river beds are to be found in all countries (see Fig. x-3). They often afford the only permanent water for homes or villages in semi-arid areas and in oases in the desert, and may support animals as well as the human population, especially in the dry season. It is a remarkable experience in the dry season to see herds of thousands of head of stock coming and going all day long and every day at important wells or well fields, and the observer is struck both by the remarkable yields of such wells in arid regions and by the immense size of the herds in what appears to be poor grazing conditions. The high yields are normally the result of rare combinations of hydrological conditions – particularly rapid absorption of run-off and an impervious stratum at moderate depth.

The term 'well' is often used for a groundwater structure excavated by hand, and such a well usually ranges from about 3 to 6 feet in diameter, whereas a bore-hole, or bored well, is drilled by a drilling machine, and is usually 6 to 8 inches in diameter, but may be as much as 12 inches or more.

In most countries the local people have been accustomed for many generations to dig their own wells, and these vary greatly in construction from place to place according to the nature of the enclosing rocks. The water is often raised in skin containers with leather thongs, usually by hand and sometimes with the aid of stock, either hauling direct or working mechanical gear. Often the shafts are of small diameter, about 20 inches, just enough for one man to work in with difficulty, and there may be scores of them, old and new, within a few acres; others are immense crumbling excavations reaching down to the water table, as in the gypsum wells of northern Kenya; others follow a zigzag course downwards along the weathered joints of solid rocks, and the water is passed up hand over hand in small buckets, while yet others are great funnel-shaped excavations in loose sand

Fig x-3. Water-hole at Campina Grande, Brazil. (*Photo UNICEF, from 'Unesco Courier'*)

30 feet or more in depth, in which the diggers are not infrequently trapped. But modern drilling and pumping equipment enables wells to be carried down to far greater depths in hard or soft rocks and to yield much larger supplies.

Those arid and semi-arid regions which have no perennial or seasonal rivers or groundwater supplies depend in part for their domestic water supply on rainwater. The character of rainfall in arid climates presents both advantages and difficulties in storage, for, whereas the great majority of rainy days produce small amounts of rain each, there occurs practically every year a small number of days with rainfall of great intensity and considerable amounts, often compressed within a few hours. The same pattern holds good for the most extremely arid areas; only here the intervals between heavy rainfalls can be spaced over years.

To catch this rain and store it various types of cisterns are used. A common type found in the Near East formed the sole supply of water for many cities of this region until a generation or two ago. This cistern is generally located underneath the interior courtyard of the building. It has a narrow opening in its roof or ceiling which is normally covered. The receptacle of the cistern is bottle- or rather flask-shaped and lined with mortar to make it impervious. The courtyard, often the roof of the house as well and sometimes even the adjoining street, function as catchment area. The water is conducted into the cistern either by a number of channels or through the opening in its roof. These cisterns are of very different sizes, from that of a normal room which would supply the needs of a family at pre-modern levels of consumption to the huge public cistern found in many cities. According to observations of Fisher and Dubertret in the Lebanon, 'it seems that an Arab peasant farmer can live and irrigate his lands, under Mediterranean conditions, from a total water supply of about 2½ cubic metres per head per annum. So with an annual rainfall of, say, 200 millimetres, catchment from an area of 13 to 15 square metres per person could be enough to sustain settlement, which is rather surprising' [11]. The city of Jerusalem had in 1921 about 7,000 cisterns with a total capacity estimated at between 0·45 and 1·6 million cubic metres [12, 13]. If properly kept, these cisterns have the great advantage over open reservoirs, pools or tanks – the latter much used in semi-arid India – in that they involve no evaporation losses and can keep the water clean and unpolluted. Every year there occurs a sufficient number of days with intensive rainfall to permit the cisterns to be filled.

In fully arid areas a different and very interesting type of cistern is found. Whereas the conventional cistern is designed to catch rainwater, this cistern is designed to catch floodwater. This permits it not only to store the rain which falls on its roof as the immediate catchment, but to store part of the water of a small hydrologic catchment. Floodwater cisterns have apparently a considerable dispersal throughout the Near East: they are called *harabe* by the Bedouin or *maagurah* in Hebrew, and often number many hundreds in regions where they occur.

Whereas the rainwater cistern has its inlet at the top, floodwater cisterns have the inlet in the side wall, being excavated sideways into a slope. The water filling it comes from the wadi-bed alongside which it is located, or from a canal draining the slope above or a small side-valley nearby. The floodwater cistern can have considerable dimensions, and sometimes a whole series of them is excavated in the low part of a slope. Their form is rectangular, resembling a room. Often a flight of steps leads down into them, permitting walking down to the water level as it recedes with the advancing season. Some of these floodwater cisterns reach dimensions of 10 by 14 metres, having a depth of 6–7 metres; this would give them a storage capacity of over 800 cubic metres. Groups of them, one adjoining the other, provided storage for multiples of this figure. The main cavity being inside the mountain, the floodwater cistern shares with the rainwater cistern the benefit of low evaporation losses. Maximum possible use is made of existing rock conditions to facilitate quarrying and reduce losses by seepage. The floodwater cisterns in the arid Negev of Israel, for instance, are nearly without exception located at the geological contact of a hard limestone overlying a soft chalk of much reduced permeability. Quarrying in the chalk was easy, whereas the limestone formed a strong and stable roof. In large cisterns a broad square pillar was left standing to support the roof. To minimise sedimentation, many floodwater cisterns have a settling basin at the entrance, the bottom of which is at the same level as that of the cistern, and lower than the threshold of the opening connecting the two [14]. To permit some of the sediment to settle outside the actual cistern is most important indeed, as the nature of the short but strong floods of the desert loads them with a large amount of sediment. In this respect the floodwater cistern is superior in design to the common rainwater cistern into which is washed all the mobile material lying on the roof.

Floodwater cisterns are apparently of considerable age. In the Negev of Israel they predate the Arab Bedouin who use some of them today. Most *maaguroth* there were choked with debris and it was only on the initiative – and order – of the British Mandatory authorities that a certain number of them, but by no means all, were cleared of debris and made fit for use again.

Both rain- and flood-water cisterns are found in the Negev, and there is no reason to suggest that they were constructed at different periods. Both obviously must be dated to the time when the Negev prospered, i.e. the late first millennium B.C. and the greater part of the first millennium A.D. But there is a distinct difference in their distribution. In general, the rainwater cistern is limited to cities and such other few and small settlements as there are. By contrast, floodwater cisterns are absent in cities, but are widely dispersed throughout the area, often at many kilometres' distance from the nearest site of settlement.

The rainwater cistern can be constructed on – or rather under – any type of ground. If the house or settlement which it is to serve is located on hard rock, its excavation is difficult and expensive. The floodwater cistern must be located next to a wadi bed or on the lower part of a slope which is to serve as its catchment. To construct a large cistern at relatively low cost requires it to be located at the contact of a hard cap-rock and a soft, easily workable and impervious reservoir rock. Only rarely, therefore, will it be possible to provide a city with floodwater cisterns. It appears that cisterns were prepared in large numbers to serve the caravans, which were the immediate reason that brought about the *mise en valeur* of certain deserts, e.g. the Negev Desert in Nabataean times.

IRRIGATION AND STOCK WATER SUPPLY

Whereas the use of wells or cisterns for water supply for stock or for small irrigated fields in addition to domestic use is a common but simple and limited technical operation, the use of dams, aqueducts, pipelines and qanats (see

Chapter v) involve sophisticated techniques which have, however, been in use for many centuries, even millennia. Cisterns have been dealt with at some length as they are basic to sedentary and nomadic occupation of arid lands, and qanats are notable for their high degree of adaptation to the arid environment, making it possible to deliver water over a distance of kilometres. Both installations are characteristic of populations at a low economic level, more specifically with low-priced labour.

Societies operating at a higher economic level, both in the past and certainly in the present, arrange for their water supply by more conventional though more expensive methods. Such methods involving canals, aqueducts and pipelines are applied equally in humid, semi-arid and arid areas. Their water may be derived from springs, rivers, reservoirs or bores. These are signs of advanced civilisation, and their operation and maintenance demand a steady and efficient administration. They require a strong regime as well, as no area can rely on a water supply which is not properly secured. Many times in arid lands the breakdown of long-distance water supply has followed closely upon the deterioration of a political system, spelling doom for the area concerned.

D. H. K. AMIRAN

References

[1] MEIGS, P., cited by H. L. SHANTZ, in G. F. White (Ed.), *The Future of Arid Lands*, Washington, 1956, p. 5.

[2] BOWMAN, E., *Desert Trails of Atacama*, New York: Amer. Geogr. Soc., Spec. Publ. No. 5, 1924, p. 75.

[3] KEDAR, Y. 'Water and soil from the desert: some ancient agricultural achievements in the Central Negev', *Geogr. Journ.*, Vol. 123, 1957, pp. 179–87.

[4] EVENARI, M., L. SHANAN, N. TADMOR and Y. AHARONI, 'Ancient agriculture in the Negev: archaeological studies and experimental farms show how agriculture was possible in Israel's famous desert', *Science*, Vol. 133, No. 3457, 1961, pp. 979–96.

[5] SPENCER, B. and F. J. GILLEN, *The Arunta*, London, 1927.

[6] THOMSON, D. F., 'The Bindibu expedition: exploration among the desert aborigines of Western Australia', *Geogr. Journ.*, Vol. 128, 1962, pp. 1–14, 143–57, 262–78; ref. pp. 11, 154–5.

[7] GOLDBY, F., C. S. HICKS, W. J. O'CONNOR and D. A. SINCLAIR, 'A comparison of the skin temperature and skin circulation of naked white and Australian aboriginals exposed to similar environmental changes', *Australian Journ. Exper. Biol. and Med. Sci.*, Vol. 16, Pt 1, 1938.
This is one in a long series of papers reporting on physiological investigations carried out by Sir C. Stanton Hicks and his associates on various

central Australian tribesmen. These were published in the *Australian Journal of Experimental Biology and Medical Sciences*, Vols. 8–16, 1931–8.

[8] WULSIN, F. R., 'Adaptations to climate among non-European peoples', in L. H. Newburgh (Ed.), *Physiology of Heat Regulation and the Science of Clothing*, Philadelphia, 1949, pp. 3–69 (see esp. pp. 32–5).

[9] DEBENHAM, F., 'The Kalahari today', *Geogr. Journ.*, Vol. 118, 1952, pp. 12–23.

[10] CHAPTAL, L., 'La lutte contre la sécheresse: La captation de la vapeur d'eau atmosphérique', *La Nature*, Vol. 60, 1932, pp. 449–54. See also the report by J. Gottmann in *Geographical Review*, Vol. 32, 1942, pp. 660–1, from which this quotation is taken.

[11] FISHER, W. B., 'Problems of modern Libya', *Geogr. Journ.*, Vol. 119, 1953, pp. 183–99, ref. p. 185, note 2.

[12] ROSENAU, M. J. and C. F. WILINSKY, 'A sanitary survey of Palestine', in *Reports of the experts submitted to the Joint Palestine Survey Commission*, Boston, 1928 (see p. 689).

[13] MASSEY, W. T., 'The Jerusalem water-supply', *Palestine Exploration Fund*, Quarterly Statement, 1918, pp. 172–5.

[14] For illustrations see D. H. K. AMIRAN and Y. KEDAR, 'Techniques of ancient agriculture in the Negev of Israel', *Comptes rendus, XVIII Congrès Int. de Géographie, Rio de Janeiro, 1956*, Vol. I, 1959, pp. 207–17, Figs. 6–8.

CHAPTER XI

Man in Arid Lands:
II – Patterns of Occupance

Apart from the isolated and unstable occupance related to mineral exploitation and trade routes, there are three main types of arid zone occupance: oasis settlement, sedentary settlement of semi-arid lands, and nomadic occupance which may be found in both semi-arid and arid areas.

Oases

Permanent settlements – villages or towns – are found in arid areas as oases only; they may be natural or created, and although the word 'oasis' is generally used for fertile areas of crop growing, it is applicable also to mining or industrial towns located in arid areas and supported by some special productivity or service.

Oases often depend upon springs, and in northern Africa the springs frequently mark the natural outflow of artesian supplies; sometimes wells have been put down to tap the artesian flows; other oases depend on surface or near-surface water, and in every case it is the supply of water that is the critical factor.

The oases vary in size and importance from a small cluster of palms to great palmeries occupying broad valley floors like that of Tafilalet in Morocco, and settlement occurs either in clusters or in a string-like pattern, depending on the nature of the source of water.

The very nature of the oasis located in the empty desert makes it vitally different from conventional settlement patterns of the humid zone. Whereas a city normally is the central settlement of a smaller or larger hinterland (or *umland*) and forms part of a pyramid of settlements as described above, desert settlement lacks pyramidal structure, a hinterland, and a balanced geographic and economic base.

Isolation is marked deep on oasis settlements. They depend on their own products and a few supplies brought from other areas, necessarily a small amount, as the long distances make transportation expensive. Recurrently throughout history, oases on important trade routes have been politically controlled by forces on either side of the desert, for whom it was vital to secure the use of the staging point of the oasis. Often too it was not a power at one of the terminals of a desert route which exercised political and economic control over it, but the nomads who controlled and operated the routes. The people of an oasis were more often subjected to outside rule than independent.

A careful oasis farmer should make well-planned and economic use of water on fields which are often mere plots. True, this is not always the case and in arid lands no less than in other climatic regions, water often is used in a very wasteful manner.

The need to make the most of the limited water and soil available often brings about minifundia as the prevalent type of land holdings. In contrast to humid lands, where one often sees a gradual decrease in intensity of land-use from the immediate site of a village to the fringes of the village lands, most oases display an abrupt transition from the intense, sometimes lush, vegetation of the oasis to the barren desert. More often than not this border is as sharp a line as there occurs in nature, only a very few metres wide.

A few observations from the oases of Pica and Toconao in the Atacama Desert of Chile may illustrate some salient points. Both oases are located at the foot of the Andes, at their border with the *pampas*.[1] Pica, at an altitude of 1,350 metres, is located on a sandy fan, Toconao, at 2,500 metres, between a low cliff of soft volcanic tuff covering the foot of the broad slope leading to the Andean range and the Salar de Atacama, a salt basin 48 by 86 km across, at an altitude slightly above 2,200 metres.

Pica, with a population of about 2,000, cultivates an area of about 80 hectares, operated by 477 farmers. Small wonder, therefore, that some holdings amount to no more than 80 square metres! Two hectares is considered an extremely large holding. At Toconao as well most holdings are of 100 square metres and less. Pica grows excellent juicy oranges which enjoy national fame, lemons, and some mango, guava and avocado. An average annual yield (in million pieces) from the 80 hectares might be: 24 oranges, 30 lemons, 5 mangoes.

The water available to Pica[2] from springs and subterranean galleries is collected in open tanks – called *cocha* – measuring e.g. 30 × 30 metres and $1-1\frac{1}{2}$ metres deep; such a cocha is filled by a socavón (qanat) in 12 hours. The largest tank holds 1,000 cubic metres.

The orchards are intensively irrigated, the use of water being allotted by time. The largest farm, of two hectares, has the use of the 1,000 cubic metre tank once every 14 days for 24 hours. To utilise the water to its maximum both at Pica and at Toconao no growth but the trees is allowed in the orchards, all undergrowth

---

[1] 'Pampas' is used in Chile for the longitudinal flight of broad intermontane basins between the coastal Cordillera and the Cordillera de los Andes. Much of the Chilean pampas is an extreme desert, in striking contrast to the grassland pampas of Argentina.

[2] According to Bowman [1] 155 l./sec, of which 119 come from springs and 36 from qanats. The figure quoted to the present author in 1959 was only half this amount.

including grass being carefully removed. The orchards are divided into small plots, which are flooded for irrigation. The tree thus stands on clean sandy soil to which is applied manure brought from a distant valley near the coast.

Toconao derives its water from a spring and two little brooks draining the high plateau above it. The main small gorge is most intensively cultivated, again with irrigated orchards or rather fruit gardens, under as 'clean' cultivation as at Pica. The main tree here is the fig, with some citrus and some poplar. Minifundia are the rule, as stated above. Not infrequently a large fig tree belongs to several owners.

Certain brands of fruit, often of high quality or special taste or aroma, are one common crop of oases. Valley oases in addition grow certain staple foods, and often fodder crops as well. It is often pathetic to observe how farmers in remote spring or valley oases still continue to grow alfalfa or other fodder crops catering for the mules, horses or camels of caravans or armies which years ago went over to motorised transport! But such is the conservatism of many an oasis population that it strives to make a living out of an antiquated way of life, changing but slowly with the times. The influence of isolation is manifest.

This illustrates another characteristic trait of oasis economy. Their physical potential often is greater than their economic potential, especially in most isolated spots.

Although in Algeria and elsewhere in northern Africa great new date-palm oases have been established around high-yielding artesian bores with the object mainly of date production, many of the old historic oases have greatly diminished in importance owing to changing economic and social circumstances, as a result of changing markets and new forms of transport by desert vehicles and aircraft.

Sedentary occupance of semi-arid lands

From the human point of view, desert proper is the area lacking rainfall, except at a few spots or except on very exceptional occasions which may cause more damage than benefit. The semi-arid area by contrast is not at all lacking in water resources, but these are generally of limited availability both geographically and seasonally. Here certainly is none of the physical marginality of life characteristic of the desert proper, but man inhabiting semi-arid lands has to allow for the limitations, seasonal or general, of water supply, and it is here, therefore, that most of the water installations mentioned before have been installed since ancient times – river irrigation, pipelines, aqueducts, qanats and cisterns. The semi-arid area has the cumulative benefit of an advantageous climate with high heat rates combined with a long duration of sunshine, and the possibility of ensuring an adequate and reliable water supply, either natural or by water engineering. In semi-arid lands it depends, therefore, on man's wisdom and planning whether or not he suffers from drought. Obviously rainfall fluctuates considerably over shorter and longer periods and with it the recharge of groundwater resources. Water consumption for essential purposes should, therefore, be geared to average minimum recharge values to prevent its permanent deterioration during low recharge years. Water consumed in excess of this amount should be used for inessential uses only, providing a bonus in good supply years, without its in-availability causing calamities in meagre ones.

As a transition and border area semi-arid lands have experienced very considerable fluctuations in occupance. Some of these were the result of changes in political stability. Others were created by an over-optimistic expansion of cultivation during wet stages of the very minor, recent fluctuations of climate. A return to medium conditions, and worse, dry stages, of the fluctuations often forces the abandonment of some rash 'pioneer settlement'. North America has had some spectacular experiences of this kind, including the notorious 'dust bowl' [2].

In the semi-arid region water is regularly available, though often in limited amounts and with sometimes excessive periodical fluctuations. Still, as potential evapotranspiration rates are considerably in excess of precipitation and very often of the total amount of available water, semi-arid lands have generally more land available than water to cultivate it. This is another strong contrast to truly arid areas where water and soil are equally at a premium.

The seasonality of rainfall and climate in general makes its mark on life in semi-arid lands. At the end of the dry season everyone keenly anticipates the first rain – to bring fresh, moist air instead of the dust of the preceding weeks, to refill the cisterns where they are used, and to assure the farmer of a good new crop year. Rainfall, adequate and of proper distribution, is the keynote to life in semi-arid lands under dry-farming practices. Any great delay in the beginning of the rainfall season is therefore a matter of grave concern, watched with growing anxiety even in the most technically advanced societies of today. For the farmer, including the modern market-oriented irrigation farmer, a proper distribution of rainfall is no less vital than an adequate amount. Bad distribution makes prolonged irrigation necessary, thereby raising production costs and often making a crop economically submarginal, and rain at the wrong time may seriously interfere with the drying of fruits, even with the quality of the crop, and with the control of pests.

The semi-arid environment

To counterbalance its environmental disadvantages, especially the limitations of available water, the semi-arid zone has certain advantageous features for the farmer as well as for recreational and various other uses. Especially where its landscape is manifold, as in areas with variegated relief and in maritime lands, its pleasant climate might be a very positive factor. This contrasts favourably with the oppressive moist heat of many wet tropical areas, the drizzle and dreariness of many mid-latitude lands, the long, hard winter months of snow climates, and the intense glare and scorching heat of the desert with its invigorating but often bitterly cold nights.

This attractive environment has, therefore, been the scene of notable civilisations and developments in the past, and one such area at least can claim at present the second highest rate of population increase among the major regions of the world. This is south-west Asia, with an annual population increase of 2·6 per cent, a figure considerably in excess of the world average of 1·8, and exceeded only by Central America, whose population increases by 2·7 per cent per annum.

These advantages of the semi-arid area not only permit such spectacular rates of population growth in certain regions of it, but attract people from the adjoining humid or arid areas. Together, these can exert considerable population

TABLE XI - I *Population density in selected semi-arid areas (density per square kilometre)*

Country	Province/district	Population density ca 1960	Percentage increase of population density 1920–1960
Chile	Aconcagua	13·7	24·6
Argentina	Tucuman	35·4	172
USA	California	38·6	454
Spain	Sevilla	88·1	97·4
	Alicante	121·4	38·3
	Albacete	25·0	28·9
	Ciudad Real	29·6	35·2
	Badajoz	38·5	29·6
	Caceres	27·3	28·2
	Zaragoza	38·3	39·3
Italy	Apulia	176	50·5
	Calabria	136	39·3
	Sicily	183	24·5
Greece	Crete	55	33·5
	Iraklion	81	76·8
	Lassithi	39	21·9

pressures on semi-arid countries, and have often done so both in the past and the present. Whereas the desert represents a stable environment, conditions in the semi-arid area are much more delicate. There is a multitude of reasons for this. The seasonality of climate, especially rainfall; the delicate balance of some semi-arid vegetations and soils and the comparative shallowness of some of the latter; the limited water supply; the strong regional contrasts within the semi-arid area and with the adjoining areas; the periodical interference with natural equilibrium by large-scale agricultural and other land-use (this factor, active in many areas, is of particular impact in areas of transitional climate and most prominent in semi-arid lands); population pressure; and, with some populations, low levels of education which do not permit them to appreciate fully the influence of technological changes.

From the point of view of conservation of a balanced environment, the semi-arid zone is, therefore, a problem area. Examples of this could be drawn from many countries. They would include: destructive uses of vegetative cover including destruction of forest, soil erosion, depletion and sometimes exhaustion of water resources. Every one of these phenomena is a sign of human maladjustment to a semi-arid environment. Very often this has not been effected by stable populations of long residence, well adapted to the region, but either by people who moved into it comparatively recently or, especially, by nomadic herdsmen who roam through semi-arid lands, damaging the land and its vegetation – natural and cultural – and sometimes looting the inhabitants. This general picture is too common in the semi-arid lands from Morocco to Mongolia to need substantiation here. But it may usefully be illustrated by the case of the semi-arid transition zone between the desert north and the Mediterranean centre of Chile: the Norte Chico.

The Norte Chico of Chile

The Norte Chico or the *small* north – so called in contrast to the Norte Grande or the *great* north (the Atacama Desert) – is a typical semi-arid transition area. In distinct contrast to the Norte Grande which 'boasts' such record stations as Arica and Iquique with average rainfalls of 0·6[3] and 1·9 mm respectively, including, of course, many years with no rainfall at all, the Norte Chico (located approximately between 30° and 32° of latitude) receives winter rainfall of 100–260 mm, the amounts increasing southward. Unlike the many areas devoid of soil and others with saline soils in the Norte Grande, the Norte Chico has basically a coherent cover of good soil. Remnants of flora prove that most of this land was originally covered by forest.

The nearby Andes Mountains provide the valleys even at this northerly latitude with a free flow of water of good quality. Being descendants of Mediterranean immigrants, the valley farmers of, for example, the Elqui or Guatulame valleys utilise this in most intensive high-class farming. The water in these deep valleys is diverted far upstream in open canals, sometimes two or even three at different levels one above the other. These lead the water for many kilometres downstream, in isolated instances as far as 60 km! Sometimes a side valley is crossed by means of a pipe-syphon. The water so provided is used on terraces built on the steep slopes in the best tradition of terrace cultivation, that method so characteristic of mountain agriculture in semi-arid lands. These terraces,

[3] Bowman [1, p. 40] states in connection with the absolute uselessnes of averages in the desert: 'There is no such thing as a normal desert rainfall.' This 'average' for Arica is strongly supported by 10 mm of rain, all of which fell in January 1918. See E. Almeyda Arroyo: *Pluviometria de las zones del desierto y las estepas calidas de Chile*, Santiago, n.d. (1950), p. 85.

Another notable rainfall, of 5 mm, was reported as having occurred at Arica on 24–25 August 1959, causing considerable damage to houses which are entirely unadapted to such downpours. A few of the adobe houses even collapsed.

meticulously built and maintained, may extend vertically over 200 metres of the slope.

The crops grown on them are equally typical of high-quality semi-arid agriculture. The Elqui Valley grows grapes almost exclusively. Some of them are made into pisco, a strong drink known and consumed all over Chile; the rest are dried for sultanas. The deep valley, its slopes inclined at 20–35°, heats up a good deal in summer, creating excellent conditions for high-quality grapes of high alcohol potential. The first-ranking irrigated crop of the Guatulame Valley is tomatoes, which here at 31° latitude ripen very early in the season. First they are shipped to Santiago, the capital, where they fetch excellent prices and from there are marketed as far south as Puerto Montt. When with the advance of the season Santiago gets cheaper tomatoes from central Chile, the Guatulame farmers market their tomatoes to the cities of the Norte Grande – a lucrative regional arrangement.

Speciality crops of outstanding quality or out-of-season crops are the logical land-use in the special natural conditions of semi-arid lands.

As mentioned before, the Norte Chico once was forested. Most of its forest cover was cut down many years ago, apparently to be sold as mine props to the mining area of the north. In the deforested areas second- and third-generation successor bush established itself. As these lands quickly became of marginal value man introduced goats, one of the most doubtful blessings inflicted by the conquistadores upon South America. The goats rapidly grazed down the land they were unleashed upon to a state of barrenness. As a result the soil, of good chemical composition, suffered considerable deterioration of texture, and severe gully erosion developed. This forced the husbandman into additional marginal areas, accelerating the spread of soil erosion.

The overall picture of the Norte Chico is of valleys blessed with running streams, many of them sustaining an irrigation agriculture of the highest standard, and marginal, semi-pastoral uplands blighted by soil erosion, often the worse the higher up one goes. It is a picture repeated in many other semi-arid lands.

Villages and houses in semi-arid lands

Obviously, it is difficult to generalise about types of settlements and houses in semi-arid lands, but a number of general traits can be isolated.

With few exceptions settlement is in villages, dispersal of rural settlement being rare. Among the reasons for this are the localised nature of water supply and the need for centralised administration in all matters of irrigation. Other reasons operative in certain areas are the limitation of good agricultural land and consequently the desire to concentrate the village area in order to spare the maximum amount of good land for cultivation; or insecurity not infrequently encountered in countries at the border of permanent settlement. Villages of this latter type are often densely nucleated, were sometimes walled, and may occupy a conveniently defensible site on a hill-top or spur.

The grouped buildings of an Australian station homestead or an American ranch are characteristic of cattle- and sheep-raising country. They house communities and are in some ways comparable with villages, although family groups are few, and many of the hands are itinerant or temporary employees. Such small specialised and predominantly male communities have their own special problems.

The rural house itself is generally much less elaborate than that in cool humid lands, as under conditions of semi-arid climate much labour and some storage can be effected outdoors. Whereas in many desert oases houses are built of adobe and may sometimes suffer severe damage when one of the intensive desert rainfalls occurs, adobe if used in semi-arid areas is given a strong protective plaster. Wherever available, stone is preferred for its stability as well as for its insulating powers against summer heat and winter cold, but there is great variation between different countries and regions. Wood, or galvanised iron on wooden framing, is common in Australia, where the run-off from iron roofs is generally collected in tanks to provide drinking water.

The same characteristics apply to the urban house as well, especially before its design was standardised according to general architectural patterns. In urban houses some additional features are common. Often the house is built around a central *patio*, a little garden or courtyard, sometimes even with a small pool and a few vines or fruit trees. The outer walls of the house may look bleak and uninviting, but inside all the rooms open on to this courtyard, the focus of relaxation and family life.

The design of the roof and sometimes of the whole upper storey often facilitates the reduction of heat and sometimes ensures maximum ventilation (Fig XI-1). A common design in many Near Eastern countries is slight cupola form, to draw the hot air above the living level under the cupola, which provides in addition a larger radiation surface than a flat roof would do. The very high rooms (4 metres and even more) found in some semi-arid countries are another attempt to make houses comfortable in summer, although they create considerable discomfort in winter, being difficult to heat. In many countries the upper storey consists of a frame only without walls – a kind of elevated roof (Fig XI-2). Sometimes there are only beams holding the roof, sometimes lattice work forms the wall, permitting good ventilation whilst preserving privacy, and various other designs are met with. To this breezy upper storey man retires for his afternoon rest and often sleeps here on summer nights. An arrangement of this kind is most desirable indeed, as it takes some hours for the interior of the house to cool off and the midday heat is preserved inside the house for hours. During the late afternoon and evening life inside the house is thus less comfortable than life outside. The inhabitants of the majority of houses which have no such open upper storey sleep on the flat roof on hot nights.

Fig XI-1. Houses at Naharayim in the semi-arid section of the Jordan Valley. Note the protruding eaves and the elevated central part of the roof ensuring proper ventilation. (*Photo S. J. Schweig*)

Fig XI-2. Typical urban house at Iquique, a coastal town in the northern desert of Chile. Note the upper storey built without walls. (*Photo D. H. K. Amiran*)

The brejos in the Nordeste of Brazil

The semi-arid regions surveyed so far are all located in the lower mid-latitudes outside the tropics. There are, however, semi-arid regions in the tropics as well. Here the result of semi-aridity is apt to have much severer effects, as it occurs at a consistently high temperature level throughout the year with attendant high potential evapotranspiration. One finds in these areas, therefore, severe semi-arid to arid conditions at amounts of rainfall which if available to one of the semi-arid areas dealt with before would place it comfortably outside the semi-arid region.

As semi-aridity is intensified in the tropics, its alleviations stand out in striking contrast. Here, therefore, we find proper oasis complexes in semi-arid environments. This is exemplified in the dry north-east of Brazil, an area where aridity is found within five or six degrees from the equator (Fig XI-3).

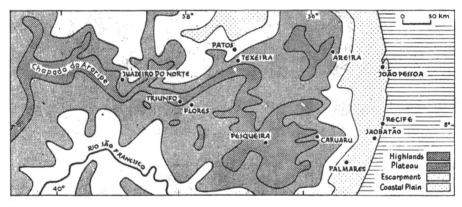

Fig XI-3. A section of the Polígono das Sêcas and their brejos in north-eastern Brazil. (*N. Z. Baer*)

After traversing the moist, narrow coastal plain from the Atlantic and ascending the even moister escarpments, one comes upon the extensive rolling plateaux of upland Brazil. Proceeding inland, the humid character of the land gradually changes to one of pronounced semi-aridity. This is the northern sertão merging into the caátinga, the dry, leafless, thorny forest of the arid tropical uplands.

It is generally topography which introduces modifications into this semi-arid environment by orographic increase of moisture and rainfall. These moister areas which might either be found in a mountain area or linked to one of the cuesta escarpments of upland Brazil or in a valley deriving water from it, are proper oases in their semi-arid environment. They are known in Brazil as *brejos*, literally virgin land, indicating their continuing fertility and their occupance subsequent to that of the villages and major towns serving the routes of the semi-arid areas. These towns are supplied from the brejos above them.

The brejo is a real oasis in every respect. It has natural forest vegetation and is intensively cultivated, reaching high population densities. Crops grown here include coffee, sugar cane, cotton and a large variety of fruit such as guava, banana, mango, oranges, anona (*fruta da pinha*) and pomegranates, as well as staple manioc for flour, corn, etc. Like the oasis of the desert, the typical brejo has a large number of small farmers, each cultivating a very small unit of land [3, 4].

Nomadic occupance of arid lands

In both arid and semi-arid lands one finds nomadic populations. Their numbers, status and importance have changed through the centuries. They include very primitive as well as quite advanced populations, the latter usually with some quite wealthy members. They roam over smaller or wider areas in the region they inhabit, mobile and always travelling light. Economically the keynote is to live off the land with a minimum of investment given in return (see Figs XI-4, 5).

The wide-ranging hunting nomads are usually keen and skilful. Early nomadic populations may have extinguished some varieties of fauna and severely reduced others in numbers, but most probably they only accelerated a natural process of animal depopulation brought about by increasing aridity (see p. 136).

Thomson in his recent survey of the primitive naked Bindibu people of central Australia [5] stresses once again their amazing skill in supporting themselves in a very arid environment. But by the nature of the desert most of its nomad people go hungry at times, gorging themselves on amazing quantities of food when it is plentiful. The central Australians set out early in the morning 'on a food-gathering quest, before the hot sun and wind obliterate the tracks of game. Sometimes each family hunted as a separate unit, at other times the women and men in separate parties – the women harvesting vegetable foods, lizards or small mammals.' Their amazing skill in tracking is attested by Debenham for the Bushmen of South Africa thus [6]:

When I suggested that trees would give vantage-points for seeing the game they were after, they became quite merry, saying that the last thing they wanted was to see the game or be seen by it – 'The whole story of the game is written in the sand at our feet, so why this silly climbing of trees ?' Their ability at spooring animals and interpreting every mark on the sand equals if it does not excel the skill of the Central Australian Blackfellows, and is quite incredible to witness.

Thomson found preserved vegetable food, some of it desiccated and carefully stored in tree caches or on top of some of the brushwood shelters of the Bindibu.

The discovery of a reserve of prepared and desiccated vegetable food stored in this way was of much interest, particularly in a drought year and in face of the belief that is widely held that the Australian aborigines live from hand to mouth and make no attempt to conserve food. The nature of this material and the way it was stored showed that the desert people were not only well supplied with food, but that they were living a stable existence as food-gatherers in spite of the long drought.

Fig XI-4. Tent of Bedouin, Negev, Israel. (*Photo D. Nir*)

Fig XI-5. Spring pasture, north of Ouargla, Libya. (*Photo Petit*)

The limited supply base of these and desert people at higher levels of civilisation brings about habits of family limitation in many nomad populations of the desert. This, of course, is influenced and modified by general social habits and religious tradition. Many people of this category practice both abortion and infanticide to prevent an increase of population beyond the resource base of their environment [7]. The habit of killing one of a pair of twins at birth, or even both, may be of particular significance, although this habit is met with in many non-arid populations too.

Even for nomads at higher levels of civilisation, who are graziers as well as hunters, life is full of risk and stress. Apart from the need to move the tribal encampment at frequent intervals to provide pasture for the livestock and water for man and beast, climate is a stringent and dangerous factor. This is well illustrated by the Bedouin of the Near East. He has to accommodate himself to intense solar radiation, high evaporation rates, scorching heat by day and cold by night. In winter sub-zero (Centigrade) temperatures are by no means infrequent in many deserts, but even then midday temperatures rise to 30°C (86°F) and above. The high winds of the desert are another source of environmental stress and irritation, the latter especially near dune areas where they easily whip up sand. It is in this respect that the traditional Bedouin mounted on his camel is much better adapted to his environment than his modern cousin or modern man in a motor-car. On camelback, but not in a car, one is often above most of the sand blown about by desert winds.

The Bedouin is, therefore, much better adapted to his environment than naked primitive man, for he wears clothes that protect him against the injurious effects of intense solar radiation in the dry desert air with its clear skies. Loose-fitting garments, like the *abbaya* of the Bedouin, meet this requirement (see Chapter XVII). The more well-to-do Bedouin uses a light-coloured (generally off-white) *abbaya* of cotton for wear in the daytime. On cool winter days, on most evenings and at night he wears a brown or black *abbaya* woven of goats' wool and quite heavy in weight. This is worn over light cotton underwear. If he has to perform hard manual labour by day, such as drawing water from a well, he strips to his underpants. Only the very poor Bedouin has to make do with one *abbaya*. The cool night being a greater inconvenience to him than the hot day, he chooses the heavy dark *abbaya* as his sole over-garment. Anyhow, tradition restricts his work and activity during the warmest hours of the day to a minimum.

Rain is another mixed blessing for the Bedouin. He needs it to obtain good pasture and he therefore 'follows the rain' wherever it falls. Most of the time desert pastures are incredibly poor and one somehow admires the Bedouin camel or goat and sheep that contrives to live on such pasture. But one has to see the dense grass and flower carpet of the outer fringe of semi-arid areas after a rainfall. The brilliant, many-coloured, exquisite flowers are often of incredible density for natural growth. This is one of the most elating and beautiful sights the desert has to offer. But the same rains make life miserable for the Bedouin, soaking him and his belongings in their semi-arid outdoor life. Often rivulets run between his tents or even through them and he may well be immobilised when the desert surface turns quagmire. One of the rare cloudbursts or protracted heavy rains may even imply actual danger for him, for men and animals may be drowned in a desert flood. The Bedouin, therefore, will never camp in a wadi bed, aware of the mortal danger of a flash flood.

In distinct contrast to sedentary villages in semi-arid lands which often seek high ground for their site, the typical Bedouin encampment is tucked away in some fold of the ground or a small side valley on the slope of a hill. There are reasons for this: protection from wind, and concealment from the sight of roaming Bedouin of other tribes and other even less welcome trespassers in the area. In contrast to the nucleated village, the Bedouin encampment is much more widely spread, sometimes a hundred metres or more between one tent and the other. An encampment of 20-30 tents may well be spread over a square kilometre. This habit of dispersing the encampment often remains with the Bedouin if he takes to living in villages with houses. It is generally easy to distinguish a Bedouin village at first sight, by the very unvillage-like, wide spacing of the houses [8].

Whereas one of the major considerations in village location is water supply, the Bedouin sites his encampment first and foremost for pasture. He is used to bringing water from a distant well, a few hundred metres and very often some kilometres away. Formerly water was lugged in leather bottles and bags; today the ubiquitous jerrycan is the common container. To meet a group of boys with donkeys (or camels), each one loaded with two jerrycans, on their way to the well or back is a common sight. Strangely enough, this habit of bringing the water to the place of residence persists with those Bedouin who take up permanent residence in houses in Bedouin villages today. They even often build their village without considering water supply on the spot; water is brought on donkeys in traditional fashion over distances of a few kilometres.

Changes in desert transportation
and their influence on nomad economy

The whole way of life of the Bedouin and other nomadic populations of the desert is based on the camel and other livestock. This was in response to a general need for camels, mainly as beasts of burden for the large caravans traversing the deserts. The long intracontinental desert routes, such as the ancient silk road connecting Turkestan with northern China or the trans-Saharan routes, required great numbers of camels, including relays and replacements, for smooth functioning. At least in the case of the trans-Saharan caravans up to a thousand camels were required for the larger caravans bringing Negro slaves from the lands south of the desert to the southern shorelands of the Mediterranean and the Middle East, and salt across the Sahara from north to south (see Chapter x). The crossing required three to four months, including long rests at oases and pasture grounds. Other caravan routes were considerably shorter and less difficult, but some of them carried even heavier traffic including some of the pilgrim caravans. Desert supply routes which carried considerably less traffic but were of vital regional importance existed in both the Americas as well.

As long as non-motorised traffic in the desert persisted in such volume, those who handled it enjoyed a key position. They provided the means of transportation, the men handling the animals, the guides who had the vital knowledge of routes and watering-places; very often they supplied food and travel equipment as well and they knew where to obtain pasture for the animals. The *masters of the desert* thus held a position based on all the privileges of monopoly. They not only levied heavy charges for their services but knew how to extract a regular fee from the traveller for the favour of not being robbed. Gradually the Bedouin and similar nomads of Asia and Africa imposed their rule on the villagers (the *fellaheen*) who had the misfortune to be their neighbours. These either paid regular levies to them or lived under permanent threat of plunder. The sheikhs of the major Bedouin tribes were the kings of the desert and its fringe lands.

This situation, which prevailed in many desert areas until the beginning of the present century, is now history. The advent of motorised travel, by passenger car and truck, followed in more recent years by aircraft, has irrevocably changed the traditional pattern.

Instead of travelling laboriously and sometimes dangerously for weeks on end, goods and passengers are moved at far less expense in a few days, sometimes in a matter of hours. They are delivered in much better condition too than goods rocked for weeks on camelback. The increasing range of the stages of travel has made motorised travel much more independent of the localised services which were vital for caravan traffic. No less revolutionary for the desert was the introduction of air travel, both for goods and especially for passengers. Today the Sahara is crossed daily by several score scheduled flights in each direction, in a few hours' flying time. Ground supplies on the way are no longer needed.

All this has made the caravan animals of the Bedouin or similar nomads obsolete; the Bedouin himself, one may say, has become obsolescent in mid-twentieth-century society. Today the camel, once the proud *ship of the desert*, is mainly a source of meat supply. The Bedouin with his keen natural intelligence has not been slow to adapt to the changed conditions, and this to a considerably greater degree than is appreciated by people living far from deserts who find it hard to part with the time-honoured Hollywood and Lawrence ideal of the noble desert nomad living a life of hardship and adventure. Today the nomad of the desert and its borderlands is gradually becoming part of modern society, as a labourer in oilfields, a truck driver, an industrial plant worker or a building hand. In short, the Bedouin today is in many regions a wage labourer. Possibly his brother or sons still take the family's camels and other animals to the traditional pastures. For obvious reasons this adaptation to modern conditions is farther advanced with nomads in semi-arid areas who live close to industrial towns, than with those of the desert proper. These often form a welcome reservoir of unskilled labour available at the lower range of wage scales. To the nomad whose standard of life is still quite simple and cheap these rates are often real income of a level neither he nor his fathers dreamed of.

D. H. K. AMIRAN

References

[1] BOWMAN, I., *Desert Trails of Atacama*, New York: Amer. Geogr. Soc., Spec. Publ. No. 5, 1924, p. 21.

[2] — 'Our expanding and contracting "desert" ', *Geogr. Rev.*, Vol. 25, 1935, pp. 43–61.

[3] LACERDA DE MELO, M., *Nord-est*, Livret-guide No. 7, XVIII Congrès Int. de Géographie, Brésil, 1956, Rio de Janeiro, 1956, pp. 115–24, 151–6, 206–14.

[4] — and M. CORREIA DEL ANDRADE, 'Um brejo de Pernambuco: região de Camocim São Félix', *Boletim Carioca de Geografia*, Vol. XIII, No. 1–2, 1960, pp. 5–45.

[5] THOMSON, D. F., 'The Bindibu expedition: exploration among the desert aborigines of Western Australia', *Geogr. Journ.*, Vol. 128, 1962, pp. 1–14, 143–57, 262–78; ref. pp. 12, 155.

[6] DEBENHAM, F., 'The Kalahari today', *Geogr. Journ.*, Vol. 118, 1952, p. 18.

[7] MURDOCK, G. P., *Our Primitive Contemporaries*, New York, 1934. See p. 34 for Aranda of central Australia, p. 494 for Nama Hottentots of south-west Africa.

[8] AMIRAN, D. H. K. and Y. BEN-ARIEH, 'Sedentarization of Beduin in Israel', *Israel Exploration Journ.*, Vol. 13, 1963, pp. 161–81.

ADDITIONAL REFERENCES

UNESCO, *Nomades et Nomadisme au Sahara* (in French only), Unesco, Paris, 1963 (Arid Zone Research XIX).

— *Land Use in Semi-arid Mediterranean Climates*, Unesco/IGU Symp., Iraklion, Greece, Unesco, Paris, 1964 (Arid Zone Research XXVI).

CHAPTER XII

Irrigation in Arid Lands

In arid regions successful growth of crops depends on irrigation. In semi-arid regions irrigation may greatly improve or extend the area of crop production, and in generally humid regions that suffer limited and occasional periods of drought, supplementary irrigation may be highly profitable. There is a corresponding variation in equipment: large channels are needed where irrigation has to meet the entire water requirements of a crop; narrow pipes suffice to deliver the small quantities required to supplement moderately high rainfall. In arid and semi-arid regions the area of land usually far exceeds that which can be irrigated with the available supply of water, so it is important to choose for irrigation the most suitable land. Once that choice has been made there follows a large investment of money and labour, so that it becomes important to keep the chosen area productive. Negatively, the objectives are to avoid bad land and waste of water and to avoid collapse of the enterprise from waterlogging and salinity, weeds, pests and diseases. To achieve these objectives there are needed professional skills and, above all, a stable and coherent social structure so that good agricultural practices may be made known and may be consistently applied throughout the area.

There is a wide range in methods of irrigation. Women, donkeys and camels lift water from wells in the Wadi Hadhramat; the creaking water wheel actuated by oxen is a common sight from Egypt to India and so is the well sweep, a counterpoised lever enabling the fellah to lift water from the source, to be tipped into an irrigation channel. From Arabia to Pakistan qanats are widely used to supply water for irrigation (see Chapter x) and water is also diverted from seasonal streams. Further east one finds many ingenious devices utilising the bamboo. Modern aids include power-driven pumps installed on river banks, on barges, or deep within wells, even those of narrow bore. Earthen and masonry dams permit storages of water which may then be distributed by gravity flow by canals or through pipes. Spray or overhead irrigation is successful in many places even though costs of installing and maintaining equipment are high.

Public health

Introducing irrigation into an arid region greatly increases the density of population. Because of this, benefits to agriculture are unhappily offset by increased incidence of disease, mainly due to inadequate arrangements for disposal of human excreta. From this cause arise bacterial infections producing bacillary dysentery and cholera; protozoal infections producing amoebic dysentery; infections by parasitic worms including *Ascaria* (round worm), *Ankylostoma* (hookworm) and *Schistosoma*, the vector of bilharzia. A great educational and administrative effort is needed to convince people that these diseases can be controlled by improved sanitation and are not a necessary consequence of irrigation. There is also need for research on methods of sanitation other than water-borne sewage, which should not be regarded as a complete solution of the problem. Irrigation may also provide breeding places for mosquitoes that spread malaria and other diseases. Nearly complete control has been established in various places by co-ordinated use of appropriate measures, attacking larval and mature forms of the mosquitoes and eliminating sources of infection. Here again there is need for a closely knit and well-informed community so that measures of control may be uniformly effective.

Choice of land for irrigation

Arid soils within reach of a projected irrigation system have mostly been deposited by wind or water and their properties largely depend on a long complex history of transport and deposition. If, in such an area, one is to choose the soils best suited for irrigation, it is helpful to discover as much as possible about this long history. Examination of aerial photographs supported by study of soils in the field will reveal intelligible patterns in the distribution of the soils and it may be possible to decide which parts, if any, are suited for development (see Chapter VI).

There is little need for an elaborate survey when the enterprise is small and concerns a few farmers only. They may simply decide that some area looks promising and try their luck.

There is little need for elaborate survey where turbulent seasonable floods pour down channels that are normally dry.

The floods are caused by heavy rain falling on uplands poorly protected against erosion by their sparse cover of vegetation. Stones and boulders are rolled down the stream bed and the floodwaters carry a heavy load of sand, silt and clay: their natural tendency is to build up alluvial deposits until these reach a height that swings subsequent floodwaters towards a new course. Then the process is repeated: new sediments gradually raise the land level and eventually the floodwaters are again deflected. A sort of natural land rotation is thus set up: splendid crops may be obtained from newly watered land free from the usual weeds, pests and disease. In these circumstances the greater part of the alluvial fan is capable of producing good crops when the water gets there: a single heavy watering may suffice to carry a cotton crop from germination to maturity; less heavy watering may suffice for a satisfactory crop of grain and fodder.

This occurs for example on the alluvial fan of the River Gash, whose seasonal floods pour down from the bare hills of Eritrea.

Because the floodwaters under such conditions carry a heavy load of suspended solids, storage in a dam is not practicable. Irrigation works of considerable

Fig XII-1. After a single heavy watering these widely spaced plants of Egyptian type cotton, free from pests and disease, will bring to maturity more than 100 bolls per tree. The site is an alluvial fan about 35 miles east of Aden

ingenuity are designed to divert a part of the floodwater which can be usefully spread: it is usually not desirable to divert a major part of the flood, since this might simply cut a new course throughout the alluvial fan and perhaps escape uselessly to the sea. The local people have usually learned where to find modest assured supplies of drinking water for themselves and their animals. Here and there within favoured parts of the alluvial fan or from the underflow in a dry river bed there may be good prospects of pumping large quantities of groundwater for urban use, as at Aden, or for irrigation of small orchards and vegetable gardens. As a rule permanent supplies of groundwater are so rare and so valuable when found that they may not be available for irrigation on any wide scale.

On the other hand, where there is a large and assured supply of water, such as the Nile flood, domestic and urban needs are more readily met and there is a vast surplus that will flow into the sea unless used for irrigation. Here it is necessary to seek out the most suitable soils and to discover the best means of irrigation, including works for storage and distribution of water and time-tables for its effective use. In general, when choosing land for irrigation, it is desirable to find areas of convenient location and extent, having a gentle slope, unlikely to suffer from floods or from poor drainage or from groundwater at shallow depth, free from obstructions such as gullies, sand dunes or boulders, having soils that are reasonably uniform with depth and of medium texture (neither coarse sand nor heavy clay), sufficiently deep, sufficiently fine-grained to retain enough water to last a crop from one irrigation to the next, that are of low salt content, only moderately alkaline and free from toxic substances.

These criteria should be used to guard against disastrous mistakes by bringing to notice any severely adverse condition that cannot be put right with such technical resources as are now available. For example, if an area is small and far from water there may be no present possibility of bringing it under irrigation because the estimated cost of taking water to it is too great. Until some new technique shows promise of overcoming this difficulty the area must be treated as non-irrigable land. Irrigation by gravity-flow through watering channels and over the ground surface is difficult to maintain on slopes exceeding 2°, and very difficult on slopes exceeding 4°. Still steeper slopes must be excluded even though the difficulties could be surmounted by making terraces and channels such as have been constructed through the age-long industry and skill of Chinese farmers or, perhaps, by installation of spray equipment. The exclusion is not absolute but is a warning that severe obstacles will be encountered or that an expensive installation is required.

Costs of production must be considered in relation to possible markets for produce. A hundred years of experience in India has led to the view that central government should bear the whole cost of installation in expectation of adequate indirect returns. These benefits will not accrue unless the project is technically sound. Drainage may be poor on very flat or low-lying tracts and it may be difficult to prevent the water table from rising to within 6 feet (180 cm) of the soil surface, especially if the rise is promoted by inflow of seepage from higher land receiving irrigation or from leaky canals. In a particular area it may or may not be easy to decide whether difficulties of this kind are likely to become acute. On the other hand, sand dunes, gullies or boulders may be an obvious hindrance to even distribution of water and to mechanical cultivation.

If the land configuration is generally favourable, the soil surveyor should, by study of aerial photographs and surface indications, try to divide the area into parts each of which looks moderately uniform and may or may not be so. Taking each part in turn he will study its soil and, in particular, the soil's capacity to store water for plants. The assessment is difficult (a) because of local variations which cannot be thoroughly explored, (b) because the criteria are vague and incoherent. Thus, it is far from obvious how to measure the available depth of a soil or its permeability to water; nor, if depth and permeability were measured, is it obvious how the two measurements could be combined. Of course, there are ways of getting round these difficulties, for example by deciding whether soil is slightly or severely defective in respect to the various appropriate criteria, and then totting up the bad marks to give a combined assessment. However, if this assessment is to be even moderately reliable the surveyor needs technical training, experience, good judgement and integrity.[1]

The soil surveyor, examining in the field a vertical section of soil exposed in the side of a gully or a sample pit, must decide what depth of soil will be available for crop roots, that is, what depth of soil will have enough water and enough air to permit healthy growth of roots. Entry of water depends mainly on the presence of root holes, cracks, and other more or less vertical passage ways or pores, that are sufficiently stable not to collapse when the soil is wetted. Thus, attention is directed to the spaces between the solid particles of soil rather than to the particles themselves. Downward movement of water may be checked by some more or less impermeable layer or by an abrupt change in texture of the soil, for example, when sand overlies clay or when clay overlies dry sand. The soil surveyor is hoping to find a good depth of fairly uniform, permeable, well-drained soil. In a well-drained soil, water that drains away by gravity or that evaporates or is absorbed by plant roots, leaves spaces permitting entry of air and escape of carbon dioxide given off by roots and by other organisms in the soil. This movement of gases is essential for healthy growth of roots and will occur provided the soil has plenty of interconnected pores wide enough to be seen by the naked eye or under a hand lens. Soils differ in the depth that is well drained and crops also differ, some requiring a good depth of well-drained soil and others being less sensitive to defective aeration. Soils differ widely in the amounts of water they can retain, a clay soil retaining much more water than a sand. Of the water thus retained a part only is available for the use of plants, the remainder being more firmly held by the soil, from which it can be removed, however, by heating (usually to 105°C). If the soil is weighed before and after heating the change in weight corresponds to the amount of water originally present in the moist soil and now driven off in the form of vapour. The moisture content can then be calculated and might be, for example, 32 g of water per 100 g of oven-dry soil. Of this 32 g perhaps only 10 g are available for use of crops, the remaining 22 g being too firmly held. However, in relation to irrigation, quantities of water are more conveniently described in terms of depth. Thus, a clay soil may hold 4 inches depth of available water in 2 feet depth of soil whereas a sandy loam may need to be 5 feet deep to hold 4 inches depth of available water (clay may hold 10 cm depth of available water in 60 cm depth of soil; sandy loam may

[1] Good instructions are contained in some 500 pages of the *Soil Survey Manual* of the United States Department of Agriculture [1]. Another useful book is *Irrigated Soils* by Thorne and Peterson [2]. See also 'Relations between Water and Soil' by T. J. Marshall [3]. In the present chapter a broad outline must suffice.

hold 10 cm depth of available water in 150 cm depth of soil). These figures are mentioned because they are the quantities of available water that suffice, in normal practice, to carry a crop without injury during the two-week period between waterings. When irrigation is by gravity flow it is inconvenient to water more often than this, and it is difficult to apply evenly less than 4 inches depth. Storage capacity for 4 inches depth of available water is therefore a first consideration when examining soil with a view to irrigation by gravity flow.

Fig XII-2 illustrates changes in total moisture content and in available moisture content of a heavy clay soil, about 100 miles south of Khartoum in the Sudan, before (May), during (August) and after (October, November) summer rains amounting to about 16 inches. The shaded part of the diagram roughly indicates

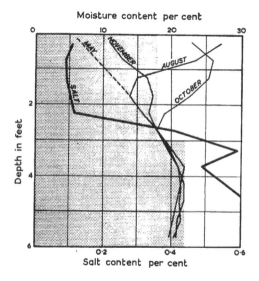

Fig XII-2. Changes in moisture content of heavy clay soil near Wad Medani in the Sudan. The shaded area represents moisture too firmly held by the clay to be available for growth of the grain crop, sorghum

water so firmly held by the soil that it is not available to the plant, in this case, sorghum. In May, before the rains, the top foot of soil had dried out far below the wilting point of any plant. Rains of July and August made up this deficit and at the time of sampling in August the top foot of soil held 1·1 inches depth of available water. With continued rainfall water penetrated into the second foot also, and at the time of sampling in October there was 1·0 inch depth of available water in the top foot and 1·0 in the second foot, but only 0·1 inch depth in the third foot. Of course during the rains the soil had held more water than this and the plants had used it: the curves for August and October merely show the situation when the soil surface was dry enough to make sampling convenient. By November the plants had used all the available water and the soil was drying out again under the influence of warm sunshine and dry winds. Depths of water represented in the diagrams were as follows:

Total water content of soil expressed in inches depth

	May	Aug.	Oct.	Nov.
Top foot	1·2	3·9	3·8	1·8
Second	2·0	2·1	3·8	2·8
Third	3·0	3·0	3·2	3·0
Fourth	3·5	3·5	3·5	3·5

Water available for growth of sorghum

	May	Aug.	Oct.	Nov.
Top foot	—	1·1	1·0	—
Second	—	—	1·0	—
Third	—	—	0·1	—
Fourth	—	—	—	—

In heavy clay soil such as this an irrigation of some inches depth increases the amount of available water in the top 2, 3 or 4 feet depth of soil, the changes in moisture content resembling those shown in the diagram but extending further to the right. The soil surveyor is likely to have a separate classification for extremely porous soils and for heavy clays. Porous sands are liable to require excessive amounts of water, most of which is lost by seepage, creating troubles elsewhere. Such soils may be excluded from irrigation unless water can be put on by spray. Heavy clays require special consideration because they present cultivation problems and are suited to a less wide range of crops.

Soluble salts compete with the crop for the water in the soil; the competition becomes severe when the salts constitute more than 0·2 per cent of the soil. The assessment is made easily and accurately enough by measuring the electrical conductivity of a solution obtained by treating a sample of soil with an appropriate amount of water. Salts are not a permanent feature of the soil: they may be carried to the surface if there is an upward movement of water through the soil; they may be washed away downwards where drainage is good and where sufficient irrigation water is applied. If the irrigation water itself tastes salty, it may give trouble because the water evaporates, leaving the salts behind to restrict or prevent plant growth. Common features of salty areas are: their native vegetation restricted to salt-tolerant plants; patchy occurrence of salts; patchy growth of crops under irrigation. Soils that have been in contact with soluble salts (for example, soils that have been in contact with sea water) are likely to contain a rather high proportion of sodium loosely combined with clay and a correspondingly low proportion of loosely combined calcium. The sodium and calcium thus loosely combined are called exchangeable because they are readily replaced when the soil is brought into contact with some other salty solution, for example, a solution of ammonium acetate as commonly used in analysis. When the proportion of exchangeable sodium is high, the soil is likely to be unproductive, strongly alkaline, sticky and impermeable when wet, very hard when dry. These adverse features cannot always be corrected at economic cost and it may be best to leave such soils out of cultivation, or possibly, to use them for fish ponds.

(Further comments by the present writer on selection of land for irrigation are contained in the FAO pamphlet 'Using Salty Land' [4] and in Chapter 7 of the *Unesco Guide Book to Research Data for Arid Zone Development* [5].)

LAND CLASSIFICATION FOR IRRIGATION

Although the soil surveyor has to make many technical judgements on matters that are obscure and uncertain, it is obligatory for him to reach and record on a map a classification of soils that is definite and easily understood. For example here are definitions of irrigability classes as given by R. G. Thomas and J. G. Thompson [6]:

Class A.1   *Eminently suitable* for irrigation and capable of sustained high productivity. Soils are deep, medium textured and free draining; available water capacity is good. Topography is even and slopes are gentle. Provision for disposal of surplus surface water will be relatively simple and inexpensive.

Class A.2   *Suitable* for irrigation and capable of sustained productivity. Soils are moderately deep and free draining; available water capacity is adequate. Topography is moderately favourable, but more work may be necessary in land preparations and provision for disposal of surplus surface water than in Class A.1.

Class B   *Approaching marginal suitability* for irrigation owing to reduced permeability, shallowness of soil, moderate but correctable salinity, or unfavourable topography. Adequate productivity is attainable with good management and maximum efficiency in the use of irrigation water, but risks are greater and special care is necessary. Corrective measures such as leaching and gypsum applications, field drainage for the disposal of surplus surface water and sub-surface drainage to control the water table, may be necessary.

Class C   *Of very restricted suitability* for irrigation owing to severe limitations in permeability, depth of soil, or topography, or owing to high salinity. Except in special circumstances, irrigation of this class is restricted to relatively permanent crops such as pastures or orchards.

Class D   *Unsuitable* for normal irrigation owing to excessive limitations in soils or topography, but suitable in some cases for irrigation of paddy rice or selected pastures.

Class S   *Excessively pervious sands* with inadequate available water capacity. Irrigation of this class is, under normal conditions, unlikely to be economic because of high water-losses, the necessity for frequent irrigations, the almost certain rapid rise in the water table, and low inherent fertility.

Thomas and Thompson also describe the technical criteria they employ to arrive at this classification. However, instead of following their account further we may consider a closely analogous system of land classification illustrated in Fig XII-3. This is a simplified version of part of a detailed map made according to the specifications of the United States Department of the Interior, *Bureau of*

Reclamation Manual [7]. The map includes a canal running along a slight ridge and a drain running along a shallow valley. The symbols on the map indicate

ABOVE THE LINE: Land class (1, 2 and 3 arable; 4 limited arable; 6 non-arable) and deficiencies (s of soil; t of topography; d of drainage). *Class 1 arable* is confined to lands that are highly suitable for irrigation farming, being capable of producing sustained and relatively high yields of a wide range of climatically adapted crops at reasonable cost. *Class 2 arable* comprises lands of moderate

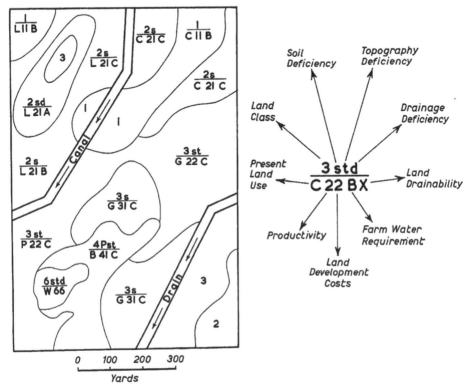

Fig XII-3. Simplified version of part of a detailed soil map prepared according to specifications of the United States Department of the Interior, *Bureau of Reclamation Manual,* Vol. V

suitability for irrigation farming, being measurably lower than Class 1 in productive capacity, adapted to a somewhat narrower range of crops, more expensive to prepare for irrigation or more costly to farm. They may have a lower available moisture capacity as indicated by coarse texture or limited soil depth, they may be only slowly permeable or moderately saline. They may have an uneven surface or steeper slopes; farm drainage may be required at moderate cost. *Class 3 arable:* Lands approaching marginality for irrigation because of more extreme deficiencies in soil, topographic or drainage characteristics. *Class 4 limited arable:* Lands with one or more excessive deficiencies but having limited use. The nature of the limitation is indicated (4P pasture; 4R rice; 4S sprinkler irrigation, etc.). *Class 6 non-arable:* Lands not suited for irrigation, being steep, rough, etc., or having soils of very coarse or very fine texture, shallow soils, soils with inadequate drainage, etc., etc. The symbols on the map indicate

BELOW THE LINE: present land use (C irrigated crops; P irrigated pasture; L non-irrigated crops; G non-irrigated grassland; B brush or timber; H suburban; W waste), productivity (numbers corresponding to land classes), development costs (numbers corresponding to land classes), farm water requirement (A low; B medium; C high) and land drainability (X good; Y restricted; Z poor).

In the illustration an area of Class 2 land in the top left-hand corner has been downgraded because of a soil deficiency. It includes a slight depression of oval shape of which the bottom is Class 3 but the margin is Class 2, having moderate deficiency of soil and drainage. Here the farm water requirement is low, A, instead of being medium, B, or high, C. Between the canal and the drain there is some Class 1 land with no deficiencies and a considerable area of Class 3 land. Part of the Class 3 land has deficiencies of soil and topography which correspond to lower productivity (grade 2) and higher development costs (grade 2). In another part severe deficiencies of soil reduce productivity to grade 3 and the land is Class 3 although topography is satisfactory so that costs of development would be low (grade 1). In all this Class 3 land the water requirement would be high, C. The 4P land class is indicated for an area in which shallow soil and uneven surface constitute such severe deficiencies of soil and topography that the land, if irrigated, should be used only for permanent pasture. In an adjacent small area of Class 6 land deficiencies are so acute that the land must be left waste. Its low productivity and high cost of development correspond to land Class 6.

There are other systems of classification and other symbols that differ from those represented in this illustration. No matter which system is used the soil surveyor is under an obligation to reach and record definite opinions upon which further action can be based. Here, for example, is B. Anderson's [8] general description of soils of the Usangu plain in Tanganyika which is dry for a large part of the year:

In drawing up a classification of soils for this survey it was necessary both that the soil groups distinguished should be recognisable on the aerial photos which were the basis of the survey and also that the groups should as far as possible be a guide to the irrigation engineer and agriculturalist.

In the classification adopted the primary division is into alluvial and non-alluvial soils. The alluvial soils are then subdivided on their texture. This produced reasonably-sized units for the sandier soils but left a large block of alluvial clay which it was considered desirable to subdivide further. Here a conflict arose between the needs of the agriculturalist and those of the surveyor since properties of the clay which will probably prove of most importance to the agriculturalist could not always be recognised on the aerial photos. For reasons explained below, when the clays are described in detail, the clays were divided into five groups. Three of these indicate the intensity of flooding, under present conditions, and a fourth is for alkaline clays, as far as these can be recognised in the vegetation. A fifth group includes some small areas of clay with a very hard surface.

In places the pattern of the soils was too complex for the different textural groups to be shown on the map scale of 1 : 125,000 (often it was too complex to be distinguished on the photos at a scale of 1 : 30,000) and it was necessary to map such areas as complexes. Three complexes were used, one for predominantly sandy soils, one for predominantly heavy soils, and a third where both sandy and heavy soils occurred together. This last group will be difficult for the irrigation engineer to evaluate; it includes substantial proportions of both sandy soils and of clays but it is impossible to indicate what is the proportion of each, valuable though this information would be.

The non-alluvial soils have been shown in much less detail since most of them are of no value for irrigation.

This quotation from a report on a reconnaissance survey illustrates the importance of basing classification on what can be seen in aerial photos and on obvious features of the landscape. More detailed studies can be made later when and where required.

The soil surveyor is usually under pressure to produce a more favourable report than the facts warrant. He will more sturdily withstand such pressure if he bases his system of classification on those features of the area that are obviously of decisive importance, for example: angle of slope in a hilly area; particle size in an expanse of alluvial deposits; depth to water table in a low-lying area. Then those important people who seek to challenge his findings will have to admit that at least in some respects the soil map is correct. The surveyor will of course need to take into consideration other features and factors that are less readily explained. There he will do well to admit that he does not claim 100 per cent accuracy but offers just such an opinion as his observations and experience permit. The cost of soil survey, whether reconnaissance or detailed, is small in relation to expenditures involved in setting up an irrigation scheme. The soil map displaying some indisputable features of the territory is likely to prevent waste of time and effort even though there remain many matters that are uncertain.

SUBSOIL WATER

Almost always far too little is known about subsoil water, also called groundwater. The prime purpose of irrigation is to spread water evenly over the land. This is rarely achieved without serious losses in transmission along canals and without excessive applications to some fields. Seepage from canals and fields adds to whatever subsoil water had accumulated by natural processes, including infiltration of rainfall and infiltration from rivers or the sea. An example of seasonal changes in level of subsoil water is provided by observations made in alluvial lands near Wadi Halfa on the River Nile. Houses had been damaged by rise of subsoil water and it was sought to ascertain whether this was due to heightening of the Aswan dam or to seepage from a canal. Observations represented in Fig XII-4 showed that seasonal changes in level of subsoil water are clearly related to changes of water level in the river due to the summer flood or to changes in storage level in the Aswan dam. The turbulent waters of the Nile in flood do not always follow precisely the same course but, where there is a bend, may bank up more steeply in one year than another. The changes in river levels are faithfully followed, though on a reduced scale, by changes in the level of subsoil water.

The effects become less clear at greater distances from the river bank and, in this location, they were much diminished at a distance of 1,000 metres. At a greater distance from the river and nearer a canal, seasonal changes were observed that could plausibly be ascribed to leakage from the canal, because there was a

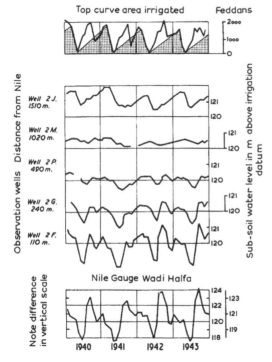

Fig XII-4. Seasonal changes in level of subsoil water at Dabeira, near Wadi Halfa, compared with changes in level of the Nile. The effects are inconspicuous at distances exceeding 1,000 metres from the river bank but the effect of seepage from a canal becomes evident

steady increase in level of water while the canal was in use. This increase did not correspond to changes in the area of land irrigated so that, in this location, seepage from overwatered fields appeared insignificant. The effect of seepage from the canal would have been less readily observed had the canal been in use throughout the year; the level of subsoil water near the canal might then have remained constant, although it might have shown a gradient, higher near the source of seepage and lower where the subsoil water escapes as it slowly moves towards the sea. Seepage from a canal may be directly estimated by blocking successive small stretches and noting how quickly the water soaks away. There are many ways of cutting down seepage from canals and they all cost money: more resolute action to lessen such losses may be expected as people become aware of the value of water. There are some practical difficulties in measuring levels of subsoil water: small boys are much inclined to drop stones down the observation pipes, and their fathers are inclined to remove the pipes altogether, either for their private use or because they fear there is some plan to steal their land or inflict some other injustice upon them.

Changes in level of subsoil water are not always slow. In an alluvial fan forming part of the coastal plain some 35 miles east of Aden water levels in some wells quickly rose several metres when partial control of violent floodwaters permitted them to be more widely spread. The alluvial fan, fed by floods during

Fig XII-5. Salt accumulation in surface soil caused by a high level of subsoil water has prevented growth of cotton in this part of an alluvial fan about 35 miles east of Aden

the monsoon rains, consists of an intricate pattern of deposits varying in particle size from clay, silt and sand to stones and boulders. Evidently the engineering works admitted floodwater to some permeable deposits overlying less permeable deposits, the new addition greatly exceeding the customary outflow after earlier rare floodings and slow infiltration. Because of the high level of subsoil water a part of the alluvial fan has become waterlogged and saline and is incapable of producing a crop. However, on the whole, the new measures of water control have greatly increased agricultural production from this alluvial fan.

Pumping of subsoil water may be a most effective means of lowering levels of subsoil water and remedying the damage consequent on waterlogging and salinity. Pumping is particularly attractive when there is prospect of using the pumped water for irrigation. Tube wells are widely and successfully used for this purpose in some parts of India and the United States. However, where alluvial deposits adjoin the sea, the subsoil water usually consists of an upper layer of fresh water resting on a lower layer of sea water. A serious difficulty arises when much fresh water is pumped out of the upper layer, for it may be replaced by an incursion of sea water. Fortunately, there are situations in which this process can be reversed: floodwaters suitably spread over a permeable alluvial formation may gradually build up an overlying layer of fresh water driving some of the sea water back into the sea. Quite elaborate investigations may be needed to discover what is happening. Dr W. L. Balls, having a life-long interest in agriculture of the Nile delta, came to the conclusion [9] that a damaging rise in level of subsoil water had occurred there and needed to be corrected. His observations, however, were not entirely convincing. In general it is necessary to be aware that subsoil water may rise; to keep a watch for signs of this trouble and, if it occurs, to consider how it may be checked either by more efficient distribution of irrigation water, by changing to spray irrigation, by digging a diversion ditch, or by getting rid of excess water by open ditches, tile drains or by pumping.

In the western United States [10] pumping substantial quantities of water from deep wells has been an effective method of lowering the water table, thereby preventing or curing damage to land by salinity. In some humid areas it is possible to apply fresh water at high tide and to permit outflow of drainage water at low tide: there may be similar opportunities in some arid areas, though not in those around the Mediterranean.

AGRICULTURAL PRACTICE

Water for irrigation is likely to be plentiful during some months and scarce at other times. There are seasonal changes also in the need for water. In contrast to equatorial regions, arid regions usually experience a hot summer and cold winter, this being mainly determined by seasonal changes in incidence of radiation from the sun. This also largely determines the rate of evaporation from an open surface of water and the water requirement of actively growing plants. Plants are affected by other climatic factors: clearly some crop plants grow best in summer and others in winter, this being determined by their diverse physiological characteristics that can be changed only to a limited degree by purposive selection and breeding. Crop growth is limited also by changes in the prevalence of pests: irrigation that makes possible introduction of a crop automatically offers a new opportunity to its familiar pests. Thus in Egypt increasing severity of boll-worm attack obliged farmers to sow more closely and earlier: they then found it profitable to apply nitrogenous fertilisers that were not needed by the former widely spaced and late sown crop. Pests or disease may rule out a crop, but efficient means of control are continually coming on the market so that the farmers have a widening choice of crops. Of great practical importance is the possibility of producing early crops to catch a lucrative market. Much money and skill are needed to introduce and maintain irrigation farming, so it is urgently necessary to produce crops of good quality in high yield. This demands a high standard in agricultural practice and in marketing. The main emphasis should be on production of high value crops for export. However, rotation of crops may be possible and desirable, or it may be enforced by seasonal changes in climate and availability of irrigation water. This is likely to offer good scope for grain or fodder crops destined mainly for home consumption. Crops suited to high temperatures, such as cotton, sugar cane and dates, may be grown to compete with artificial fibres, sugar beet and fruits of more temperate regions.

AGRICULTURAL DISCIPLINE

There is no generally agreed body of laws and regulations controlling the use of water for irrigation. The need for such regulation is so apparent that small groups of co-operating farmers have separately developed and have willingly observed quite complicated codes of behaviour. Difficulties arise, of course, when separate groups have conflicting interests, as may happen when they all wish to draw water from a limited supply. Thus there has been a long dispute between India and Pakistan about sharing the waters of the Punjab. Similarly there have been disputes between Egypt and the Sudan about sharing the waters of the Nile. It would be absurd to pretend that such disputes can be avoided, but the situation is usually made less tense when discussed on a purely technical basis by irrigation engineers, who can usually agree on some appropriate procedure for allocating the available supply. By further technical discussion further sub-division may follow until, with luck, an agreed supply is directed to a group of neighbouring farmers able to recognise their mutual interest in making the best use of it. They may agree to take water at the approved times and in the approved amounts, to sow approved seed at the time and rate recommended, to weed early and thoroughly, to harvest cleanly and to uproot the crop in due season. Failure to observe these practices injures not only the individual farmer but his neighbours also; irrigation implies a uniformly high standard of farming; losses are severe if this is not kept up.

Water may be needed not only for irrigation but also for watering stock, for domestic or industrial use, for fish ponds or for recreation. When choices have to be made between many different ways of developing water resources in the hope of obtaining optimum benefits, the problem calls for elaborate mathematical study (as described, for example, by Hall and Buras [11]). Commonly, however, decisions are reached by more rough-and-ready methods and sometimes in a hurry under pressure of some adverse situation. Farmers should continually strive for increasing economy in use of irrigation water, so that their claim for a share may not be easily challenged on the score of waste and extravagance.

Irrigation farming is a highly skilled competitive business of absorbing interest to alert and industrious men. There is a continuing need for a variety of technical services, testing and supplying good seed, persuasively insisting on the good practices that help to secure profitable yields, alert to discover and overcome

Fig XII-6. Aerial view of irrigation channels in the Griffith area, New South Wales. (*Photo by courtesy CSIRO Irrigation Research Laboratory, Griffith*)

pests and disease, thoroughly informed about progress and mishaps in other agricultural countries. It should not be supposed that present techniques are near perfection: it is far more likely that we have yet much to learn about distribution and use of irrigation water.

HERBERT GREENE

References

[1] US Dept. Agriculture Handbook No. 18, *Soil Survey Manual*, Washington, 1951.
[2] THORNE, D. W. and H. B. PETERSON, *Irrigated Soils, their Fertility and Management* (2nd ed.), London, 1949.

[3] MARSHALL, T. J., 'Relations between water and soil', *Tech. Comm. No. 50, Commonwealth Bureau of Soils*, Harpenden, UK, 1959.

[4] GREENE, H., 'Using salty land', *FAO Agricultural Studies*, No. 3 (2nd ed.), FAO, Rome, 1953.

[5] — 'Soil Resources', Chap. 7 of *Guide Book to Research Data for Arid Zone Development*, Unesco, Paris, 1957.

[6] THOMAS, R. G. and J. G. THOMPSON, 'The classification and assessment of soils for irrigation in Southern Rhodesia', *Proc. Third Inter-African Soils Conference*, Vol. I, Publ. No. 50, Commission for Technical Co-operation in Africa South of the Sahara, 1959, pp. 345–50.

[7] US Dept. of the Interior, *Bureau of Reclamation Manual*, Vol. V, Irrigated Land Use Part 2, Land Classification, Rel. 29, 1953.

[8] ANDERSON, B., 'A soil reconnaissance survey of the main irrigable areas of the Rufiji Basin', Vol. VII of *The Rufiji Basin, Tanganyika*, FAO, Rome, 1960.

[9] BALLS, W. L., *The Yields of a Crop, based on an Analysis of Cotton-growing by Irrigation in Egypt*, London, 1953.

[10] RICHARDS, L. A. (Ed.), 'Diagnosis and improvement of saline and alkali soils', *US Dept. Agric. Handbook*, No. 60 (2nd ed.), Washington, 1954.

[11] HALL, W. A. and N. BURAS, 'The dynamic programming approach to water-resources development', *J. Geophys. Res.*, Vol. 66, 1961, pp. 517–20.

CHAPTER XIII

Small-scale Industry and Crafts in Arid Regions

This chapter is devoted essentially to the analysis of artisanship in the arid zone of the Old World. It is, in fact, only there that the complex ways of life (the contrast between pastoral nomads and sedentary workers) and the raised technical level have combined to permit different types of development of craftsmanship to appear, and thus to necessitate an explanation.

Types of craftsmanship

DEPENDENCY OF NOMADISM ON THE SEDENTARY ARTISAN

In the schism between nomads and people of sedentary occupation which dominates the ways of life in arid regions, the inferiority of the nomad in the domain of artisanship would appear to be inherent. The materials of greater or less complexity and the more or less cumbersome installations that are required by the artisan for his work are scarcely compatible with the mobility and constant shifting of nomads, who find difficulty, moreover, in stocking up with raw materials, even those of little weight, or with the products of their work other than those intended for their immediate personal use.

One must, then, have a town in the desert. The normal type of development of handicrafts is characterised by the existence of urban centres, serving to group together the great majority of artisans, and working for the nomads. This type is well represented in the Arab lands of the Middle East. Let us consider the very complete list drawn up by A. de Boucheman [1] of materials used in the life of the Sba'a Bedouin of the Syrian desert, in which list care has been taken to note whether the objects are of nomadic or of sedentary origin.

Cloth for articles of clothing is always acquired from sedentary workers, and almost all the clothing made up by them. In the tribe, certain women are able at best to make shirts or men's pants, the women's clothing being always made in the towns. The only exceptions are certain pieces of leather work – sandals (but these traditionally are worn very rarely, as for burglary) – or belts of leather or plaited thongs, the work of the Bedouin or of the Sleyb caste of hunters.

Harness for horses is always made in the towns, and generally that of the camel also. Objects made of leather or tissue – halter and girth, drapes that protect the front of a palanquin – are more often made by the nomads, but saddle

Fig XIII-I. Nomadic woman of the Larbaa tribe in the Laghouat region of Algeria weaving a tent wall out of goat and camel wool. The tent itself is black. (*Photo Jean Despois*)

rugs of wool with a short close pile are woven by sedentary town dwellers in Palmyra, while the nomads content themselves with making much rougher rugs.

For woodworking (generally in tamarisk) the division of labour is similar. All the complex pieces for saddles, for the structure and ornament of palanquins and for pack-saddles are made by townsfolk; the nomads make little but the support of the palanquin and the framework for water-packs, of which the cover is made by sedentary workers. As for tentage, the strips of woven goathair of the tent itself are sometimes made by the nomads and sometimes bought ready-made in the towns by those who do not own goats, while the pegs and poles generally come from the towns (Fig XIII-I). Occasionally, tents may be made by sedentary artisans with wool supplied by certain nomads, as for example thirty years ago at Safad in north Palestine [2]. Among household chattels, the utensils for making coffee, and also cooking utensils, are entirely supplied by sedentary artisans, but leather bottles and buckets and portable drinking-bags for camels are of Bedouin manufacture. Pulleys, axles and metallic supports are purchased from workers in towns. The leash, hood and shackles used in falconry are made within the tribe, but perches are obtained from sedentary workers. Knapsacks, small bags and infants' hammocks are also woven, in general, by the Bedouin, and, finally, these people often make rudimentary violins for themselves – one-string fiddles made

Fig XIII-2. Coppersmith's shop, Isfahan, Iran. The work is done by sedentary craftsmen. (*Photo L. Trikojus*)

of skin covering a rectangular frame of small wooden boards, which is their only musical instrument.

This enumeration, which might appear to be tedious, is in fact significant, as it reveals that nomads rarely make objects for themselves, either wholly or almost wholly, except those made of leather. For fabrics, a kind of division of labour is established: clothing and good quality cloths come from townsfolk, while coarser materials – the fabric of tents and woven stuff for bags – are most often made among the tribe. An analogous division is found for rush and reed mats, which are sometimes made by sedentary workers and sometimes by nomads, as for instance in Palestine north of the region of Huleh, before the partition of the country [2, p. 123]. It is essentially the same with basket-work. On the other hand, the making of wooden objects is predominantly by sedentary workers, and all the more delicate pieces are their work. Moreover, their superiority is absolute for all metallic objects and for ceramics (see Fig XIII-2). The craftsmanship of the nomads is thus limited to the most rudimentary modifications of primary materials in leather, wool or hair provided by the breeding of herds. The fact that in the Near and Middle East most of the clothing is made of cottons explains the predominance of the sedentary worker in the clothing trade.

THE NOMADIC ARTISAN

1. *Diffused crafts*

The subordination of the nomad with regard to the sedentary artisan is made possible only by the existence of an exchangeable currency. Let us recall, for example, that at the end of the nineteenth century the average price of the wooden framework of a yurt among the Turkmens of the steppes south-east of the Caspian attained the equivalent of a well-bred camel, and at certain periods had probably reached the price of three or four camels [3]. This suggests a pastoral activity or social supremacy that permitted the levying of a more or less disguised tax or tribute. Even in the Near and Middle East where dependence on centres of sedentary work is pushed to its maximum, it has been noted that it is especially the richest Bedouin who provision themselves from the artisans in the oases, while the less well-to-do more often make their own necessities. This social stratification is largely based, for clothing and tent-stuffs, on the quality of the products, which, we have seen, is much higher among the sedentary craftsmen [4]. Even if at regular intervals one buys the products of their crafts from town workers, day-to-day repairs must of necessity be carried out in the tribe, without the need for travelling to a commercial centre. Even for arms, which are very complex articles bought almost entirely from sedentary workers, the Rw'ala gun-smiths are at least accustomed to repair them, although in fact they rarely make new ones [5].

Thus there develops of necessity a type, very different from the preceding, where craftsmanship is largely diffused through the nomadic tribe. In this type of diffused craftsmanship, each person makes anything and everything for the daily needs. The proportion of objects made from primary materials furnished from the flocks, leather or wool and hair, rises considerably, and in the ultimate, almost attained by the nomads of eastern Tibet [6], the extent of sedentary craftsmanship is very much reduced. It is limited to work with certain pelts particularly suited for delicate work, as for example boots, while all the nomad women are capable of preparing, cutting and sewing the skins for more ordinary uses. It includes certainly all articles made of metal or clay, but these are not widespread among the nomads, except for metallic jewellery made chiefly by Chinese craftsmen. In the Sahara, where the materials of nomadic life are more complex and where ceramics have penetrated into tribal life, the tribes themselves have been initiated into their manufacture. Although, among the Tuareg, ceramics are generally made by potters sedentarised in the interior of the Tuareg zone of influence, nomadic potters exist in the Azaouagh [7], and in the Arab-speaking northern Sahara pottery-making by the nomad women is general. They nevertheless limit themselves to modelling pottery for kitchen utensils; pottery made on the wheel, for carrying water, remains the prerogative of sedentarised workers.

It seems, then, that neither at this moment of time, nor recently, can one find nomads living uniquely on their own craftsmanship and totally independent of settled workers, at least in the Sahara and the Middle East. Perhaps the sole remaining people who fully support themselves as to their needs in hand-made articles are the few tribes of Australian aborigines who still remain out of touch with sources of manufactured articles. Such a tribe, the Bindibu, lives in a state of nakedness over 300 miles across desert and semi-desert from the nearest town (Alice Springs), and 225 miles from the nearest homestead. Nomadic hunters and gatherers, they have few possessions except their weapons, their religious regalia of decorated stone or wood, and a few simple wooden vessels and string carrying bags [8]. Between such primitive people, still Neolithic, and the nomadic herdsmen of the Middle East with their well-developed endemic cultures, are many gradations, some strongly influenced today by tourism, and organised sedentary craftwork for sale to visitors, as with the Navajo Indians and those Australian aborigines who have settled in missions or on cattle stations.

In diffused nomadic craftsmanship the only division of work is, in general, of sexual origin, and is often a matter of an elemental determinism related to the nature of the predominant occupations of the sexes. Thus among the Toubous [9] the men make objects of wood and leather (tent-poles, wooden plates and ladles, bottles and bags of leather, tent material) which are connected especially with travel, while the women make household articles – basket-work, panniers and matting. In the same way, needlework is reserved for the men, who wear sewn clothing, while the women wear draperies. These distinctions are, however, not absolute; they carry no tabu, no notion of rigorous interdiction. Each Toubou makes saddlebag or pannier as occasion arises. Elsewhere it is common to see nomadic man spinning or knitting, doubtless to occupy his long periods of inactivity.

The determinism may, however, be more subtle: specialisation may be different among the nomads from among settled peoples. For example, in the northern Sahara the pottery modelled by the nomads is the work of women, while that turned on a wheel is made by male sedentary workers [10, 11]. A very complex case is afforded by rugs [12]. Today in North Africa rugs are woven by men in Algeria, notably at Djebel Amour, and, on the contrary, by women in Morocco. It may readily be understood why the *reggam*, the master weaver, must be a man. He is the sole person acquainted with the quite complicated techniques of calculation which are necessary. The woman, who goes out infrequently, does not master them. The man alone has the necessary authority to be the creator and organiser of the work. It is probable that also in the Orient weaving was originally in the hands of the man. The transference of the technique of carpet-making to the women is, without doubt, an expression of sedentarisation, or of the influence of city merchants who distribute the work of the tribes, impose designs and ensure

a certain standardisation of products for commercial purposes. It has as a consequence a certain sterilisation of motifs and inspiration. The handing-over to
women of the work of carpet-making is consequently general among sedentarised
workers. At Mzab, for example, the women weave carpets while the men weave
burnous [13].

The diffusion of craftsmanship among the whole nomadic group, and the
absence of specialisation other than between the sexes, can definitely be correlated with the obviously poor level of their products. The richness of the
material life of the Bedouin in the Near East cannot be understood, as we have
seen, except by recognising their marked dependence on neighbouring sedentary
workers. There is, however, another consideration, that of the division of work
within the tribe, and the existence of artisan castes.

2. Artisan castes

The development of specialised craftsmanship among the nomads rapidly results
in spectacular effects on their social organisation. The driving force behind this
is a scornful attitude towards artisans, which is general in pastoral and nomadic
societies [14]. In contrast with the occupation which above all is noble, that of
the management of herds, manual work appears as a base or even servile occupation. Whatever may be the springs of this derogatory view, its results are always
translated into the formation of a veritable artisan caste-system, involving more
or less rigorous segregation. Such castes are widespread throughout the southern
Sahara [15]. It is working with iron, which always involves narrow specialisation
even outside the arid zone and among sedentary workers, which is somewhat a
subject of mysticism and arouses a rather fearful respect for the smiths, that
gives rise to the commonest specialised groups. Next in order is ceramics, which
by its very nature is foreign to the nomadic way of life. Among the Tuareg of the
Hoggar, who are thought of as being in a primitive condition, the artisan caste
of the Inadan is essentially that of blacksmiths [16], while pottery is confined to
black women, who are sedentary but not in the caste-system [17]. Among the
Toubous, in a more evolved stage, the Azzas constitute a caste of hunters and
artisans, paying tribute in kind to their suzerains, with whom marriage is
rigorously forbidden. The men are smiths, working with iron, copper and silver,
while the women are potters. It is the same in the Azaouagh, where there exists,
in addition to the Inadan smiths and woodworkers, the Ikanawen caste of potters,
in which the women bake the pottery while the men collect suitable clay [7]. It
is probable that artisan caste of this type, or more or less narrow specialisations
of certain tribes, existed in central Asia among the Mongols before peace under
Gengis Khan led to the development of sedentary nuclei and a related dependence of the nomads upon these. The Secret History of the Mongols mentions
smiths and carpenters in the twelfth century, permitting one to read between the
lines a real division of labour in a nomadic society whose trade exchanges with
the world outside were reduced to a minimum [18]. Among the nomadic Turks

of the Middle East, in Iran and Anatolia, there are many tribal names of a professional character, for example *demirciler* (blacksmith), *kuyumcular* (goldsmiths), which are also preserved in Anatolian place-names and which surely indicate a formerly analogous situation [19].

The ethnic origin of these artisan castes appears to be varied – it does not necessarily reside in inferior or subject people. Although in the Sahara it is for the most part coloured people (for example the Ibarogan among the southern Tuareg), they may be white, as with the Inadan of the Hoggar, who appear to constitute a complete amalgamation. Although their common origin, according to legend, was in the Adrar of Iforas, it is possible that there are among them descendants of the Jews of Tamentit at Touat, who were dispersed at the end of the fifteenth century, and also perhaps of Sudanese elements [16]. Thus it is certainly possible in many instances to see here a recent addition to the nomadic society, newcomers who are sedentary artisans drawn to the tribe by the lure of gain, or constrained to embrace the nomadic life by the vicissitudes of history, or by the decay or destruction of their sedentary habitats. The fact that these people have the character of strangers may help to explain the phenomenon of caste, and also that of rigorous endogamy. Obvious examples are afforded in the Near East by the frequent presence of Gipsies in this role, in the bosom of the nomadic Turks of southern Anatolia [19], or of Iranian nomads in the Fars [20], or of the semi-nomadic Arabs of northern Palestine a quarter of a century ago – the Nawar, who made especially copper utensils and coffee-mills [2].

The integration of such new arrivals with the life of the nomads could only have been progressive. Its divers repercussions could develop only slowly. The Touareg of the Hoggar afford a typical case of the process, sufficiently recent, it would seem, to permit it to be analysed. The artisans, the Inadan, are still small in numbers, and annexed only recently to the tribes. It is still possible to recognise traces of a not-far-distant past, when work with diverse materials, but notably wood, was still considered by the Touareg women to be a noble occupation [16]. The Inadan are exclusively blacksmiths in origin; but their presence soon modifies the conditions of every kind of craft work, which becomes more and more restricted to them. The Touareg have acquired the habit of applying to them not only for metalwork, but for all kinds of other things. During the past two generations craft specialisation appears to have been actively in progress, while diffuse artisanship sensibly declined.

Principles of the geographical distribution of fundamental types of artisanship

As we have seen, the essential contrast is that between two types of deserts: on the one hand deserts with urban oases which constitute dense concentrations of artisans, with nomads largely dependent on the sedentary social system; on the other, deserts without or with poor and purely agricultural oases, where the nomads are in large measure forced to undertake their own crafts. Geographically, within the desert zone of the Old World, the Near East, broadly regarded, stands out as being of the first type, the Sahara almost entirely of the second [21].

It is only necessary, for instance, to compare the catalogue of crafts of the poorest of the Middle Eastern oases, say Soukhné, described by A. de Boucheman [22], with that of all the oases of the Fezzan, given by J. Despois [23]. On the one hand there is a prodigious variety of small trades serving the nomads, on the other an almost non-existent industry. Central Asia belongs in an intermediate category. All of High Asia – Mongolia and Tibet – is characterised by great ergological poverty of the nomads and by their considerable independence with respect to the oases, which are generally underdeveloped or of recent development (the origins of Ulan-Bator, capital of Mongolia, go back only to the seventeenth century), while the great cities of lower central Asia have at all times dominated the life of the steppes and deserts which surround them. The problem of the distribution of types of industry in the arid zone is related firstly to the presence of oases, next to the prosperity of the oases themselves. It poses one of the most fundamental questions of human geography in the arid zone.

Firstly, one may seek for natural causes for these contrasts. The absence, or mediocrity, of oases in the whole of High Asia may be explained largely by the essential concurrence of the activities of the nomads and the sedentary peoples in the same spheres at the same periods of the year. The pastoral territories of the nomads correspond in summer largely to the altitudinal belt where rain-cropping is possible only at this time of the year, and in winter the pastures lie in the piedmont areas, where irrigation is attempted. This coincidence has paralysed the development of agriculture in all the regions of nomadic preponderance in High Asia.

By way of contrast, the hot desert climate of the Asian lowlands, with winter rains, results in an almost absolute dissociation of the nomadic territories, which are forced to great altitudes in summer, and isolated in winter among the dune systems some distance from the piedmont zone where sedentary agriculture prospers. This was the fundamental factor influencing the possibility of the development of urban manufacturing centres in central lower Asia [24].

This deterministic interpretation, the partial truth of which cannot be doubted, must, however, be modified for the greater part of the Saharan–Sind region by factors of another kind. If the Saharan oases have not been able to play a role comparable with those of Arabia, it is because they have often been destroyed by the ravages of nomads, as in the Fezzan, for example, where trans-Saharan commerce has never succeeded in re-establishing prosperity, having been constantly menaced and periodically ruined since the great nomadic developments based on the camel took place [23]. Most of the fixed centres, at least in their present structure, are of relatively recent historical development due to progressive recolonisation by nomads who are gradually becoming sedentarised. In this contrasting situation, social relationships between nomads and sedentarised people do not seem to be involved. At a time of bellicose nomadism they lie almost always within the compass of a marked suzerainty of the nomads with respect to the inhabitants of the oases: even the urban centres of the Arabic-Syrian desert, and indeed of the margins of the fertile crescent, do not escape this rule. Is it perhaps simply the smaller size of the Arabian desert, with its peripheral centres always more or less within reach of the nomads, in contrast with the immensity of the Sahara where access to the border zones is often impossible, that affects the situation? Or must one seek the basis in the prosperity of the commerce of caravans, always more highly developed in the Near and Middle East than across the Sahara, and requiring the support of the oases?

The reason for the differences may, however, equally be sought among the nomads themselves. The capacity for craft work of any group of nomads depends strictly upon their level of civilisation, their accumulated traditions and techniques. It may be said that the character of pastoral nomadic life is a 'facies of convergence' which unites people of different origins in a common mould imposed by natural conditions – ancient hunters promoted little by little to a more elevated stage as a result of the domestication of their animals, or sedentary people more or less driven and constrained to bury themselves in the desert [25, 26]. The craft-complex of the nomads will necessarily bear the mark of their origins. Thus it is that the absence of ceramic crafts among the majority of nomads has been seen as the proof that their origin lay in the groups of 'marginal hunters' [27], living on the periphery of sedentary centres affected by the Neolithic revolution, but not themselves having experienced this. It is probable that the introduction of ceramic work among the Tuareg is a relatively recent event, perhaps connected with the introduction of millet deriving from the Sudanese zone into the range of their foodstuffs [17]. Similarly, the ergological poverty of the nomads of the Am-Do has been seen as a proof of their origin from technically retarded hunters [6]. It would, then, be quite understandable that the nomads of the Near East, who undoubtedly spring from the most ancient of sedentary centres, should have taken quite different traditions with them into the desert.

Finally, it would not be too much to seek in religious causes the explanation of certain deficiencies and successes which do not seem to be entirely related to the factors already indicated. Industry, and the urban life which it generally requires, enjoys a most privileged place in the Islamic world [28], in contradistinction to nomadic life with its constant occasions of impurity and impiety. One cannot but be struck with the coincidence of the prosperity of the life of urban craftsmen in those deserts and regions where Islamic culture is most ancient and most deeply entrenched. The artisan caste triumphs among the Toubous, whose Islamisation scarcely goes back to the nineteenth century. This appears to be largely the imprint of a pre-Islamic mental attitude and economic structure. And in the eastern parts of central Asia it may perhaps be necessary to seek the explanation of the economic mediocrity of certain centres less in the influence of the distribution of summer rains than in the essentially monastic character of those centres within the Lamaistic domain? On the other hand, the flourishing of cities in the deserts of the Near East appears to be inseparable from that marked dignity which has always accompanied these centres of origination of the Islamic religion.

The complexity of the questions raised by the nuances of variation shown in the diffusion of craftwork in the arid zone of the ancient world may thus be seen. All the great human problems of these regions are involved, through the medium of this extremely revealing test of the general level of traditional activity in the desert. But will not this significance diminish markedly in the face of the transformations which must come and which are, in fact, already affecting them?

Contemporary changes in industry in the arid zone

What is, then, and what will be the fate of the artisan with the irruption into the arid zone of the products of major industry? The rather primitive craftsman would appear to be condemned to vanish very shortly. General demographic and economic conditions do not appear to require his survival. It is known that in many underdeveloped countries (for example, India) the persistence and even the growth of crafts is an expression of the weak development of modern industry combined with a stultified agricultural outlook and intense demographic pressure, which constantly throw on the labour market those who only thus can find work for themselves without considerable capital investment. Nothing like this is possible in the arid zone, and the general sparseness of population rules out such consideration. In fact, in many cases, decay is rapid. The picture of the crafts at Tamentit (Touat) drawn in 1961 permits the point to be made [11]. Traditional modelled pottery is quickly replaced by enamel-ware. Blacksmiths and carpenters no longer live in the oases. Work with leather still subsists among the nomads but is nothing more than an accessory occupation, no longer affording a living for a man, among the sedentary workers. Nevertheless, two categories of callings resist the change and even prosper. The jewellery trade and silver work are expanding along with the enrichment of the oases and the growth in circulation of money, connected with the increased wealth of the Sahara and the penetration of modern economy. On the other hand, the trades finding openings through tourism remain sufficiently remunerative. This is the case with basket-making, and with turned pottery, made by men, and strictly decorative and lacking in practical use, which has taken the place of the traditional utilitarian pottery modelled by women, which is in the process of disappearing. Studies permitting one to determine to what extent such an example may be generalised are lacking. In any case, modern industry can compete only with difficulty with the products of traditional craftsmanship for certain objects adapted to the conditions of desert life, for example, woollen burnous. Thus it is that the textile trade of Souf, formerly devoted in part to luxury which cannot be compared with the products of European industry, are turning more and more today to the making of these coarse woollens [29, p. 96].

However, not everything is disadvantageous in the penetration of modern economy. Before the triumph of the products of major industry, a transition period may generally be defined, in which the trades prosper, linked with the much greater facility of transportation which opens new markets, and with the pacification and suppression of raiders, which eliminates major losses. This development is in many ways comparable with that of pastoral nomadism in the early years of the revival of security, when the herds left the fortified villages and went out into the desert. It certainly is only an ephemeral effect; but in the Sahara it has been sufficient to extend the range of manufactured products quite markedly, many dating only from the last century (that of French penetration), and in any case to completely transform the conditions of supply of primary materials. Thus in Gourara the weaving of *dokkali*, tapestries sold to caravaniers or to the merchants of Timimoun, has its origin in the weaving of pieces of rough cloth sold to the Touareg. It was only in 1851 that colour was added at Timimoun; but it was not until 1909 that the type most widespread today appeared at Fatis, based on a model brought in from Gabes. The wool comes today from the Saharan Atlas (Geryville), the cotton from Algiers or Oran, and chemical dyes are also imported. 1,000–1,500 pieces are made each year by 350 craftsmen. There is here a development of craftwork which, in essentials, goes back to the last half-century [30, pp. 143–5].

A remarkable example of recent expansion, or rather of tenacious resistance based on the continual creation of new markets, is afforded by a type of mineral extraction, the extraction of salt, which is widespread in the southern Sahara and which, in its methods, belongs to the traditional small-scale industry based on the artisan. It is well known that West Africa lacks salt and that it normally derives its supplies from the Sahara, where the great salt-caravans, the *azalai*, move regularly towards Taoudenni or Bilma, giving rise to a large and constant flow. The conditions of work in the salt mines are so extremely severe that labour cannot generally be assured except by a system of slavery or captivity, to the advantage of the carriers. With growing competition of marine salt coming from the coast, these social and economic conditions permitted it to be foreseen about 1920 by enlightened observers that the application of liberal regulations on the liberty of the individual, and the progressive organisation of commercial contacts for the distribution of marine salt, would tend, with only a brief delay, to the ruin of the rock-salt industry. But this has not occurred. Certainly the limit of the zone of influence has receded little by little towards the north, to the advantage of marine salt, but quite slowly. This movement has been represented cartographically in the Chad region [31, pp. 123–30]. The diminution in total value of the exploitation of salt mines with reference to 1929 as base has been estimated at 50 per cent. But the salt of Borkou still sold in 1958 up to the north-east shores of Lake Chad, after having reached Maraoua in the Cameroons thirty years earlier. The salt from Demi in the Ounianga, which reached Fort Archambault in 1952, by 1958 attained the line of Ati-Abéiché in the Ouadai. A transition zone existed, where the marine salt sold 15 to 20 per cent cheaper than the rock-salt, but where this latter still continued to find a market parallel with the former. Force of habit alone explains this anomaly, which was not founded in the Chad country, a zone of commercial individualism related to the tribal anarchy of the Toubous, nor was it due to any powerful organisation comparable to the great *azalais* of the western Sahara. But the intervention of these social factors has created, in this latter example, conditions for a striking expansion. Taoudenni had 50 miners in 1922, 246 in 1957. Production was raised from more than 23,000 cakes of salt in 1901 to 160,000 in 1958. The commerce of the *azalais* has not ceased growing. The explanation of this phenomenon must be sought in sociology rather than in economics. For the young carriers of the tribe of the Berabiche, transport in the *azalai* is the traditional proof of virility. Despite the disappearance of all compulsion, applicants to become miners are numerous among the debtor workmen of Timbuctu, who go to the salt mines for one or more seasons to pay off their debts, but return there as a matter of habit. Finally, the machinery for commerce in rock-salt is present throughout West Africa, with its distribution network, and lives on through repute and advertisement. The African market increases continally and there is a place in it for the two types of salt. Thus it may be explained that trade in rock-salt from the desert has been able to grow in absolute value even though it has lost ground relative to sea-salt [32].

There is no doubt that the future of the artisan and his work in the arid zone has surprises in store, and it is probable that there will be some 'fireworks' before the inevitable decline. The profound singularity of the ways of life and of human environments in the desert will doubtless in the long run express itself again in this sphere by archaisms which have not ceased to arise and to astonish us.

X. DE PLANHOL
(*Translated from the French by E. S. Hills*)

References

[1] DE BOUCHEMAN, A., 'Matériel de la vie bédouine recueilli dans le désert de Syrie (tribu des Arabes Sba'a)', *Docum. études orient. Inst. franç., Damascus*, Vol. III, 1935.

[2] ASHKENAZI, T., *Tribus semi-nomades de la Palestine du Nord*, Paris, 1938, pp. 118, 123, 124.

[3] DŽIKIEV, A., 'Poselenija i žilišča turkmenov jugovostočnogo poberez' ja kaspÿskogo morja XIX–XX V.V.', *Sovietskaja etnografija*, Vol. 4, 1957, pp. 76–90.

[4] DICKSON, H. R. P., *The Arab of the Desert*, London, 1949, p. 73.

[5] MUSIL, A., *The Manners and Customs of the Rw'ala Bedouins*, New York, 1928, p. 281.

[6] HERMANNS, M., *Die Nomaden von Tibet*, Vienna, 1949, pp. 79–84.

[7] NICOLAS, F., 'Les industries de protection chez les Twareg de l'Azawagh', *Hesperis*, Vol. 25, 1938, pp. 43–84.

[8] THOMPSON, D. F., 'The Bindibu Expedition: exploration among the desert aborigines of Western Australia', *Geogr. Journ.*, Vol. 128, 1962, pp. 1–4, 143–57, 262–78.

[9] CHAPELLE, J., *Nomades noirs du Sahara*, Paris, 1958.

[10] JEST, C., 'La poterie dans le Sahara oriental (Souf, Touggourt, Ouargla)', *Trav. Inst. Rech. Sahariennes, Univ. Algiers*, Vol. 19, 1960, pp. 105–18.

[11] RIO, CNE., 'L'artisanat à Tamentit', *Trav. Inst. Rech. Sahariennes, Univ. Algiers*, Vol. 20, 1961, pp. 135–83.

[12] GOLVIN, L., *Les tapis algériens*, Algiers, 1953, pp. 605.

[13] MERCIER, M., *La civilisation urbaine au Mzab*, Algiers, 1922.

[14] CLÉMENT, P., 'Le forgeron en Afrique noire. Quelques attitudes du groupe à son égard', *Rev. Géogr. humaine et ethnol.*, No. 2, 1948, pp. 35–58.

[15] CAPOT-REY, R., *Le Sahara français*, Paris, 1953, p. 226.

[16] LHÔTE, H., *Les Touaregs du Hoggar* (2nd ed.), Paris, 1955, pp. 317–23.

[17] — 'La poterie dans l'Ahaggar', *Trav. Inst. Rech. Sahariennes, Univ. Algiers*, Vol. 4, 1947, pp. 145–54.

[18] VLADIMIRTSOV, B., *Le régime social des Mongols*, French transl., Paris, 1948, p. 53.

[19] DE PLANHOL, X., *De la plaine pamphylienne aux lacs pisidiens, nomadisme et vie paysanne*, Paris, 1958, pp. 107, 272–3.

[20] BARTH, F., *Nomads of South Persia*, London, 1961, pp. 91–2.

[21] MONTAGNE, R., *La civilisation du désert*, Paris, 1947, p. 42.

[22] DE BOUCHEMAN, A., 'Une petite cité caravanière, Soukhné', *Docum. études orient. Inst. franç.*, Damascus, Vol. VI, 1939, pp. 16–85.

[23] DESPOIS, J., 'Géographie humaine du Fezzan', *Inst. Rech. Sahariennes*, Mission Fezzan, Vol. 3, Univ. Algiers, 1946, pp. 177.

[24] RAKITNIKOV, A. N., 'Nekotorye osobennosti istoričeskoj geografii zemledelija i životnovodstva v Srednej Azii', *Voprosy geografii*, No. 57, *Istoričeskaja geografija*, 1960, pp. 71–90.

[25] DE PLANHOL, X., 'Genèse et diffusion du nomadisme pastoral dans l'ancien monde', *Rev. Géogr. de l'Est*, Vol. I, 1961, pp. 291–7.

[26] VON WISSMANN, H., Article Badw., *Encycloped. of Islam* (2nd ed.), Leyden, 1959.

[27] WERTH, E., 'Zur Verbreitung und Enstehung des Hirtennomadentums', *Inst. f. Menscheit u. Menscheitskunde*, Vol. 16, 1956.

[28] DE PLANHOL, X., *Le monde islamique, essai de géographie religieuse*, Paris, 1957.

[29] BATAILLON, C., 'Le Souf', *Mem. Inst. Rech. Sahariennes, Univ. Algiers*, No. 2, 1955, p. 96.

[30] BISSON, J., 'Le Gourara', *Mem. Inst. Rech. Sahariennes, Univ. Algiers*, No. 3, 1957.

[31] CAPOT-REY, R., 'Borkou et Ounianga,' *Mem. Inst. Rech. Sahariennes, Univ. Algiers*, No. 5, 1961.

[32] CLAUZEL, J., 'L'exploitation des salines de Taoudenni', *Monogr. régionales Inst. Rech. Sahariennes, Univ. Algiers*, No. 3, 1960.

CHAPTER XIV

Industrialisation

The making of articles by nomads and by sedentary urban dwellers, men and women skilled in traditional handicrafts concerned chiefly with the materials of everyday life, is generally on a small scale of production and is an integral part of established cultural and social patterns in arid lands, as discussed in Chapter XIII. 'Industrialisation' implies activities on a larger scale, with well-organised channels of trade, both internal and foreign. Manufacturing, mining, petroleum production, packaging and canning food, and similar organised industries that may be established in urban or rural districts in arid regions, have important economic and social effects. These latter do not, perhaps, differ in principle from the effects of industrialisation in densely populated humid regions, but in the desert and semi-desert a mine, an oilfield or a prospecting camp is truly an oasis of technology (Fig XIV-1), obvious and even incongruous in the local scene, requiring of the local inhabitants a considerable effort of assimilation, and of the newcomers a like effort of accommodation to the indigenes and to the environment.

It is true that some long-established industries have become well adjusted into the social structure of desert communities, as with the rock-salt industry of the Sahara.[1] Even in the most primitive of extant cultures, those of the Australian aboriginal tribesmen, the beginnings of industry and trade may be seen in extensive ochre mines at Umberatana in the Northern Territory, which are so large as to imply considerable antiquity, while intertribal trade was engaged in with a tough greenstone particularly suited for making axes, which was quarried in Victoria but is found hundreds of miles away in central Australia. Indeed, the mineral and stone industries have, in arid as in other regions and certainly for some thousands of years since the dawn of urban civilisations, been so highly organised as to have afforded a prototype for industrial development. Plans of the workings of mines, drawn on papyrus, are preserved from the time of Seti I in the nineteenth dynasty in Egypt (1300 B.C.) (Fig XIV-2), and it is clear that apart from the manual labour done by slaves, the overseers and technical staff

[1] Chapter XIII.

were capable and responsible men, working within a well-organised industrial structure. In the circumstances of today it is perhaps chiefly the abolition of forced labour that constitutes a fundamental difference from the conditions of former centuries or millennia, for while slave labour did involve a definite cost element, the labour was directed to the job, however unwillingly, whereas today

Fig XIV-I. Aramco Exploration trailer camp in the 'Sand mountains' of the Rub'al-Khali, Saudi Arabia. Tents are for the Saudi Arab escort. Sand-tyred trucks on right; gasoline, water tank trucks left and centre. (*Print by courtesy Petroleum Information Bureau, Melbourne*)

some inducement is required to obtain a work force under conditions of hard living in deserts and semi-deserts.

In most countries it has been the hope of considerable gain that has led to industrial developments in rural arid environments, and the enterprises have accordingly been limited in type. All costs are high, except for land values in non-irrigated areas, and these factors have led to the characteristics of large industry in dry lands – the enterprises are each large, generally widely separated, surrounded by inhospitable terrain, and, to a considerable degree, they remain as 'foreign' elements among the local agricultural or pastoral inhabitants. The situation differs in large urban centres in arid regions, where in fact a management cadre, a work force and other facilities (power, water, transport) are often already available, or may more readily be created. Given adequate technical education, manufacturing industries may find a favourable environment there, while in non-irrigated rural areas, mineral and oil exploitation are the chief industries that have been established. Irrigated lands afford special conditions, where the packaging and canning of foodstuffs are undertaken in factories, especially in large highly productive regions. Koenig [1] and others have argued

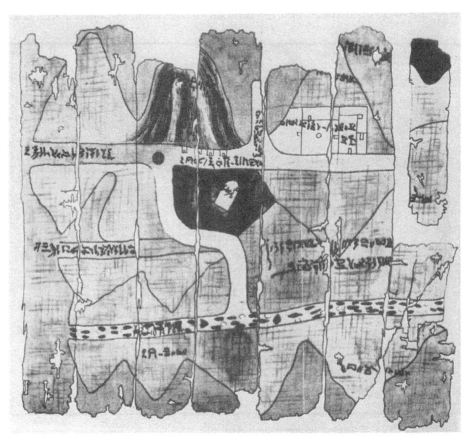

Fig XIV-2. Ancient mine plan of the time of Seti I (XIXth Dynasty, 1350–1200 B.C.) in Egypt. Turin Papyrus. (*After Hume, 'Geol. of Egypt', Vol. II, Cairo, 1937*)

that instead of further expanding the area of irrigated land, any additional water that may become available would be put to better use by establishing manufacturing in the cities of irrigated oases, and, indeed, such notions are often attractive to governments seeking ways of decentralising industry, to provide avenues of employment, other than farming, for growing rural populations. The economics of such schemes is, however, generally disadvantageous due to high overhead costs for transport and power, and the social benefits must be weighed against this in deciding the issue. The imponderables are many, since even ways of properly estimating the benefits of irrigation have not been fully studied. Many such issues are in these circumstances subject to strong political and group pressures, and tax-remissions, subsidies and other forms of government-sponsored aid are commonly used.

Of late years new influences have added to the industrial potential of arid regions. The tourist industry, aided by modern means of transportation and by air-conditioning, has increased, partly in response to growing populations which have led to overcrowding of traditional resorts, and partly because many people enjoy the wide open spaces and the company of the people who live there. The

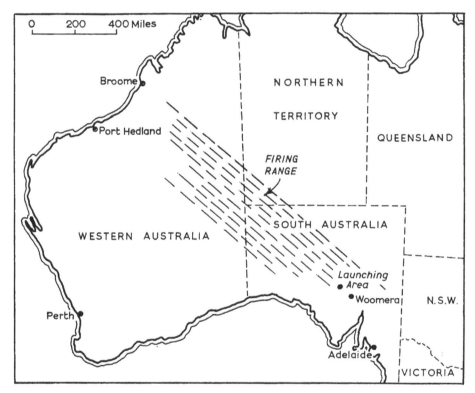

Fig XIV-3. The Woomera rocket-range, Australia, showing the great area involved in the operation

unusual, even bizarre vegetation, rock formations and animal life of deserts are also attractions. It is difficult to foresee all the consequences of the world population explosion which will double the number of inhabitants of the world by the year 2,000; but it is obvious that open spaces will increasingly be at a premium and reduced hours of work and longer vacations all add to the requirement for recreational space. The arid zone offers these as no other region does. As well as tourism, commuting between rural areas and the towns is already leading to the growth of settlement in arid areas.

Further new developments connected with modern technology have already to some extent been introduced and give a pointer to things to come. Atomic power stations, research institutes and testing grounds both for military and for peaceful uses, together with their satellitic workshops and townships, are often located in deserts, and thus make use of very large areas which otherwise would have little actual or potential use. While it is unlikely that atomic stations for supplying power for industry will eventually be located far from the major users because of the cost of power transmission over long distances, research and other dangerous or secret activities may well continue to be sited in arid regions. Rocketry, too, calls for experimental and testing grounds of great size, and as rockets will certainly play an increasing part in scientific as well as in military affairs, and as the dawn of a space age is upon us, the arid zone may again be expected to provide room for deployments in these operations. Already in Australia a testing range over 1,000 miles long and some hundreds of miles wide stretches across South Australia and Western Australia, in an otherwise virtually uninhabited belt of very arid land (Fig XIV-3).

Mining

Because of low rainfall, soluble salts, which in more humid regions would be dissolved, may crystallise at or near the surface, either in the soil or in salt lakes and pans. Nitrates are formed from the decay of organic matter. In India, efflorescences and incrustations of potassium nitrate are formed in the soil, but sodium nitrate, which was formerly of very great importance in the manufacture of nitrogenous fertiliser, occurs in the nitrate desert of Chile and, before the manufacture of artificial nitrogenous fertilisers, was widely exported (see Chapter x). Sodium nitrate is also a constituent of the Esna Shales, especially along the Nile Valley in Egypt, and the shale has been applied as a fertiliser by the fellahin for generations, although the sodium is deleterious.

Among the salts that crystallise in dry lake beds, sodium chloride (common salt) and gypsum are the commonest, and both afford opportunities for industrial development. Salt may be 'harvested' annually, as in the Mallee of south-eastern Australia, where it is scraped into heaps after the winter, but where the saline water is more voluminous (often being derived from saline springs in the lake bed), salt pans and works may be established. Gypsum is generally worked from the deposits of former lakes, rather than from existing lake beds. Gypsum is valuable as an additive to heavy clay soils to improve their texture, and also for making plaster of Paris and a range of building materials, notably plaster-sheeting. Among other useful salts are alunite (potassium aluminium sulphate), borates, sodium carbonate, and iodides.

Rock-salt, which is found in very thick beds in the geological succession inter-bedded with or intruded into other rocks, was formed in lagoons and other enclosed waters along sea coasts under arid conditions in the long geological past. Many of the important deposits are situated in parts of North America, Great Britain and Europe that now enjoy humid climates, although it is true that other deposits occur in arid lands in Africa, in the oilfield belt of Iran and Iraq, and elsewhere. The present climate of areas where ancient salt deposits occur bears

Fig XIV-4a and b. Pits and mounds made in mining opal at White Cliffs, New South Wales. The field is now abandoned. (*Photo E. S. Hills*)

no relation to the occurrence of salt, although, to be sure, the salt in arid regions may come up to the surface and even form salt glaciers, because of the low rainfall.

The requirements of human beings and animals for salt are such that in many regions it has a high value and is extensively traded (see Chapter XIII).

Apart from the deposits in salt lakes, and salt encrustations in the soil (which are found only in arid regions), the one mineral which is peculiar to deserts is opal, the production of which is increasing in Australia. Opal is formed by the precipitation of silica gel from groundwater percolating through the joints and along the bedding in sedimentary rocks, and the subsequent partial dehydration of the gel. Ground and water temperatures must be high, and there should be little or no dilution by rain. The chief Australian fields are now those of Coober Pedy in South Australia, where opals valued at about £A600,000 were obtained in 1960. The fields at White Cliffs, east of Broken Hill, NSW (Fig XIV-4), are now exhausted. Guano, a deposit of the excreta and bodies of sea birds, is collected for use as fertiliser along the coastal islands adjacent to the desert of Peru where, again, the low rainfall permits the deposits to accumulate. Some 40–150 feet of guano have been stripped away but new deposits are still forming [2].

As to all other minerals and metallic ores, there is in fact no reason based on geological or geophysical principles why they should not occur in deserts as much as in other climatic regions. Mineral deposits in arid and semi-arid areas do have certain special features due to the considerable depth to the water table, and surface and near-surface ores are modified due to groundwater conditions. Fossil soils of older periods may be preserved, notably lateritic soils which in places are found to be high in alumina and thus constitute bauxite, the ore of aluminium. Some reconstitution of the originally iron-rich laterite probably goes on in most such examples. With regard to fossil fuels, it is a little strange that deserts have proved to be rather poor in coal, but extremely rich in petroleum (Fig XIV-5). Indeed, the value of mineral and petroleum production from arid lands is strikingly great, and the volume of production is such that it greatly contributes to the economy of nations and thus is a powerful influence in world trade and international affairs.

All mineral deposits are wasting assets which must eventually become worked out, despite the best efforts of management and technical advances leading to improved recovery and the economic working of lower grades of ore and deeper coal and oil. It would, however, be unwise to dwell overmuch on this, despite the many ghost towns in formerly active mining fields, because in the overall picture mineral wealth will clearly rank very large for many years to come, although economic setbacks may affect particular mines or districts. It is also true that manufacturing and indeed all industries are subject to fluctuations which can equally give rise to local reductions or even cessation of production, with consequent unemployment. Thus in the changing scene of national and world economics there is little difference between economic recession due to reduced production of nitrate in Chile, of gold or copper in Australia, of ships on the

Clyde, or of textiles in Manchester. It is clearly the responsibility of governments and industry to ensure that a proportion of the profits of flourishing industry is used to establish a variety of services and other industries which may give immediate community benefits and help to cushion the effects of recessions in any one. This is particularly true of the mineral extractive industries in arid lands.

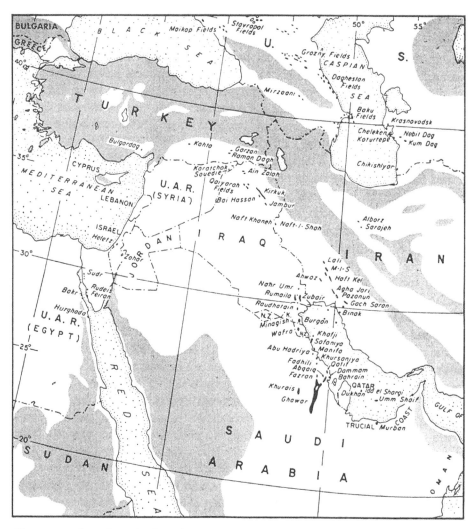

Fig XIV-5. Location of oil pools in the Middle East. Stippled areas are older, non-productive rocks. (*After Hotchkiss, 'Amer. Assoc. Petrol. Geol.', Vol. 46, 1962. Reproduced by kind permission of the Association*)

In Australia, several large towns and many smaller townships in the arid zone owe their existence entirely to mining. Kalgoorlie (population 21,759 – census of 1961) and Broken Hill (population 30,810) are the largest and oldest of these towns. Mount Isa (population 13,358) is a rapidly developing town based on copper and lead mining.

As always in arid regions, adequate water supply is a necessity. Reservoirs have been made at Broken Hill and Mount Isa, but there is no site for any large dam at Kalgoorlie and the town depends upon a pipeline constructed in the early years of this century, which brings water over 350 miles from the Helena River near Perth. In addition, the line supplies water for a million acres of farmland and 34 towns *en route*. Kalgoorlie is, as well as being primarily a mining town, an important commercial centre serving, through road and rail traffic, a large area of surrounding country.

Government support for gold mining is afforded in Australia chiefly through taxation concessions on company expenditure for prospecting and for development of ore reserves, sometimes also by subsidies or by technical services provided by government agencies. Dividends are tax-free, and a premium may be given over the fixed gold price in certain circumstances. An increase in the world price of gold would lead to considerably increased production and employment in the industry, so that the aim is to preserve as many working mines and as many skilled miners as possible in existing circumstances, while maintaining gold production, which yields international credits in hard currency. While very similar considerations apply to an industry such as gold mining in any region, it is true that in the Australian scene closure of mines in arid and otherwise unproductive country would result in rapid depopulation of the region, the effects of which would be felt in many ways throughout the country.

With highly profitable mining enterprises and petroleum production, the economic situation is different, and governments may derive much income from royalties and taxes, or by direct participation in dividends either through nationalisation or shareholdings. The greatest of any such benefits are undoubtedly by the revenues derived from petroleum exploitation in the Middle East. In Iraq, petroleum revenues have been applied to construct new factories, fertiliser plants and water-control works. In Iran the Quhistan land development scheme which includes hydroelectric generation and irrigation, sugar production, and a petrochemical industry, is financed from petroleum revenues, together with housing, education, transport facilities and other community benefits. The hope is that as oil production decreases in the years to come, permanent material benefits will have been derived from the general increase of productivity and potential land-use built up on oil revenues. But to be realistic, the lessons of arid land development, with many failures due to salinisation of irrigated lands or the overproduction of certain crops, must be remembered. It is with the hope, but not with certainty, of sound development that capital is invested in arid areas. Not only is wise administration called for, but the difficult decision whether to invest in already more favoured land where productivity may be increased, rather than in the arid parts, has to be made. An enlightened approach to the productive industry itself is equally important, for the amount of 'risk' capital needed to develop new ore-bodies or to discover new oilfields is considerable, and the industry must be left adequate funds to finance this.

Despoliation

It is an old and oft-repeated tale that a mining community has caused despoliation of the countryside for several miles around. Timber is required for mine
props and for fuel, both industrial and domestic. Some members of the community attempt to augment their incomes by rearing goats: land holders are
tempted to run more cattle than the land will naturally carry: prospectors and
small-scale miners establish camps and temporary settlements, and around these

Fig. xiv–6.   A large mining town in a semi-arid region. Broken Hill, N S W.
(*Photo E. S. Hills*)

the same processes go on. Since 1883 when mining claims were first pegged out
at Broken Hill (New South Wales), the countryside around has suffered from the
ravages of man and animals (Fig xiv-6). Natural fodder plants have been eaten
out, the species disappearing in succession – the more palatable first – and so on
until little fodder of value grows, and the plant communities become reduced to
an ecologically depressed state. The pressure on plants of little value to animals
being low, these spread and become stable, so that the better plants have to
compete both with the animals and with their plant competitors. Under natural
conditions such a system remains permanently debased and little can be done
on a regional scale to remedy this situation. Similar devastation of former plant
ecosystems in the nitrate desert of Chile is described in Chapter xi, and many

other instances could be cited. Where the wealth produced from the mines is great, it may be possible to establish fenced and protected areas in which, aided by artificial seeding, regeneration of a vegetative cover may take place. Aesthetic reasons alone may well justify this, as at Broken Hill (Fig xiv-7), for it will indeed prove expensive on a regional scale, and certainly not economic with regard to any eventual financial return that might be got from well-conducted pastoral pursuits. It is obviously highly desirable, when mining or oil production is to be undertaken, to prevent such despoliation in the first instance, but it will equally be very difficult to control human beings to the extent that is necessary, especially where the indigenous people have little feeling for the aesthetics of nature preservation[1] or for controlled economic exploitation either of mineral wealth or of fodder plants. Enclosed regeneration plots have indeed required armed guards to keep out nomadic pastoralists in North Africa.

The deterioration of land about a mine thus seems to be only partly, and to a minor degree, due to the mine itself. It is the miners, their families and the people who, often unemployed for periods, congregate about the mine and cause most damage. The groups concerned are, then, of two types. Firstly, the technological persons who have entered the region, know little about it and often care less, and the indigenes who, for one reason or another, are attracted there. These latter may soon become conditioned to a new life dependent on the mining community, but their reactions differ according to their circumstances, and often according to age groups. Those who, being landowners and successful in a long-established pattern, see in the mines only a source of interference with their rights, are often antagonistic. Australian pastoralists in New South Wales kept secret the discovery of gold in 1849, fearing the effects on the workers and an influx of undesirable types. Their fears were realised when the news became known in 1851. After some years during which there were certainly many rapid social and economic changes, the pastoral industry remained while the mines gradually closed down. The miners and their families, indeed, added to the wealth of the community both by gold production and by their energy in taking and developing land holdings.

In North Africa, in Arabia, Kuwait, Iran, Iraq and Pakistan, oil and gas fields have had notable economic and social effects in arid, and even in truly desert, areas. In the exploration stages, local merchants are called upon for greatly increased supplies, transport workers, guides and local workmen are in demand, and the expenditure of much money goes on in the district and in the servicing towns. If the venture becomes permanent, the social environment is affected as described, young men in particular finding employment as wage-earners, and often receiving remuneration far greater than that to which they have been accustomed. With enlightened management and government, permanent settlements and adequate housing, schools and medical services, transport and communications are established and the old ways of life can rarely be maintained

---

[1] Jarvis [3] reports that in Palestine in former years the general populace placed no value whatever on the preservation of trees, or even of pasture.

(a)

(b)

Fig. XIV–7. Regeneration areas at Broken Hill, NSW: (a) before regeneration (1936), (b) after regeneration (1962). (*Photo by courtesy Conzinc Riotihto (Aust.) Ltd*)

without some, or even major, changes, the magnitude of which in the social and political spheres is perhaps proportional to the economic gain of the enterprise.

But, apart from the rich bonanzas that oil and mineral deposits provide in favoured spots, the growth of industries in arid lands is likely to be slow, and to take place rather as a result of flanking movements and probing into the most favoured regions, than as a frontal attack on the vast emptiness of the true deserts.

Where action is called for to assist planned projects in transportation, education, administration, tourism or defence; where research, perhaps as yet not performed, offers some hope of gain through the use of cheap land, solar energy or wind power, then industry will move naturally into the desert. Perhaps, with a better appreciation of the economic advantages of the use of water for urban rather than for rural purposes, communities will, in fact, react more favourably to the arid lands than they have in the past – although the great cities that were created in the course of history – some since destroyed while others are still flourishing – remind us that such a realisation was not beyond the powers of the ancients to achieve.

E. S. HILLS

References

[1] KOENIG, L., 'The economics of water sources', *The Future of Arid Lands*, Amer. Assoc. Adv. Sci., Publ. No. 43, 1956, pp. 320–8.
[2] UNESCO, *Courier*, No. 8–9, 1955.
[3] JARVIS, C. S., *Desert and Delta*, London, 1947.

CHAPTER XV

The Use of Arid and Semi-Arid Land

Scope and definitions

The terms arid and semi-arid, as related to land-use, have been used rather loosely and with differing meanings in recent literature on the 'arid zones'. In this chapter, a more precise connotation is proposed for these terms, based upon position in relation to a particular threshold of expected or effective rainfall, a threshold which will vary from one region to another according to the growing season of plants, to the nature of the climate (summer rainfall, winter rainfall, monsoonal, etc.), and to altitude, latitude and other factors. On one side of such a threshold, the conditions are said to be arid, and extensive grazing is the only reliable form of land-use. On the more humid side of the threshold, conditions are said to be semi-arid, and crop cultivation using dry-farming techniques and drought-resistant or drought-escaping crops can be recommended. Naturally this threshold is not precisely along any fixed isohyet or the line of any given expected rainfall. There is a transition zone in which rainfall and therefore crop cultivation are quite unreliable, but into which agriculture is frequently forced under a variety of social and economic pressures.

The geographical location and extent of the arid and semi-arid regions as here defined vary widely according to the seasons, and so also does the type of land-use. For instance, in the *kharif* or monsoon season in the drier parts of India, much of the country is sub-humid to humid; in the *rabi* or winter season the greater part is semi-arid (when drought-resistant crops adapted to lower temperatures may still be grown on light rains or residual soil moisture following the growing of a monsoon crop) or arid (when crop cultivation is not possible and domestic livestock wander indiscriminately over both the cultivated and the grazing land). In the Mediterranean and Near Eastern regions, much of the land that is semi-arid in the winter season of rainfall and low evaporation becomes arid during the dry, hot summer season. Any map of the arid and semi-arid zones should take these marked seasonal variations into account.

This review of the past and present use and, unfortunately, misuse of arid and semi-arid lands and of their potentialities for future development is based primarily on those lands that lie between the Atlantic Ocean in the west and the Gangetic Plain in northern India in the east [1–3]. Digressions to east and central Africa will be made when data appear to be relevant, although realising the great differences in social and economic background. The present status and future potentialities of arid lands are primarily related to questions of economic feasibility and social and agrarian structure. The vast region defined above is sufficiently uniform in these respects to be taken as the sample region of study; it is sufficiently different in these respects from the other major arid regions in Australia and the Americas to make comparisons difficult and generalisations dangerous. Nevertheless it is clear that in those other regions the major factors that affect land-use in arid and semi-arid zones also operate, although there may be regional differences in emphasis. Overgrazing of the poor natural vegetation and exploitive methods of dry farming may be mentioned as two practices which are known to have adversely affected arid and semi-arid lands, albeit under quite different social and agrarian systems, in Australia and the United States of America, and in some regions with permanent deleterious effects. Although a detailed analysis cannot be attempted here, further information may be sought in the bibliography at the end of this chapter.

The history of land-use and of the origins of agriculture and pastoralism is synonymous with our region of study. The agricultural revolution in the Neolithic age took place in a certain area or areas in the Near East which may not have been classifiable as semi-arid at that time but which have now become so, and that not necessarily because of climatic change. The history of land-use is the history of civilisations, how and why they arose, flourished, waned and disappeared, what effect they had on the land that they had inherited from earlier civilisations, and what effect that land had on them [4–6].

The agricultural practices developed in the early days have changed little until very recently. The free-range grazing habits are still of Neolithic vintage. The land of our region probably acquired most of its present physiognomic characteristics, the type, ecological status and density of its plant cover, the extent of erosion and general land deterioration, the evidence of land sickness in the nature and reliability of the river courses, in the last two or three millennia B.C. or in the early centuries A.D. Superimposed upon the damage caused by ignorance and steadily increasing human and livestock populations was the effect of a long sequence of wars and invasions. These not only caused deterioration of the land through the passage and conflict of armies while crops were in the field, through the burning out of enemy forces in the forests and the practice of a scorched-earth policy by armies in retreat before a victorious invader. They also led to confiscation of crops without compensation, and caused disruption of the agrarian economy through increased taxation to maintain the armies or the demand for able-bodied men from the agricultural community to replenish the fighting forces, particularly after campaigns in which there was great loss of life.

All these factors have combined to cause serious, sometimes catastrophic, deterioration of the total land resource. The present vegetative cover is nothing like the original climax vegetation, the highest stage in the ecological succession that may be reached in any particular environment. The cutting of trees for timber and of trees and shrubs for fuel, the free-range grazing and browsing by excessive numbers of uncontrolled domestic animals, the indiscriminate burning of regenerating vegetation to facilitate that grazing, and the unwise cultivation of steep slopes to escape the dangers of flood and malaria in the valleys or merely because of land hunger, all these operating in varying intensities have caused serious erosion and, perhaps even more important, desiccation of the total environment (see Chapter VI).

Oliver West [7] has stated:

The correctness or otherwise of any system devised for the use of farming land should be judged not so much by its immediate economic effect, the money it puts into the pocket of the farmer or the importance of a particular product in the economy of the country, but by the effect it has on the land that is being used. It is not enough that it should be productive but that production should be sustained and consistent with preservation of the land as a permanent resource. This can only be achieved when the farming system is as closely as possible in harmony with the environment; the climate, the soil and the vegetation.

This 'land' about which we talk may be considered both from the technical and from what may be called the philosophical points of view. In the assessment of overall actual and potential resources, consideration has to be given to a number of factors, including vegetation, soil, climate, past and present utilisation, and social and agrarian structure. The classification of land which must be an essential part of any dynamic approach can be regarded as the subdivision of a discrete part of the earth's surface for use in agriculture, grazing or forestry, individually or in combination. It should therefore consider all the factors which affect the growth and reproduction of the plants upon which man and animals are to depend [8]. These factors can be grouped into three more or less independent variables:

Geomorphology
Geology
Climate

and three dependent variables which change in relation to each other and to the three independent variables, namely:

Soil
Water (surface and underground)
Vegetation.

In this sense the concept is of land representing a complex of all the factors on, above and below the earth's surface which affect man's agricultural, pastoral and forestry activities. It is possible to synthesise the data collected by specialists in the disciplines concerned with these six variables and to define land units and systems on the lines indicated in a later section of this chapter.

There is also a philosophy in land-use in arid zones, a 'mystique' among those who can appreciate the 'légerté et lucidité' of the desert environment and way of life. These enthusiasts may be found among the peoples who inhabit the desert and who regard the settled life of the cultivator as worthy only of inferior beings and slaves. There are also the many scientists, explorers and sociologists who have done their life's work in the semi-arid and arid environments of our region of study, and to whom the arguments to be raised here regarding the viability of the agrarian structure of the deserts, settlement of nomads and industrialisation will be anathema. But we must consider the problem of land-use in the arid zones from a realistic point of view, against a background of modern improved living standards, social progress, and the present and future economics of crop and animal husbandry.

It should be stated that the main emphasis in this chapter is on the problems and techniques as they apply to non-irrigated land, cultivated or otherwise. The provision of water for irrigation at once removes the main limiting factor of semi-arid land, and makes possible a type of agriculture which would call for a complete study in itself [2, 9–11]. Irrigated cropping is, however, considered in its relation to the other types of semi-arid agriculture, and particularly to its role in relieving pressures of human and animal populations from the unirrigated areas.

Economic and social aspects

Marginal land is a term applied to any area in which, for reasons of climate, soil, elevation, accessibility or distance from markets, it is a doubtful economic proposition to undertake major and costly measures of amelioration. It is probably true to say that by far the greatest area of marginal land in the world is that in which crop and animal husbandry are risky enterprises because of the lack of water. In any planning of land classification and the consequent developmental actions, the economic feasibility of these must always be borne in mind.

This applies particularly on a long-term basis. Short-term actions on which it is not possible to visualise an immediate economic return may be justified for reasons of local political significance. Some enthusiasts for the Negev still hope to revive the vanished Nabataean agriculture, with its special techniques of water conservation [12, 13]. They wish to prove that it is wrong to conclude that desert farming, in spite of the dramatic achievements in the isolated agricultural communities, is not a realistic instrument for populating the Negev. A few more villages are to be founded for border security and to supply local food needs, but agriculture, including fishing off Eilat, is expected to absorb only 5 per cent of the Negev settlers.

Short-term actions not related to the present cost and productivity of the land are otherwise justifiable only if they create a new environment, in which agricultural production can be expected to become again economic at a higher level in the foreseeable future. The expensive new irrigation schemes in Rajasthan and along the southern semi-arid fringe of the Sahara are cases in point.

Apart from this, it must seriously be questioned whether the present forms of crop and animal husbandry under dry-land conditions as discussed in this chapter can really be regarded as viable components of the national economies in our region. And in spite of any nostalgic or romantic objections to the conclusion that they are not viable, we are faced with the trends of human populations away from the free but rugged life of the desert to the more interdependent but softer and economically more secure conditions of the settled areas. Such movements are already to be seen towards the oilfields, mineral mines and new irrigation projects to the north and south of the Sahara, and towards the mineral cities of the Negev and the new irrigation areas of the Near Eastern countries. It will certainly be seen in a mass movement of the peoples from the semi-arid villages of Rajasthan into the new conditions of seasonal reliability and prosperity to be provided by the Rajasthan Canal and the vast underground reservoir of Jaisalmer.

Such movements away from the desert are inevitable and are in fact desirable up to a point, primarily for sociological reasons. It is much easier to provide housing facilities, health services and schools for settled communities than for nomadic and migratory tribes that are on the move for most or all the year, following the seasonal rains. But it is not proposed to support complete depopulation of the desert. There are those who would argue that the human and livestock populations of the desert fringes are so small in any case and the livestock products so poor in quantity and quality that development schemes are not justified. Those economists who support this view consider that technical assistance and economic investment should be concentrated on the areas that are more favourable in climate and offer the possibility of rapid increases in production, and that the desert peoples with their misguided ideas about freedom and a way of life should be left to fight their inexorable and losing battle with the environment.

It seems wrong to accept this language of defeat in the face of difficulty. By the application of the modern principles and techniques which are becoming increasingly available from the research institutes in the region, it is possible to do much for the management and improvement of arid and semi-arid land. It must be realised, however, that it is not correct to consider this type of land alone. Measures to improve arid and semi-arid land must be related to those proposed for the more favourable environments along the climatic or humidity gradients discussed below. Wherever possible, the production policies in each zone along a gradient should be integrated with those in the other zones. Some countries with a sufficient range of climatic gradients may be able to do this within their own boundaries. Countries such as those along the Sahelian zone in West Africa lie almost entirely within one climatic zone; in their case, international collaboration would be needed with countries to the south.

The drier zones along the climatic gradient in, say, West Africa are best suited for livestock production and could provide meat for the markets to the south with or without fattening and finishing in more favoured zones. In the drier zones, the production of cereals and other food crops cannot be raised much above the subsistence level. One may visualise a two-way movement, of livestock and live-stock products from the arid to the more humid, and of food grains and similar commodities from the humid to the more arid. Australia today affords a very clear example of this.

The main theme of this chapter is the reduction of pressure of human and livestock populations on the semi-arid lands, and their partial transference to more favoured areas where intensive forms of crop and animal husbandry can be practised. While still allowing for optimal utilisation of the arid and semi-arid areas of natural vegetation, this reduction of pressures will permit the vegetation to regenerate and to play more effectively its true role as a protective barrier between the desert and the valuable cultivated land towards the more humid end of the humidity gradient.

The main economic and social objective should be to discover whether the present human and livestock populations, with their expected increases, can be maintained on an adequate economic and nutritional plane. In most cases, this is not possible in the present conditions of haphazard land-use. It needs to be dis-covered whether, by the correct reclassification of the land and the application of modern principles and techniques, a new ecological balance may be achieved between man and his domestic livestock and the environment.

Climate as a controlling and limiting factor

The best potential land-use system for any specific semi-arid environment depends on the relative influence of topography, soil and climate, and the most important of these is climate [14]. In Southern Rhodesia, for example, topography is the limiting factor over 5 per cent of the country, soil over 18 per cent, and climate over 77 per cent. If, however, crop yields vary more from year to year on the same soil than they do on different soils in the same year, this indicates that climate has a greater influence than soil on a much larger area than the·77 per cent mentioned above [15].

Rainfall in eastern Africa is seasonal (summer rain, winter dry) and erratic except in favoured areas such as much of Uganda. Where there is a rainfall expectation of over 30 inches per year, there is a reasonable prospect of harvesting crops. Where the expected rainfall is less than 20 inches per year, crops should not be grown without irrigation, and the ranching of cattle on an extensive basis is the only possible form of economic land-use under dry-land conditions. The zone between 20 and 30 inches provides the best prospects for fairly intensive ranching of beef or dairy cattle. If we take a 25-inch rainfall line as dividing the pastoral areas from the lands of high potential for arable cropping, we find that in Kenya this coincides with the 4,000-foot contour, so that only the coastal strip and the south-western third of the country fall into the 'high potential' category. In Southern Rhodesia, the superior areas lie in the north-eastern third, Uganda has only limited dry areas in the north, but Tanzania is dry at heart, with wetter oases towards the periphery [16].

If one regards the rainfall threshold for the successful growing of grain (the lowest rainfall-demanding crop) in East Africa as 30 inches per annum with a specific degree of reliability, half the land area of East Africa should be classified for pastoral use only (two-thirds of Kenya, one-third of Tanzania and limited areas in Uganda). In more than half the remainder, ecological conditions are of an unreliable and marginal nature, highly susceptible to relatively minor climatic fluctuations. Three-quarters of the population of Kenya live in one-quarter of the country, which is surrounded and interlaced by pastoral areas which suffer not only from bad grazing management, but also from climatic fluctuations [17].

This classification of land based on the primary criterion of rainfall reliability should be carried out in all projects involving land planning and adjustment of use in the arid and semi-arid zones. In this way, it would be possible to define and map the climatic zones along the humidity gradients from deserts to humid lands of high potential, as has been done in a general way for western, eastern and central Africa in Tables xv-1 and 2, and in Figs xv-1 and 2. A similar zonation along humidity or rainfall gradients for India is presented in Table xv-3 and Fig xv-5.

It should be noted, however, that the 25-inch rainfall line found to be the border between the arid and the semi-arid (the pastoral and the cultivable) lands in East Africa may not be generally applicable. This is a summer rainfall region in which crop growth coincides with a period of high evaporation. The borderline will probably be found to be at a lower rainfall level for winter growing crops in a winter rainfall environment. In the most markedly 'Mediterranean' part of the Australian continent, the south-western area of Western Australia, the most significant features of the agricultural environment are the sharply defined wet winters and dry summers, the converse of eastern and central Africa (Fig xv-3). At Perth, for example, 87 per cent of the rain is received between May and October. In this part of Australia, more rain may fall during the winter than can be utilised by plants, and on some soils serious waterlogging may occur. During the summer, on the other hand, there is such a moisture deficiency that shallow-rooted plants are unable to persist without irrigation. This climatic pattern has led to the widespread use of annual rather than perennial sown pastures, and of plants which produce maximum growth and complete their life cycles during the short period when soil moisture is available. But winter rain, falling at a time of low evaporation, is more effective than summer rain; 20 inches of rain concentrated in the cooler part of the year is more effective than the same amount concentrated in the six summer months.

From these considerations have developed the ideas of 'effective rainfall' and 'the length of growing season' where it is seasonal drought and not cold which is the limiting factor (Fig xv-4). Under non-irrigated conditions, the growing season is the period over which, on the average, effective rain may be expected. A map prepared on this basis for East Africa has been published by the East Africa Royal Commission [17]. Fig xv-4, from the *Official Year Book of Western Australia*, shows the length of the effective rainfall season in different parts of

TABLE XV - I *Ecoclimatic gradient in western and western equatorial regions of Africa*

Zones	Approximate rainfall range, cm	Dry season, months	Vegetation (Keay, 1959)	Forest formations	Chief tree species	Grass cover types (Rattray, 1960)
Saharan			Desert	—	—	*Aristida* (12); desert conditions
Sub-Saharan	100–250		Sub-desert steppe; tropical types	—	—	*Aristida* (13, 15) associated with steppe
Sahelian ..	250–600	9	Wooded steppe with abundant *Acacia* and *Commiphora*	Shrub and thorny savannah, Sahel types	*Acacia* *Commiphora*	*Cenchrus* (5, 6, 7) ass. with savannah
Sudanian	600–1,250	6–8	Woodlands, savannahs, undifferentiated, relatively dry types	Open woodland (tree savannah), Sudanese types	*Anogeissus* *Sclerocarya* *Balanites* *Prosopis* *Butyrospermum* *Adansonia* *Bombax*	*Andropogon* (3, 5, 6) ass. with savannah
Guinean	above 1,250	3–6	Woodlands; a. Northern areas with abundant *Isoberlinia doka* and *I. dalzielii* b. Southern areas, undifferentiated, relatively moist types	Seasonal (deciduous) forest, Guinean types	*Isoberlinia* *Berlinia* *Uapaca* *Lophira* *Brachystegia*	*Hyparrhenia* (22, 31, 33, 34, 35, 36, 37) ass. with savannah
Guinea Equatorial	above 1,800	short	Moist forest at low and medium altitudes	Closed rain forest	*Khaya* *Entandrophragma* *Lovoa* *Piptadenia* *Lophira* *Mytragina* *Sacrocephalus* *Aucoumea* *Triplochiton* *Tarretia* *Chlorophora* *Terminalia*	*Pennisetum* (5, 6, 7) ass. with woodland

(see Fig xv-1)

Main types of land-use	Main crops	Livestock	Countries	Zones
Nomadic grazing	—	Some camels	Mauritania, Mali, Niger, Chad, Sudan	Saharan
Nomadic grazing	—	Camels, sheep and some cattle	Mauritania, Mali, Niger, Chad, Sudan	Sub-Saharan
Semi-nomadic grazing with beginnings of semi-arid cultivation. Minor tree products and protection against desiccation	Sorghum and millet, irr. rice	Cattle and sheep	Senegal, Mauritania, Mali, Upper Volta, Niger, Chad, Sudan	Sahelian
Semi-nomadic grazing and arable cultivation with fallows generally of medium to long duration. Woodland utilisation for minor products	Sorghum, millet, groundnuts, yams, maize, irr. rice	Cattle with some sheep and goats	Senegal, Mali, Upper Volta, Niger, Northern Nigeria, Chad, Sudan	Sudanian
Arable cultivation with fallows generally of medium to long duration. Forest utilisation for local consumption. Forest plantations	Sorghum, millet, groundnuts, cassava, yams, maize, irr. rice	Cattle depending upon absence of tsetse, goats and some sheep	Senegal, Gambia, Portuguese Guinea, Guinea, Mali, Ivory Coast, Upper Volta, Ghana, Togo, Dahomey, Nigeria, Cameroon, Chad, Central African Rep., Sudan, Congo (Brazzaville), Congo (Leopoldville)	Guinean
Arable cultivation with fallows generally of relatively long duration. Permanent tree crops for export. Forest utilisation for export and local consumption (natural stands and plantations)	Upland rice, plantains and bananas, yams, taro, maize, oil palm, cacao, rubber, robusta coffee	Goats, with cattle in forest–savannah mosaic on coast	Sierra Leone, Liberia, Ivory Coast, Ghana, Nigeria, Cameroon, Rio Muni, Gabon, Congo (Brazzaville), Cabinda, Congo (Leopoldville)	Guinea Equatorial

Zones	Vegetation (Keay)	Forest formations	Chief tree species	Grass cover types (Rattray, 1960)	
Eastern Equatorial desert and sub-desert	Desert. Sub-desert steppe; tropical types	—	—	*Chrysopogon* (5); desert conditions	
Eastern Equatorial savannah	Wooded steppe with abundant *Acacia* and *Commiphora*	*Acacia–Commiphora* woodland (Sahelian Zone in West Africa)	*Acacia Commiphora Dobera Salvadora Boscia Phyllanthus Euphorbia*	*Chloris* (1) ass. with tree steppe *Chrysopogon* (1, 2, 3, 4, 6, 6A) ass. with tree steppe	
Eastern Coastal	Coastal forest–savannah mosaic			*Panicum* (1, 2) ass. with woodland *Hyparrhenia* (5) ass. with woodland	
Lake Victoria	Forest-savannah mosaic			*Eragrostis* (9) ass. with savannah *Hyparrhenia* (25, 26, 27) ass. with savannah *Hyparrhenia* (23) ass. with woodland	*Panicum* (2, 3) ass. with savannah *Themeda* (12, 15) ass. with savannah
Eastern and Central Plateau	a. Woodlands of various types with *Brachystegia* and *Julbernardia*, etc.	Miombo and Mopane woodlands	*Brachystegia Isoberlinia Julbernardia Oxytenanthera Pterocarpus Colophospermum*	*Aristida* (4) ass. with woodland *Aristida* (5, 6, 7) ass. with savannah *Eragrostis* (6, 7, 8) ass. with savannah *Heteropogon* (1) ass. with savannah	*Loudetia* (2) ass. with grassland *Loudetia* (1, 3, 4) *Loudetia* (5, 6, 7) ass. with savannah *Pennisetum* (1, 2); grassland
	b. *Acacia, Combretum* and *Terminalia* in savannah woodlands			*Hyparrhenia* (3, 5, 6, 10, 15, 17, 23) ass. with woodland *Hyparrhenia* (13, 14, 17B, 24); grassland	*Themeda* (12, 13) ass. with savannah *Themeda* (8, 9, 10, 11); grassland
	c. *Cryptosepalum pseudotaxus, Baikiaea, Marquesia,* and *Guibourtia coleosperma* on Kalahari sands, and			*Hyparrhenia* (4, 5A, 9, 12, 15A, 16, 17A, 19, 21); ass. with savannah	
	d. *Colophospermum mopane* abundant				
Eastern African Highlands	Montane grassland and forest			*Heteropogon* (2) ass. with savannah *Hyparrhenia* (30) ass. with savannah *Pennisetum* (1, 2) ass. with woodland	*Pennisetum* (3) ass. with savannah *Setaria* (3) ass. with woodland *Themeda* (16) ass. with savannah

Fig xv-2)

Main types of land-use	Main crops	Livestock	Countries	Zones
Nomadic grazing	—	Camels, sheep and goats – some cattle } Cattle, sheep and goats	Somalia, Ethiopia N. Territory of Kenya	Eastern Equatorial desert and sub-desert
Nomadic grazing and cultivation generally subsidiary to herding. Minor tree products and protection against desiccation, wildlife	Millet, sorghum	Cattle, sheep and goats	Somalia, Ethiopia, Sudan, Kenya, Tanzania	Eastern Equatorial savannah
Arable cultivation with varying periods of fallow	Rice, cassava, coconut	Few cattle and goats	Somalia, Kenya, Tanzania, Mozambique	Eastern Coastal
Grazing and arable cultivation with variable but generally short fallow periods. Tree crops mainly coffee	Plantains, bananas finger millet, sorghum, robusta coffee, cotton	Cattle and goats	Kenya, Uganda, Tanzania	Lake Victoria
a. Arable cultivation with variable but generally long fallow periods. European farming and tobacco planting	a. Finger millet, sorghum, cassava, maize, flue-cured and Turkish tobacco. Limited sown pastures in European farming	a. Few livestock of all classes – limited by tsetse	Tanzania, Congo (Leop.), Angola, Zambia, S. Rhodesia, Malawi, Mozambique	Eastern and Central Plateau
b. Grazing and arable cultivation with variable but generally short fallow periods. European maize, cattle and mixed farming	b. Maize, sorghum, sown pastures in European farming	b. Cattle with some sheep and goats depending on absence of tsetse		
c. Grazing and arable cultivation with variable but generally long fallow periods on the sands, and short fallow periods and permanent cultivation on associated alluvial and peat soils	c. Bulrush millet, cassava, maize, finger millet	c. Cattle with some sheep and goats depending on absence of tsetse		
d. Ranching and grazing in tsetse-free areas. Arable cultivation mainly on associated alluvial and other soils.	d. Maize, sorghum, bulrush millet	d. Cattle with some sheep and goats		
Grazing and arable cultivation with short fallow periods and permanent cultivation of bananas and plantains in moist forest zones. European cattle ranching, mixed farming and coffee planting	Maize, sorghum, plantains and bananas, coffee, pyrethrum, tea, limited sown pasture	Cattle predominant, goats and some sheep	Ethiopia, Kenya, Rwanda, Burundi, Tanzania, Congo (Leop.), Malawi, S. Rhodesia	Eastern African Highlands
Woodland utilisation for local consumption. Forest plantations				

Zones	Approximate rainfall range, cm	Dry season, months	Vegetation	Forest formation	Chief tree species
Arid	30 and below	10–11	Desert	Scrub	a. *Acacia jacquemontii* *Calligonum polygonoides* *Calotropis procera*
Semi-arid	30–60	7–9	Thorn forests	Open scrub woodland	(i) *Prosopis spicigera* *Salvadora oleoides* *Acacia leucophloea* *Acacia arabica* *Capparis decidua* *Balanites aegyptiaca* *Zizyphus nummularia*
					(ii) *Acacia senegal* *Anogeissus pendula* *Cordia rothii*
					(iii) *Acacia leucophloea* *Acacia arabica* *Acacia latronum* *Capparis* and *Zizyphus* spp.
					(iv) *Acacia planifrons* *Acacia latronum* *Albizzia amara*
				Savannah woodland	(v) *Pterocarpus marsupium* *Anogeissus latifolia* *Acacia latronum* *Albizzia amara*
Dry sub-humid	60–120	6–8	Woodlands, savannahs	Open dry deciduous types (i) Dry teak	*Tectona grandis* *Adina cordifolia* *Emblica officinalis*
				(ii) Dry *sal*	*Shorea robusta* *Pterocarpus marsupium* *Terminalia tomentosa* *Anogeissus latifolia*
Moist sub-humid	120–160	5–8	Open woodlands with eastern areas covered by *sal* and western areas by teak and *Terminalia*	Seasonal forests with mixed ever-green and deci-duous species (i) Moist teak	*Tectona grandis* *Terminalia crenulatae* *Terminalia paniculata* *Terminalia bellerica* *Dalbergia latifolia* *Lagerstroemia lanceolata* *Grewia tiliaefolia*
				(ii) Moist *sal*	*Shorea robusta* *Lagerstroemia parviflora* *Dillenia pentagyna* *Pterocarpus marsupium* *Syzygium cumini* *Schima wallichii* *Terminalia tomentosa*
Per-humid	160–200	2–6	Moist forest at low and medium altitudes	Evergreen *seasonal* forest (western form)	*Actinodaphne angustifolia* *Olea dioica* *Pouteria tomentosa*

Grass cover	Main types of land-use	Main crops	Livestock	Indian State
Lasiurus sindicus–Cenchrus spp.*–Panicum* spp. (*P. turgidum* and *P. antidotale*)	Nomadic and semi-nomadic grazing, dry farming	Millets, sorghum, barley	Camels, sheep, goats, cattle and donkeys	Rajasthan
Cenchrus–Dichanthium annulatum, Eremopogon foveolatus, Eleusine compressa–Dactyloctenium sindicum, Aristida	Nomadic and semi-nomadic grazing, intensive dry farming, utilisation of trees and shrubs for grazing, also cultivation with well irrigation, dairy industry (northern India), fallows of 3 to 7 years' duration	Lesser millets, (Eleusine) millets, sorghum, barley, cotton, oil seeds, some wheat	Camels, sheep, goats, cattle, some buffaloes, donkeys (note: goats are for mohair and mutton)	Rajasthan, Punjab, Gujarat, Saurashtra, Maharashtra and Madhya Pradesh
Cymbopogon jwarancusa–Sehima nervosum	Intensive dry farming, multiple cropping, free grazing, utilisation of forests and minor products for fuel and timber, fallows of 2 to 3 years (southern India)	As above, besides groundnuts, cotton, irrigated rice, sugar cane, chillies, wheat and pulses	Sheep, cattle, goats, some buffaloes	Maharashtra, Mysore, Andhra, Madras
Dichanthium annulatum–Themeda quadrivalvis, Pseudanthistiria heteroclita				
Heteropogon contortus, Aristida redacta, Aristida funiculata				
Cymbopogon coloratus–Themeda triandra				
Cymbopogon coloratus–Heteropogon contortus–Themeda triandra, Iseilema laxum, Chrysopogon montanus, Aristida spp.	Dry and irrigation farming, seasonal fallows, horticultural crops, utilisation of forest products, forest grazing for local consumption and export	Millets, sorghum, maize, cotton, tobacco, oil seeds, wheat, pulses, mangoes, citrus	Cattle, buffaloes, goats, some sheep, horses, donkeys	Punjab, Madhya Pradesh, Gujarat, Maharashtra, Andhra, Madras, Uttar Pradesh
Chrysopogon montanus Bothriochloa intermedia Eulaliopsis binata				
Pseudanthistiria heteroclita Themeda quadrivalvis Dichanthium annulatum (in plains)	Some shifting cultivation, forest utilisation on scientific basis, arable cultivation with seasonal fallows, forest grazing, horticultural crops	Rice, jute, sugar cane, pulses, tobacco, mangoes, bananas	Cattle, buffaloes, goats, some sheep	Maharashtra, Mysore, Orissa, Bihar, U.P. and Bengal
Arundinella pumila Sehima nervosum (on the hills)				
Imperata cylindrica, Narenga porphyricoma, Saccharum spontaneum, Themeda arundinacea, Microstegium ciliatum				
Arundinella–Cymbopogon	Shifting cultivation, intensive cultivation, short fallows, forest utilisation	Jute, rice, pulses, wheat, sugar cane, cinchona, rubber (cashewnut in W. Ghats), coffee (Coorg)	Cattle, buffaloes, goats	Maharashtra, Mysore, Kerala, Assam and Bengal

TABLE XV - 3   *Ecoclimatic gradients in India (continued)*

Zones	Approximate rainfall range, cm	Dry season, months	Vegetation	Forest formation	Chief tree species
				(eastern form)	*Cinnamomum cecidodaphne* *Amoora wallichii* *Mesua ferrea* *Altingia excelsa* *Manlietia insignis* *Artocarpus chaplasha* *Eugenia* spp. *Tetrameles nudiflora* *Stereospermum chelonioides*
Wet	300–400	2–4	Wet forests at medium altitudes; optimum type	Closed rain forest (western form)	*Poeciloneuron indicum* *Palaquium ellipticum* *Mesua ferrea* *Dipterocarpus indicus* *Hopea parviflora* *Cullenia excelsa* *Vateria indica* *Calophyllum elatum*
	200–300	2–3	Wet forests at medium altitudes; optimum type	(eastern form)	*Dipterocarpus pilosus* *Artocarpus chaplasha* *Artocarpus heterophyllus* *Mesua ferrea* *Cinnamomum cecidodaphne* *Altingia excelsa*

Grass cover	Main types of land use	Main crops	Livestock	Indian State
Imperata–Saccharum–Themeda				
Alpinia–Phragmites–Saccharum				
No grass in the undisturbed forest	Shifting cultivation, spice gardens, forest utilisation for timber and forest products, plantation crops, horti-cultural crops	Hill rice, tapioca, cardamum, pepper, tea, rubber, bananas, jack fruits, coconuts in the coast	Cattle, buffaloes, goats, pigs, horses	Mysore, Kerala, Assam
No grass in the undisturbed forest				

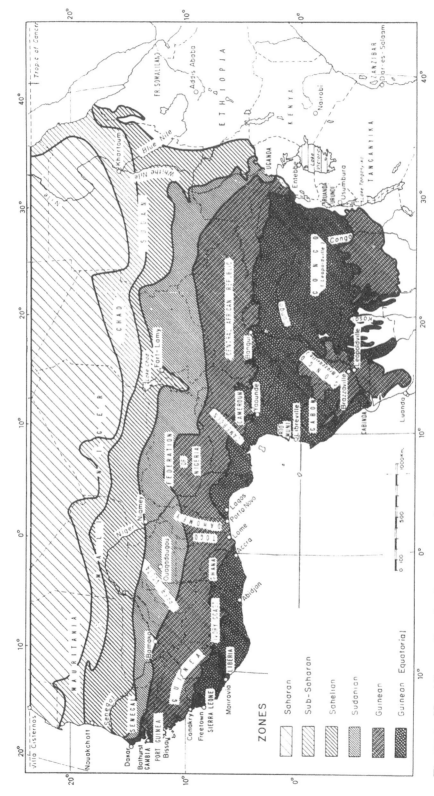

Fig XV-I. Humidity gradient in the western and western equatorial region of Africa. (*FAO Conference Document, 1961*) – see Table XV-I

the agricultural region of that State. It may be seen that at Albany soil moisture is adequate for 9 months of the year, while at Southern Cross the corresponding period is only 3 months. Owing to more uniform distribution of rain over the year, Esperance with an annual rainfall of only 26 inches has a longer growing

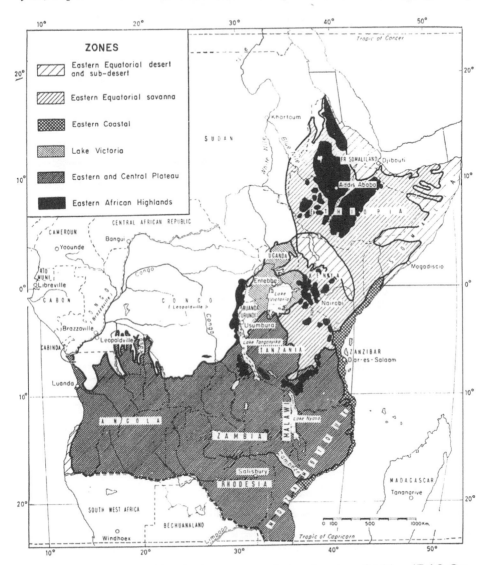

Fig xv-2. Humidity gradient in the eastern and central region of Africa. (*FAO Conference Document, 1961*) – see Table xv-2

season than Perth, where the rainfall is 35 inches [18]. There is a great need for similar analyses and mapping to be done for the Mediterranean and Near Eastern regions of comparable climate.

The degree and frequency of climatic fluctuations are of great significance to the user of semi-arid land [19]. The plant ecologist and agriculturist are probably correct in considering four broad types of alterations in climate which vary in their biological and economic significance:

(i) major changes, into or out of ice ages or pluvials,
(ii) minor changes which persist for 100 to 300 years,
(iii) variations or trends which are experienced for 10 to 50 years, and
(iv) changes induced by the action of man.

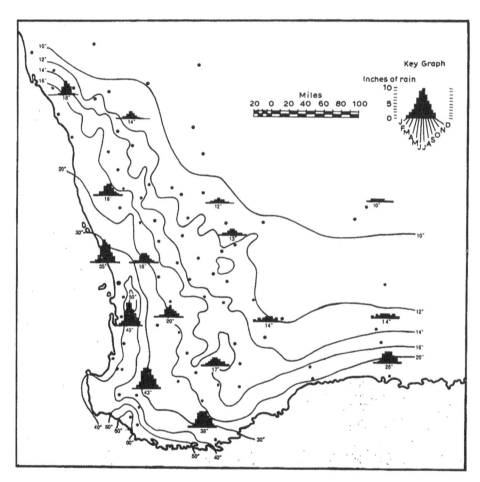

Fig XV-3. The agricultural region of Western Australia has a comparatively reliable rainfall concentrated mainly in the winter. Isohyets of yearly average totals and the seasonal distribution at a number of towns are shown here [18]

Only types (ii) and (iii) will be considered here; type (i) is not relevant and type (iv) is discussed elsewhere.

The second type of climatic change calls for the archaeologist to use his own specialised criteria and methods of deduction and correlation to decide, for example, whether the fall and regeneration of primitive communities or a change in the type of farming or of the crops that were grown could be connected with climatic change. The agricultural historian would produce evidence of the effect of these minor changes on land-use, agriculture and the production of and trade in commodities. There is no doubt that climatic changes or fluctuations over periods of 100 to 300 years, if of sufficient magnitude, must have had a profound effect on the economic life and products of the people. Many of the crops of the desert fringes are grown perilously near the lower threshold of their moisture

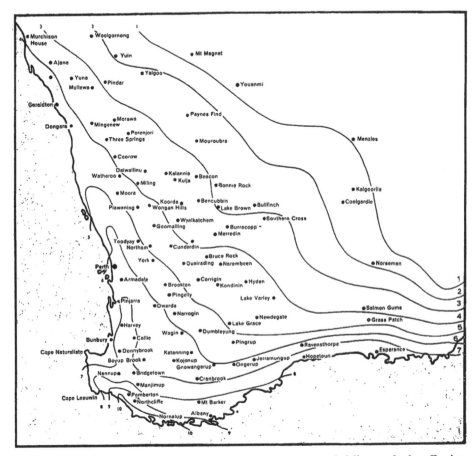

Fig xv-4. The number of months during which the average rainfall exceeds the effective rainfall is indicated. The length of this period – which may be called the 'growth season' – is an important determinant of the agricultural pattern [18]

requirement. If the rainfall and/or the available water resources should have fallen below that threshold value and remained below it for a long period, large-scale abandonment of land and movements of peoples would have inevitably followed, to start a chain reaction of migration and conflict along the humidity gradient [20].

The main emphasis in the present context should, however, be placed on the third type of alteration of climate, the variations or trends which are experienced for periods between 10 and 50 years, the two or three drought years in ten that are part of the vivid experience and memory of one or two human generations. These variations are generally smaller in scale than the preceding type, with respect both to duration and meteorological parameters. Although they are more transitory in their effects than the long-term fluctuations, they may nevertheless have a marked influence on land-use, agricultural systems and crop acreage if a threshold for any of these is crossed.

This question of threshold values is very important in land planning, particularly in the difficult and critical environments of the arid zone. The distribution of land-use systems in relation to the humidity gradient described below is based on a number of thresholds, above or below which a particular practice or crop is or is not possible. It is the realm of collaboration between the agroclimatologist on the one hand, with his calculation and mapping of rainfall reliability and expectation of high or low temperatures, and the plant physiologist, agronomist and ecologist on the other. By a better knowledge of the range of tolerance of wild and cultivated plants, by a careful choice of cultivars, and the adoption of special agronomic practices, the latter specialists may do much to reduce the economic and social results of inevitable climatic fluctuations.

Land-use along humidity gradients

Let us now be more specific with regard to the systems of land-use and husbandry that may be adopted at the different stages along the climatic or humidity gradient. African examples are given in Tables xv-1 and 2 and in Figs xv-1 and 2, and Indian examples in Table xv-3 and Fig xv-5.

This gradient may also be expressed in one example as follows, with particular reference to the development as soon as possible of livestock production units showing a certain balance between natural range and cultivated fodder crop, and the maximum integration of crop and animal husbandry.

(a) In zones which have a low rainfall and where this varies widely between seasons and is distributed in any one season in an erratic manner geographically, various forms of nomadic and migratory grazing are practised. This is the only method of utilisation where the areas to be covered are so vast and the seasonal location of the grazing so variable geographically that ranching within a prescribed area is impossible. Livestock show marked seasonal fluctuation in weight and condition, and the flocks and herds suffer severe losses in extreme years.

(b) The next stage is where the rainfall becomes more reliable in annual amount and geographical distribution, but long dry seasons may still be expected. The grazing resource becomes sufficiently reliable to maintain animals of low to medium productivity on a ranching system within prescribed limits of land. This is already practised or proposed for lands on the semi-arid fringe in Chad, the Sudan and West Pakistan. The inevitable seasonal fluctuations in weight and condition may be eliminated by the feeding of conserved fodder or concentrates, depending upon availability and cost in relation to the expected return from animal produce (see Table XV-4). It is in this zone that the first cereals, such as the very drought-resistant *Pennisetum typhoides*, may be grown under dry-land conditions.

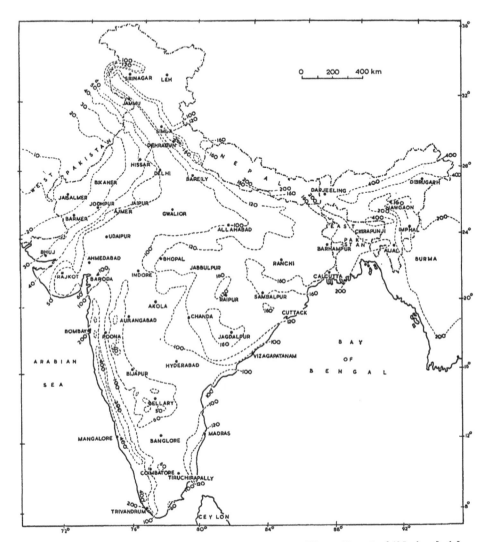

Fig XV-5. Rainfall map of India. Isohyets are in cm. (*From Map 6 of 'National Atlas of India', preliminary edition*) – see Table XV-3

(c) As the rainfall increases in annual amount and becomes more reliable, and the dry season is shorter, ranching units which incorporate a small proportion of good land for the cultivation of annual fodders become possible. There is an important threshold here; on the drier side, yields of fodder are so low that they are not worth growing and the feeding of purchased concentrates is economic; on the more humid side, yields of fodder become sufficiently high to justify their cultivation and feeding in place of purchased concentrates. The loss of weight and condition in livestock is less than in the first two zones, and calving intervals are reduced.

(d) The next stage permits a progressive increase in the cultivated fodder acreage in relation to the uncultivated grazing land, the cultivation of perennial fodders or even the sowing of pastures, and the practice of rotation with food and cash crops, but still with a considerable acreage of natural grazing land in the farming unit, or available on a commercial basis nearby. It now becomes possible to keep better animals requiring a higher plane of nutrition.

Figs xv-6 and 7 give an excellent example of the improvement in management and economic return which is possible on the more humid side of the threshold referred to in paragraph (c) above. At Matopos in Matabeleland, Southern Rhodesia, it is uneconomic to grow fodder crops under the 15-inch rainfall, and purchased concentrates or urea/molasses blocks are in order. Proceed north to the environs of Salisbury with a 33-inch rainfall and the whole situation is changed. From the former practice on a maize farm near Salisbury indicated in Fig xv-6, with grazing of the veld during the summer rainfall period, it has been possible to change (Fig xv-7) to winter grazing of the veld supplemented by winter legumes for protein, with well-managed pastures sown with superior species providing the summer grazing. Thus it has become possible to increase the annual return from beef on this 2,000-acre maize farm from £180 to £2,700 without reducing the production of maize.

(e) Finally there are the most favoured areas where rainfall in one or both of the annual cropping seasons is adequate and reliable or where irrigation facilities are available, where most of the land can be cultivated. Since these zones are usually the most densely populated, holdings now become small, and there is room at best for a short-term annual fodder crop in the characteristic mono-cultural systems.

TABLE XV - 4 *Unit price of protein in some northern Nigerian feeding stuffs* [21]

Feeding stuff	Source	Market price	Cost of lb crude protein	Cost of lb digestible protein
Eragrostis tremula hay	Maiduguri	2-lb bundle for 3d.	2s. 3d.	6s.
Mixed grass hay	Maiduguri	70-lb bundle for 5s.	1s. 7d.	5s. 2d.
Groundnut haulms	Katsina	1d. per lb	10d.	1s. 4d.
Velvet bean hay	Govt. farm	£7 per ton (long)	9d.	1s. 6d.
Silage made from *Andropogon gayanus* and *Pennisetum pedicellatum*	Collected from villages	£2 per ton	1s. 9d.	6s. 1d.
Maize–cowpea silage	Govt. farm	£3 per ton	7·5d.	1s. 7d.
Groundnut cake	Kano Oil Mill	£22 per ton	4·6d.	5·6d.
Cottonseed	Ginnery	£7 per ton	3·1d.	5·6d.

One may in a very general way say that the type of livestock production that can be maintained and hence the economic end product (if any) depend upon the place along the humidity gradient occupied by a region under investigation. The

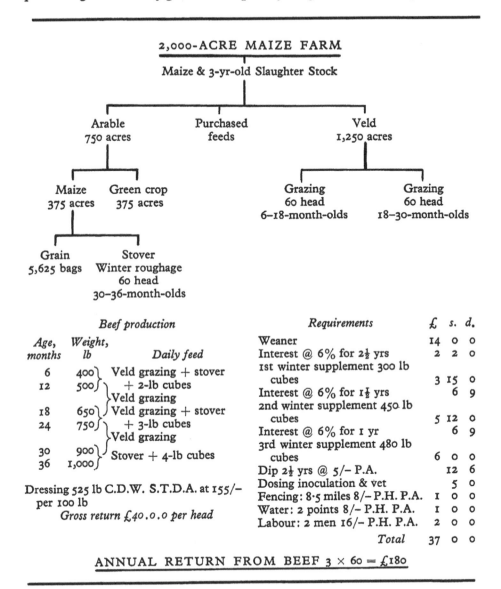

2,000-ACRE MAIZE FARM

Maize & 3-yr-old Slaughter Stock

| Arable | Purchased | Veld |
| 750 acres | feeds | 1,250 acres |

Maize	Green crop		Grazing	Grazing
375 acres	375 acres		60 head	60 head
			6–18-month-olds	18–30-month-olds

Grain — 5,625 bags

Stover — Winter roughage — 60 head — 30–36-month-olds

Beef production			*Requirements*	£	s.	d.
Age, months	*Weight, lb*	*Daily feed*	Weaner	14	0	0
			Interest @ 6% for 2½ yrs	2	2	0
6	400	Veld grazing + stover	1st winter supplement 300 lb			
12	500	+ 2-lb cubes	cubes	3	15	0
		Veld grazing	Interest @ 6% for 1½ yrs		6	9
18	650	Veld grazing + stover	2nd winter supplement 450 lb			
24	750	+ 3-lb cubes	cubes	5	12	0
		Veld grazing	Interest @ 6% for 1 yr		6	9
30	900	Stover + 4-lb cubes	3rd winter supplement 480 lb			
36	1,000		cubes	6	0	0
			Dip 2½ yrs @ 5/- P.A.		12	6
			Dosing inoculation & vet		5	0
			Fencing: 8·5 miles 8/- P.H. P.A.	1	0	0
			Water: 2 points 8/- P.H. P.A.	1	0	0
			Labour: 2 men 16/- P.H. P.A.	2	0	0
			Total	37	0	0

Dressing 525 lb C.D.W. S.T.D.A. at 155/- per 100 lb

Gross return £40.0.0 per head

ANNUAL RETURN FROM BEEF 3 × 60 = £180

Fig xv-6. The effect of a change in farming practice on a maize/beef cattle farm near Salisbury, Southern Rhodesia

production of wool, mohair, hides and skins belongs primarily to zones (*a*), (*b*) and (*c*), the production of second-quality beef to zones (*b*), (*c*) and (*d*), and the production of first-quality beef, fat lambs and cow or buffalo milk to zones (*d*) and (*e*).

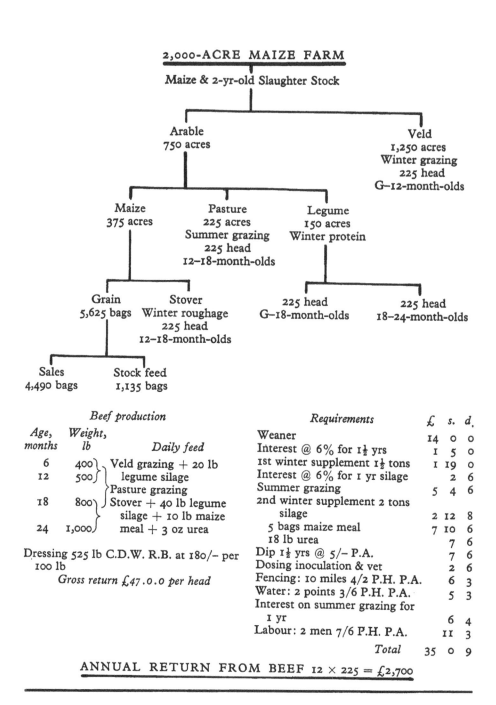

2,000-ACRE MAIZE FARM

Maize & 2-yr-old Slaughter Stock

Arable
750 acres

Veld
1,250 acres
Winter grazing
225 head
G–12-month-olds

Maize
375 acres

Pasture
225 acres
Summer grazing
225 head
12–18-month-olds

Legume
150 acres
Winter protein

Grain
5,625 bags

Stover
Winter roughage
225 head
12–18-month-olds

225 head
G–18-month-olds

225 head
18–24-month-olds

Sales
4,490 bags

Stock feed
1,135 bags

Beef production

Age, months	Weight, lb	Daily feed
6	400 ⎱	Veld grazing + 20 lb
12	500 ⎰	legume silage
		⎱ Pasture grazing
18	800 ⎱	⎰ Stover + 40 lb legume
		silage + 10 lb maize
24	1,000 ⎰	meal + 3 oz urea

Dressing 525 lb C.D.W. R.B. at 180/– per 100 lb

Gross return £47.0.0 per head

Requirements	£	s.	d.
Weaner	14	0	0
Interest @ 6% for 1½ yrs	1	5	0
1st winter supplement 1½ tons	1	19	0
Interest @ 6% for 1 yr silage		2	6
Summer grazing	5	4	6
2nd winter supplement 2 tons silage	2	12	8
5 bags maize meal	7	10	6
18 lb urea		7	6
Dip 1½ yrs @ 5/– P.A.		7	6
Dosing inoculation & vet		2	6
Fencing: 10 miles 4/2 P.H. P.A.		6	3
Water: 2 points 3/6 P.H. P.A.		5	3
Interest on summer grazing for 1 yr		6	4
Labour: 2 men 7/6 P.H. P.A.		11	3
Total	**35**	**0**	**9**

ANNUAL RETURN FROM BEEF 12 × 225 = £2,700

Fig xv-7. The effect of a change in farming practice on a maize/beef cattle farm near Salisbury, Southern Rhodesia

In the 'Agro-Ecological Survey' of Southern Rhodesia, the land has been classified in five natural regions characterised principally by climate (effective rainfall). These *natural regions* are broken up into *natural areas* which, although differentiated within a given region primarily on the basis of soil characteristics, are, however, not fundamentally soil units. Each represents a particular combination of the natural characteristics of climate, soil and relief, which finds expression agriculturally in a particular farming system best suited to it [15]. Here we again have the humidity gradient, but stated in the reverse order. This type of classification into regions on the basis of effective rainfall is highly desirable for all countries in our region of study, to indicate where potentialities for development still exist, and where a type of husbandry has been extended too far into environments of hazard. Unfortunately, meteorological data are too often inadequate.

Region I is the high-rainfall mountainous, eastern border which, because of its rather exceptional climate and highly dissected topography, has a very diversified land-use potential.

Region II enjoys a high and reliable rainfall and is suited for intensive farming based on crop production.

Region III, where the rainfall is less effective and reliable, has a greatly reduced cropping potential and is suited to semi-intensive farming, in which the main farming enterprise is livestock assisted by fodder crops, with cash cropping as a sideline.

Region IV has a fairly low rainfall, and here the main enterprise must be based upon livestock production assisted by fodder crops above a certain threshold.

Region V has a rainfall which is too low and erratic for any type of crop production, cash or fodder, and here utilisation must be based on the productivity of the veld alone.

TABLE XV - 5 *The acreage of the Natural Regions in Southern Rhodesia and their percentage contribution to the total acreage of the country*

Natural Region	Total acreage	Percentage
I	1,515,500	1·56
II	18,144,200	18·68
III	16,683,600	17·16
IV	32,020,200	32·90
V	25,805,800	26·60

Thus we find a situation common to all arid and semi-arid land, that as the rainfall increases, so does the intensity of use and the proportion of cultivated land. In Region II, productivity can be increased by the use of the plough and there is little doubt that the full potential of this region in Southern Rhodesia will not be reached until all land which could be arable is broken up and is under crops and established pasture. The importance of ploughing in increasing the productivity of the land decreases rapidly in Regions III and IV, until in Region V ploughing no longer has any importance, and use must be based on the production of the natural vegetation harvested by the grazing animal. Because established pastures or reseeded range go hand in hand with ploughing, it follows that the importance of established pastures decreases with effective rainfall, while the importance of natural pasturage rises, until in Region V natural pastures are all-important and the only source of production from the land.

Thus the correct management and conservation for the purpose of various extensive forms of animal husbandry represent the most important problems in the use of arid and semi-arid land. They will be discussed in the following parts of this chapter.

Conservation and management of vegetation

The trees, shrubs and grasses of the arid and semi-arid zones have always been free for the taking, and have suffered accordingly. This precious natural resource so well adapted to the critical environment in which it grows has been squandered. Since knowledge regarding the ecology of vegetation and its capacity to protect the land and conserve soil and water has only recently been acquired, the users of land have through the ages lived off this resource without doing anything to conserve it, nor appreciating why they should. The social and agrarian structures of the Near East have evolved and become permanent features on the understanding that this resource would continue to be freely available indefinitely. It is the responsibility of the modern land-use planner to undertake the extraordinarily difficult task of educating the people to value and conserve this resource on an individual or communal basis.

The specialist in conservation cannot on his part afford to be an unpractical idealist. He must admit that timber has to be found for building, fuel for domestic and other use, grazing and browsing for livestock, and land for the cultivation of crops. What he does say is that man and his animals are now completely out of equilibrium with their environment, and that everything possible must be done to find a new balance before it is too late, and no matter how unpalatable this may be to politicians, administrators and the users of land. Pereira [22] has stated the problem with regard particularly to the ranching areas of east, central and southern Africa.

Ranchlands of many semi-arid tropical countries are characterized by erratic rainfall distribution with violent convectional rainstorms, by high land-surface temperatures and by annual evaporation totals of two to four times the total annual rainfall. In these conditions the management of rangeland for optimum grazing and minimum soil erosion presents severe difficulties. Factors which may be critical for successful land-use include control of useless scrub and maintenance of the more valuable species for both grazing and browsing, assessment and control of stocking rates, demarcation of paddocks and organization of rotational and deferred grazing, use of fire and protection from fire. . . . Primitive pastoral tribes (in the Karamoja District of Uganda, for example), not yet completely dissuaded from inter-tribal cattle raiding, have scarcely begun to understand the problem, and are progressively destroying their environment by overgrazing. Torrential debris-laden floods repeatedly wash out roads and bridges, and then cease abruptly leaving dry water courses. Dry-season water supplies are scarce while the productivity of the grazing is critically limited by a very few inches of rainfall accepted by the soil surface and stored for use in the hot dry seasons. How can the present disastrous misuse of grazing land be corrected to give better grazing and less erosion? How much of the rainfall can be accepted and stored in the truncated subsoils which are all that remain on great areas of battered scrubland? How far will this better land-use reduce the destructive torrents? Can it prolong them into quieter streamflow for the dry season?

Only about 4 per cent of the land area of East Africa can expect to receive in 4 years out of 5 as much as 50 inches of rain a year, and it is only from this very small proportion of East Africa that perennial rivers will normally originate, as land covered with actively growing vegetation is likely to use at least this amount of water annually. Further, 55 per cent of the land is unlikely to receive 30 inches of rain in 4 years out of 5, so that in this large area perennial rivers are almost literally the life blood of the country . . . and water is one of the most valuable crops that can be harvested from this small area of high rainfall land [23].

Within our overall region of study, various sectors of the whole community of land users, past and present, have been blamed for the loss of ecological equilibrium. It must be remembered that this blame should not be placed only on the primitive cultivator who has destroyed the vegetative cover by clearing land for his crops, nor on the animal grazier who has permitted his flocks and herds to reduce the plant cover to an ecologically lower level, and who has also practised uncontrolled burning of natural grasslands or shrub communities with the intention of providing more palatable or more accessible grazing. Even before their arrival upon the scene, the hunters used fire, probably on a very large scale, to concentrate their game and to make their territories more open and suitable for future operations. The history of devegetation is, therefore, closely associated with the history of fire, the history both of the burning of natural vegetation and of man's use of fire since he developed it as a major part of his culture [24]. Long ago, the fuel was brought to the brick-kilns of Mohenjodaro by bullocks and donkeys. Now the motor trucks radiate out further and further from Riyadh and Teheran, and down into Rajasthan from the populous areas in the Punjab, to make the search for fuel more thorough and the devegetation more complete.

The definition and ecological history of desiccation call for a thorough review. One of the most serious but seldom mentioned consequences of human activity is the deterioration of microclimates to a more arid level following the disappearance of the woody plants and the reduction in the average height of the herbaceous cover. The action of wind on plants and soils becomes much more marked, temperature variations on and below the soil surface become greater, infiltration of water into the soil is affected, evaporation is more intense, and the more xerophilous species replace the more mesophilous. This deterioration in microclimates undoubtedly explains why many species, particularly the trees and shrubs, are no longer reproduced and why it takes so long for vegetation to regenerate after centuries of misuse. Perhaps even more important is the reduction in effectiveness of the desert fringe types of vegetation in acting as a protective barrier between the desert on the one hand and the more mesophilous

Figs xv-8 to 20. *Rainfall gradient on sandy soils*

Fig xv-8. Site 61/88. (Site numbers refer to the Report: *Observations sur les sols et la végétation en Mauritanie du SE et sur la bordure adjacente du Mali* by P. Audry and Ch. Rossetti, FAO, Rome, 1962.) Approximate position 15° 50′ N/9° 25′ W, Western Hodh. Annual rainfall (summer, one season only) about 500 mm. Savannah with *Andropogon gayanus* and *Hyparrhenia dissoluta* and scattered small trees (*Combretum glutinosum, Sclerocarya birrea*). These physiognomic types which contain Sudanese floristic components represent intrusions of formations belonging to more humid regions. They are often localised on flattened dunes. A red-brown soil has been identified in this sand. The annual production from the herbaceous layer may be estimated at 2,500 kg of dry matter per hectare. (*Photo Ch. Rossetti*)

types of sub-humid and humid vegetation on the other. It would be interesting to see what would happen if it were possible to regenerate the *Acacia/Commiphora* forest of the Sahelian zone, or to discontinue the lopping for fodder of *Prosopis spicigera* in Rajasthan.

Charles Rossetti, formerly of Unesco, has prepared a photographic record of the physiognomy of vegetation along a gradient in West Africa between the 15th and 20th parallel North (Figs xv-8 to 20) [25]. Most of these photographs, collected during surveys in 1959 and 1961, come from a meridian section between 7° and

Fig xv-9. Some kilometres from the site illustrated in Fig xv-8, in the same edaphic conditions; a formation resulting from the destruction of the former savannah by man for cultivation (here of *Arachis hypogaea*). The tall perennial grasses are replaced by an annual sward based on *Cenchrus biflorus*, the trees disappear and are replaced by shrubs of the same or different species (*Guiera senegalensis, Boscia,* etc.). (*Photo Ch. Rossetti*)

Fig xv-10. Same site as Fig xv-9. The physiognomic contrast with the savannah is of edaphic nature. The brown soil is a loamy sand. There is only a thin cover of annuals (*Aristida funiculata, Schoenefeldia gracilis,* etc.); patches of *Cymbopogon schoenanthus* and *C. giganteus* (perennials) indicating increased moistute in the depressions; scattered shrubs of *Boscia senegalensis* and *Acacia seyal*. In the foreground, a plant of *Combretum glutinosum*. Annual production from the herbaceous vegetation (annuals) is less than 1,000 kg of dry matter per hectare. (*Photo Ch. Rossetti*)

10° West longitude in Mauritania, but some were taken in the Adrar of the Iforas in Mali. These photographs illustrate the following conclusions reached during Rossetti's field work in this region; they show the climatic gradient from south to north, that is from the more humid to the more arid regions; certain additional illustrations have been included to show particularly striking edaphic contrasts.

Fig xv-11. Position 15° 47′ N/11° 37′ W. Under the same rainfall conditions as the sites illustrated in Figs xv-8 to 10, but on possibly more compact sand, there is a characteristic physiognomic variant: an annual sward with a short herbaceous stratum (*Cenchrus biflorus*, *Blepharis* sp., *Tribulus* sp., *Gisekia* sp.) and a shrub and tree storey of a density of about 100 plants per hectare. The photograph shows fine specimens of *Sclerocarya birrea* (left), one *Balanites aegyptiaca*, sometimes accompanied by *Sterculia setigera*. It is of course possible that the herbaceous layer is the result of the elimination of a savannah storey similar to that in Fig xv-8. Certain partial effects of closure would in fact seem to indicate that the regeneration of this savannah vegetation is extremely slow, or even non-existent under certain conditions of competition by annuals. (*Photo Ch. Rossetti*)

1. A traveller moving due south or due north on land with comparable hydrological characteristics, in this particular case on sandy soils less than 5 per cent clay plus loam, with similar topography and situations in the drainage system, can observe a succession of physiognomic vegetation types which are closely related to the distribution of certain species, especially of Gramineae.

2. It can easily be shown that this succession is characteristic of the rainfall gradient which goes in the same direction.

3. The succession is modified by three main ecological factors:

(*a*) the physiography of a given region,

(*b*) its edaphic characters in relation to physiography, and

Fig XV-12. This may well be a degraded community from which the tall tree storey has disappeared completely, and where the small trees have been replaced by a shrub storey (*Combretum aculeatum* browsed by goats, *Leptadenia pyrotechnica*, *Balanites aegyptiaca*, etc.). The botanical composition of the herbaceous vegetation is similar to that described for Fig XV-11. (*Photo Ch. Rossetti*)

Fig XV-13. Position 16° 20′ N/9° 22′ W.; Msilé plain; estimated annual rainfall: 350 mm; Site No. 61/77. Brown soil developed from sand. Short-duration annual sward (*Cenchrus biflorus*, *C. prieurii*, *Mollugo* sp., *Aristida stipoides*, etc.) with small trees and shrubs (*Acacia raddiana* and *Balanites aegyptiaca*). (*Photo Ch. Rossetti*)

Fig XV-14. The same community in a site near that in Fig XV-13, after obvious biotic interference. The tree storey has disappeared almost entirely. Steppe elements belonging to more northern regions (*Aristida pallida*, *Panicum turgidum*, etc.) have established themselves as a consequence of probable wind disturbance in the plain, due to denudation. (*Photo Ch. Rossetti*)

Fig XV-15. Position 16° 32′ N/9° 28′ W. A site slightly more to the north of the one where Figs XV-13 and 14 were photographed; decrease in rainfall between these sites is probably not significant, and the physiognomy of the vegetation shown may be considered as a valid example of an edaphic and physiographic contrast to the preceding sites. A thicket may be seen in a well-watered depression, with *Commiphora africana*, *Euphorbia balsamifera* among the small trees and shrubs, *Andropogon gayanus* (at this latitude an indicator of moisture) accompanied by annuals, particularly *Schoenefeldia gracilis*. (*Photo Ch. Rossetti*)

(c) the human factor (grazing, scrub fires, cultivation), partially determined by (a) and (b), but obviously dependent upon climatic, as well as political and ethnic factors.

In view of the total lack of methodical studies on the degradation of vegetation, the influence of biotic factors has been illustrated only by sites located very near

Fig XV-16. The belt of annual grassland, as shown in Figs XV-11 to 14, extends with variations in botanical composition over a rainfall zone between the 400- and 250-mm isohyets. Near the latter isohyet, steppe grassland appears, which may be defined as an annual grass cover interspersed with patches of narrow-leaved perennials, less than 80 cm high, often localised on areas of sandy soil where wind erosion may be clearly observed. These steppe grasslands form the transition between the annual grasslands to the south and the continuous steppe cover to the north. In the region of Timbédra (16° 25′ N/9° 02′ W; annual rainfall about 260 mm), on coarse-textured sand, was a good example of such a steppe grassland, here with scattered small trees and shrubs (*Combretum aculeatum*, *Grewia bicolor*, *Boscia senegalensis* and others), and with a herbaceous layer composed primarily of annuals (*Aristida mutabilis*, *Cenchrus biflorus*, etc.), together with patches of steppe elements (*Aristida pallida* and *Cymbopogon schoenanthus*; the latter species shows no rainfall differentiation). (*Photo Ch. Rossetti*)

each other, where it was almost entirely certain that all other factors could be excluded and that the differences observed could be attributed to these biotic factors. In no case is it possible to speak with complete assurance of a climax vegetation, since the influence of these factors cannot be entirely excluded anywhere.

In most of the countries of southern Africa, bush encroachment of the grazing lands is considered to be an economic menace as it reduces the livestock-carrying

Figs xv-17 and 18. The two photographs show the extreme contrast due to soil factors between two aspects in the same rainfall zone as the one shown in Fig xv-16 (dried lake-bed of Mahmoude; position 16° 30′ N/7° 35′ W; Site No. 61/73). Fig xv-17 shows a community with steppe affinities (*Panicum turgidum* with annuals: *Aristida hordeacea, A. mutabilis, Cenchrus biflorus, Heliotropium* sp., etc., the shrubs and small trees being *Calotropis procera, Leptadenia* sp., *Commiphora africana, Acacia raddiana*, etc.), on sand with less than 6 per cent silt + clay fraction, on which a sub-arid brown soil was identified. Fig xv-18 shows the community which appears on clay (about 30 per cent) a few hundred metres from the sand bank in Fig xv-17. Total absence of perennial grasses, few annuals. Some woody herbs (*Leucas* sp., *Momordica* sp., *Hibiscus* sp., *Abutilon* sp., *Aptosimum pumilum*) and shrubs of *Acacia ehrenbergiana* (= *A. flava*) and *Cordia rothii*. Note the contrast between the two types of colonisation on sand and clay, which are typical for these dry regions: the vegetation on clay is poorer and has a shorter growth cycle. (*Photo Ch. Rossetti*)

capacity [26]. The natural regeneration of bush is Nature's way of reconstituting an environment through the processes of ecological progression, and it is essential to find out why in terms of microclimatic and other studies, within the regenerating bush itself and also in the more mesophytic environments along the same humidity gradient. It is believed that studies of this type are planned for Southern

Fig xv-19. Dahr of Oualata, position 17° 17' N/6° 58' W; annual rainfall 100 mm; sandy plateau; Site No. 61/36. Steppe of *Panicum turgidum* with annuals (*Aristida stipoides, Gynandropsis* sp., *Heliotropium* sp., *Latipes* sp., *Sesamum* sp., etc.) and a tree and shrub storey (*Acacia ehrenbergiana, Boscia senegalensis, Commiphora africana, Maerua crassifolia*). This is the northern limit of distribution for most of the woody species and the annuals. These *Panicum turgidum* steppes cover a considerable area in north-western Africa; they are at least 1° Lat. wide. Their southern limit is not precisely governed by rainfall, but rather by the topography of the sands. The more dissected it is, the further south these steppes may be found. (*Photo Ch. Rossetti*)

Rhodesia. Perhaps as a result of them, an optimal relation between tree and grass may be worked out, one that will give maximum livestock production, but still 'in harmony with the environment' [7].

We already know a great deal about the vegetation of our region of study, which comprises three of the great plant geographical subdivisions, the Saharo-Sindian, Mediterranean and Irano-Turanian [27]. Much information has also been collected on the plant communities within these major subdivisions, on the actual or potential climax types of vegetation, and the stages of regression within each. But more data are required from basic ecological studies, and particularly with regard to the microclimates associated with the stages of regression, and the comparative capacity of these stages to conserve soil and water, and to take the sting out of desert winds.

At the practical level, it is, however, difficult to be optimistic. When so many scientists concerned with the land have no 'feeling' for natural vegetation as such and are not conscious of the inexorable and insidious advance of man-induced desiccation, one can hardly expect the graziers and fuel-cutters to give up their

Fig xv-20. Further north, *Panicum turgidum* is replaced by *Aristida pungens*; the woody vegetation becomes thinner (disappearance of certain species and reduction in the density of the remaining ones), and shrubby species such as *Calligonum comosum* and, even further north, *Cornulaca monacantha* appear.

Here, in the Aouker, with strongly undulating sands in 'aklé' dune pattern, is an almost pure steppe of *Aristida pungens* with a very scattered woody vegetation (*Calligonum, Leptadenia, Acacia raddiana*). ('Aklé' is the Mauritanian local name for a dune massif with overlapping dunes composed of usually very unstable sand.) The annual rainfall is still of the order of 90 mm (approximate position 18° 05′ N/9° 30′ W). These communities become more open as one proceeds further north; on these sand dunes *Aristida pungens* is found practically alone; it extends almost to the extreme vegetation limit in the 'desert', where sometimes only *Cornulaca* is found. (*Photo Ch. Rossetti*)

'first come, first served' philosophy, and to manage the natural vegetation for the sake of the land and not only to meet the immediate needs of economic man and his animals. The protection of limited areas by fencing or armed guards may be possible for research or demonstration purposes, as was done at Ein Arrub and Umm Safa in Palestine under the British Mandate, or at Mafraq in Jordan, but fencing or policing on the vast scale that is really necessary are quite out of the question. It may only be hoped that the combined effect of the other actions proposed in this chapter will lead to a partial retreat from the uncultivated areas. Only in this way can the pressure on this land be reduced sufficiently to permit a certain amount of natural or assisted regeneration of the plant cover to take place.

Figs XV-21 to 24. *Edaphic communities in the region with less than 150 mm rainfall*

Fig XV-21. Halophilous communities cover only limited areas in the whole West African continental Sahara, as compared with other continental deserts, and especially with the northern Sahara. These communities are found only in small closed basins with no drainage outlet. They are far more extensive in the Atlantic littoral from the 16th parallel N to Morocco, but are limited to a coastal strip varying in width from only a few hundred metres to some 15 km.

In south-western Mauritania, some halophilous communities are found on the edge of the Primary sandstone escarpment between Tagouraret (17° 30′ N/7° 30′ W) and Tichit (18° 20′ N/9° 30′ W). The site shown (in the region of Oujaf, 17° 46′ N/7° 52′ W; rainfall 80 mm) is a saltpan (*sebkha*) which is completely bare in the centre and has a shrubby fringe of *Salvadora persica* (left) and *Suaeda fruticosa* (right). On the non-saline sand in the foreground are some plants of *Panicum turgidum*. (*Photo Ch. Rossetti*)

Figs XV-22, 23 and 24. Sites No. 59/75 and following; reference position 18° 20' N/1° 20' E; region of Kidal (Adrar of the Iforas, north-eastern Mali); annual rainfall 130 mm. The characteristics of the gradient illustrated for Mauritania were also found, with small variations, in 1959 in eastern Mali. Figs XV-22 to 24 show the characteristic physiognomic gradient in a wadi in the Adrar des Iforas. In dissected terrain, Saharan wadis show three different plant habitats: the tributary gullies; one or more batha (beds) into which the gullies drain; and a flood plain where the water infiltrates rapidly because of its loss of velocity and where there is thus usually no surface flow. These three situations are physiognomically very different.

Species with a more southerly distribution often survive in the gullies. Thus Fig XV-22 shows thickets of *Balanites aegyptiaca*, *Grewia tenax* and *Salvadora persica*, separated by areas of *Andropogon gayanus*, a savannah grass which is here associated with *Dichanthium annulatum*, *Cymbopogon schoenanthus* and others.

As the gullies widen and the now more accessible sites become drier due to grazing and cutting, these floristic elements disappear. Along the wadi bed there are nearly always fairly luxuriant riparian communities. Fig XV-23 shows dense thickets of *Balanites aegyptiaca*, *Combretum aculeatum* and *Acacia scorpioides*, interrupted by dense stands of *Panicum turgidum* and other species. Away from the banks, on the compact sand of the higher ground is a poor annual grassland of *Aristida mutabilis*, associated with *Acacia raddiana*.

The batha is sometimes interrupted by intermediate flats where the stream bed disappears. In the terminal plain, with its rather deep deposits, there is a patchwork vegetation determined by the degree of wind disturbance. Fig XV-24 shows a short-duration annual grassland with small trees on sandy clay soil (loam + clay more than 15 per cent). Note the complete cover of annual grasses with a short growth cycle (*Aristida adscensionis*, *Schoenefeldia gracilis*) with herbs such as *Cassia tora*. The shrubs and trees are *Acacia ehrenbergiana* and *A. raddiana*. (Photo Ch. Rossetti)

Animal husbandry

This brief review of the major forms of land-use in the arid zone is concerned with the ecological and agricultural aspects, with only passing reference when necessary to the effects of economic, social and political pressures on nomadism[1] [28–31].

In adopting the approach of the plant and animal ecologist and the agriculturist and animal husbandman, it is very difficult to distinguish between true nomadism and the many forms of migratory grazing from relatively stable centres along relatively fixed seasonal routes which are so characteristic of the region under investigation. Since the problems which we face and the techniques for improvement that we propose are very similar in both cases, it should be understood that we are dealing with the whole problem of free-range grazing in semi-arid regions, and not necessarily with the nomadic system as such in all cases [32].

Two policies have been advocated with regard to nomadic pastoralism. In the first, a change to a sedentary way of life is considered necessary for economic and social reasons and should be achieved through the necessary investment and the use of modern techniques; in the second case, nomadic pastoralism is considered as a way of life which merits careful consideration and needs to be protected mainly for cultural reasons. In this chapter, even if the first approach may prove to be possible and desirable, the problem of nomadic pastoralism will be tackled in relation to the full utilisation of the existing natural resources and the maintenance in the desert of small communities as guardians of these assets or as a manpower reserve for further economic development.

The nomadic way of life in its combination with nomadic livestock management is a very ancient form of adaptation to an arid and semi-arid environment, which began after the Neolithic period and the domestication of plants and animals. It is now generally recognised that the agriculturist had far greater potentialities for domesticating wild animals than the hunter. The agriculturist did not prey upon the wild animals and so did not inspire them with fear; he would possess food supplies attractive to the ruminants like the ox and the sheep.

Present archaeological data indicate that domestication of the food-producing animals probably occurred in village-farming communities in the 'Hilly Flanks' area of south-western Asia [33, 34], and Reed [35] considers that cereal agriculture and the settled village antedate the domestication of all animals except the dog. Relatively intensive and successful agriculture and stock-breeding (mixed farming) societies developed in the Zagros foothills and their grassy forelands, as well as in the lower Jordan valley [36], prior to the appearance of the earliest societies of this type elsewhere, in Iran and Egypt, for example.

Reed does not accept a dramatic environmental change at the end of the Pleistocene in south-western Asia, producing 'the recurring challenge of desiccation' which Toynbee [37] regards as the main factor explaining the origins of nomadic pastoralism. With the steady accumulation of new data, we may expect

---

[1] These are dealt with elsewhere in this volume and were the primary concern of contributors to the 'Symposium on Nomads and Nomadism in the Arid Zone' published in the *International Social Science Journal* (Vol. XI, No. 4, 1959).

those who approach the problem in terms of the relatively new science of environmental archaeology to produce explanations for many of the aspects of nomadism as it exists today. And there is little doubt that the form in which nomadism exists or more correctly remains in existence today is the same as that which was characteristic of this form of human life and animal husbandry three or four thousand years ago. At least man and his animals are more or less the same but there is little doubt that the natural vegetation is not.

In these days of progress, however, of schemes for the settlement or for the industrialisation of the desert, we must look at the nomadic system of land-use with modern eyes. If we do so, we see that it is one of the major types of animal husbandry of the arid and semi-arid zones, which occur in a stratified manner in relation to the availability of water for plant growth, either from natural rainfall, floods or irrigation. The different strata may be characterised somewhat as follows, with, of course, all manner of combinations and permutations between them:

(a) Settled flocks and herds maintained for part or all of the year on artificial fenced pastures, and/or fed with cut green or conserved fodders in yards or stalls.

(b) Grazing flocks and herds maintained on natural or sown pastures on a semi-intensive system, but with a specific unit of land to which they are confined and where they must subsist throughout the year on grazing supplemented if necessary by cultivated fodder crops.

(c) Flocks and herds maintained for most or all the year on a system of free-range grazing, but following well-defined migratory routes from the communal centre to reliable seasonal pastures elsewhere when grazing at the centre is no longer available.

(d) The truly nomadic flocks and herds which have no fixed centre, but which follow seasonal routes to grazing areas, the annual availability or condition of which is governed by the occurrence of the geographically and seasonally spasmodic and unreliable rainfalls characteristic of regions in which nomadism is the only practicable method of animal husbandry.

These four major types show a marked gradation in the types and qualities of fodder or grazing that are available, and in the types and breeds of livestock. At one end we have the *prima donna* breeds of dairy cattle provided with high-quality pastures and green or conserved fodders on a year-round basis; at the other, the hardy desert breeds of livestock capable of subsisting on a grazing resource which is minimal in quantity and quality, with marked inter- and intra-seasonal variations in availability, accompanied by limited resources of drinking water.

How is the improver of the nomadic way of life and its characteristic types of domestic livestock to approach the problem of the amelioration of the arid and semi-arid environment or in some cases the elimination, total or partial, of this type of animal husbandry from the land-use system? In this connection it must be realised that some types of livestock can be maintained and some types of livestock products can be produced only on semi-arid grazing lands. Elimination of the nomadic system cannot therefore be undertaken without giving due thought to these considerations, although a change from a nomadic to a migratory system involving seasonal movements from and back to established centres would not involve too great a change in the livestock.

It is first essential to study and map the habitual seasonal movements of the nomadic, semi-nomadic and migratory types of shepherds, and to obtain information on the class and number of their animals and their seasonal fluctuations in numbers. It is then necessary to estimate as objectively as possible the carrying capacity of each of the most common types of grazing land and to map them, in order to arrive at an overall and seasonal evaluation of the forage potential within each large pastoral unit. Rapid and cheap techniques are required. The present state of much of the vegetation in the Near East would call for a reduction in number of animals to one-fifth or one-half the present number according to circumstances and at least during certain seasons. In parts of Rajasthan, village pastures have been re-allocated to Panchayat Samitis (elected village councils) on the basis of one-fifth of an acre per bovine animal, with no allocation for sheep, goats or camels. This is twenty times what the natural or reseeded grass stands can sustain and still have a desirable botanical composition. The only answer under such conditions is to cut and carry such grass hay and other fodder that it may be possible to grow and conserve, and to migrate to other areas as soon as this supply is exhausted.

The overall conclusion is that the grazing ranges badly need a rest from the intensive grazing for excessively long periods of the year to which they have been subjected for centuries, and probably at an accelerated rate in more recent times. Under existing economic and social conditions, that rest can best be provided by holding nomadic and migratory flocks and herds for, say, one month longer in the cultivated lands (to which they go at regular intervals in many parts of the area), and by bringing them back to these areas one month earlier, an action which can be regarded as one step in the direction from true free-range grazing of the nomadic type to a partially settled form of livestock husbandry and so also of human life. The stock cannot be kept in the cultivated areas proper during the growing season – they would interfere with the sowing and harvesting operations. It therefore becomes essential to establish holding centres, probably along the fringe of the cultivated zones. At these centres enough fodder reserves would be grown or assembled to maintain the livestock for these periods of one month, and for use in seasons of crisis. The centres could also be used for the provision of superior rams to the flock-keepers in transit to the desert ranges in exchange for their inferior animals, and also for veterinary treatment. The objectives of the fodder agronomist should be both to see that the fodder at these centres is produced and conserved correctly, and also to ensure that the farmers in the cultivated areas are made aware of the economic merits of growing fodder crops, and are educated in their cultivation, conservation and utilisation.

This brings us to a consideration of the basic approach to land-use in arid and semi-arid areas, in all of which grazing represents the most important activity in terms of total acreage involved, if not in production. On the one hand, we have to realise that the grazing ranges, whether of the forest or non-forest types, have suffered considerably at the hands of the nomadic and migratory graziers and their flocks and herds. On the other hand, there is practically no integration of crop and livestock husbandry in the cultivated lands – we are still back with the fundamental distinction between cultivator and shepherd that may have originated at the dawn of agriculture. The main objective in these latter areas must be intensification and the fullest integration, to use a hackneyed word, of crops and livestock, and thus to bring to the cultivated land which needs it so much the animal fertility, the droppings of millions of livestock which are at present wasted on the desert sands.

It seems to be a matter of shifting pressures – of reducing pressure on forest ranges so that they may be able to fulfil their proper functions, of reducing pressures on the non-forest ranges so that the grass and shrub cover may regenerate to a desired level; and of increasing the pressure on the cultivated land which needs livestock to increase crop yields through the supply of dung and urine and the organic matter and nitrogen of cultivated fodder grasses and legumes.

The ecologist concerned with the habitat of the desert, its climate, vegetation and soil, and with the place of man and his domestic livestock in that environment can introduce progress in two different but interrelated directions. The first is to introduce measures of improved management in the arid and semi-arid grazing lands themselves which will still make it possible to utilise this resource on the basis of conservation, and to produce the livestock products characteristic of that environment. The second is to start actions which will make it possible if not essential for the free-range graziers and their livestock gradually to rely less and less on the arid and semi-arid grazing resource, and to become more sedentary than they were before. This can be achieved in combination with continuing use of the arid and semi-arid grazing ranges by flocks and herds which move under the control of professional shepherds.

It is probable that only a combination of these approaches will achieve the fullest realisation of the production potential of these areas. Under proper management the extensive arid and semi-arid grazing lands will be able to yield a valuable annual crop of livestock which will, itself, provide the raw material for an animal industry in the intensively farmed irrigated areas. Thus, the cultivator with part of his holding devoted to forage crops will be able to obtain from the pastoral flocks and herds young animals which he can quickly fatten for the urban or export meat market, will not be tempted to overstock the grazing at his disposal. Such a development would be acceptable alike to the livestock man, the farmer in the cultivated areas and to the soil conservationist fighting desiccation due to devegetation.

Certain steps may be undertaken to promote and facilitate the transference of human and livestock pressure from grazing ranges to cultivated areas, combined with the partial acceptance of a sedentary way of life by most of the people. This would appear to be desirable for the improved welfare of man and his livestock, and for the more intensive use and increased fertility of the cultivated areas. It is necessary to establish an infrastructure which will facilitate grazing control and will help to make free-range grazing an important component part of an efficient livestock industry. Centres to be established on the fringes of the cultivated areas should provide the graziers with the superior stock, the fodder reserves and the veterinary attention which they need. If this action is combined with measures for human welfare, such as the provision of drinking water, schools, medical units, souks, etc., nomadic people may be induced to change to a migratory system of husbandry, to leave most of the families at the centre, and to send out the live-stock at appropriate seasons in the care of professional shepherds on relatively stable seasonal routes.

The problem of reliable markets must in many stituations be solved before there are adequate incentives to produce certain commodities. For example, in a society where the number of animals owned represents the prestige of the owner, for example over much of Africa south of the Sahara, and the ultimate use of the products is only to meet his own needs and those of the other members of his tribe or group, there exists no economic incentive for change. It is first necessary to create a desire for products which can be obtained in exchange for those produced. In terms of livestock production, this might mean a reduction in one class of animals for which there is no demand and an increase in another class for which a market exists or can be created. For example, in many countries lambs from range flocks of sheep are sold after weaning to agriculturists or professional feeders, who finish them for market on farm-produced fodders. The livestock producer sells the harvested forage from the range in the form of lambs, while the agriculturist sells farm fodders in the form of meat and retains part of his production as animal manure to maintain the fertility of his arable soil. This relation between range livestock husbandry and agriculture represents the desirable integration of production from arable and non-arable land.

In the operation of a production cycle based on arid and semi-arid ranges used in conjunction with farming lands, it may be necessary to consider the economics of expenditures on both classes of land. Public expenditure on both might be justified, but it is not likely that the same problem will exist in the same degree at different locations. In a situation where expenditure on range improvement would result in increased animal produce readily marketable in agricultural areas, a considerable expenditure might be justified. If, on the other hand, the managed range capacity based on livestock production were to exceed the market possibilities, it would be desirable to direct expenditures towards the improvement or expansion of arable lands through irrigation schemes, fertilisation, and the extension of improved farming practices. Analyses of the different economic criteria in different regions should be made for the improvement of high-value arable lands or the less valuable non-arable lands. When one is dealing with problems such as critical catchment areas, the encroachment of deserts, and the

depletion of potential resources, the economy of time may justify major investments in the rehabilitation of arid and semi-arid lands where measurable
economic returns cannot be anticipated for 10, 15, 25 or more years, and where
possibly no economic return can be anticipated *in situ*. It must be recognised that
market values for all land-protective (conservation) measures are still difficult to
assess.

Many developing countries in Africa and Asia maintain unproductive animals
for reasons related to sentiment, religion, prestige, bride price, or merely as an
insurance against the disastrous losses that occur from time to time in marginal
environments. These livestock are an anachronistic luxury which cannot be considered in the scientific planning of animal husbandry in relation to the economic
development of grazing and fodder resources. Land-use planners should concentrate all their efforts on making economic animal husbandry so efficient, productive and reliable that the livestock owners themselves will be induced for
economic reasons to abandon sentimental and other unproductive forms of animal
husbandry. In the meantime, these unproductive animals will continue to ruin
the natural grazing lands and the forests, contributing greatly to erosion, flooding, desiccation and land deterioration in general, but that must remain a national
problem on which no outside help or advice can be given.

Crop husbandry

It is becoming increasingly clear that the agriculturist and plant ecologist define
arid land as that land on which only extensive free-range animal husbandry is
practicable, and semi-arid land as that land on which cropping of drought-
resistant crops can be practised, with gradually increased integration of more
settled forms of animal husbandry as one progresses to more humid conditions.

The science of crop ecology in semi-arid zones is not yet sufficiently advanced
for it to be possible to produce overall conclusions and generalisations. Instead
of giving here merely a description of the existing types of crops and cropping
systems in the semi-arid zones, it has been considered preferable and more constructive to present the conclusions from two recent studies made in contrasting
environments.

The first is that by P. A. Donovan [38] on 'Crop ecology and farming systems'
in the summer rainfall environment of Southern Rhodesia which we have
already taken as one of our examples of climatic gradients. The other is the
FAO/Unesco/WMO Interagency project on agroclimatology conducted by
G. Perrin de Brichambaut and C. C. Wallén, who studied primarily the winter
rainfall arid and semi-arid zones of the Near East, with cross-reference to comparable stations in other parts of the world [39]. It is proposed to locate a new
Interagency pilot study in a different climatic environment so that the problems to
be solved would be of a different nature and call for other methods of assessment.
A tropical area with a semi-arid climate of summer rains is considered to be the
most appropriate, and it has been suggested that the study be made in Africa

south of the Sahara, on the belt bounded to the north by arid steppe and desert, and to the south by sub-humid savannah and forest. This zone is relatively homogeneous in climatic and ecological conditions and includes the main food-producing regions of several countries. The approximate extent of this new area of study would fall between the northern limit of regular dry-land farming, and the southern limit approximately where the length of the dry season exceeds that of the wet season and where precipitation occurs only during one season (tropical type of rainfall).

The Agro-Ecological Survey of Southern Rhodesia [15, 40] has defined farming systems essentially in terms of the relative importance of crops and live-stock; the productivity of an area and the economic size of holding are defined in terms of its ability to produce crops under the soil and climatic conditions prevailing. For example (and here we have a great contrast with the Mediterranean, Saharan and Near Eastern regions), in the area where tobacco growing is the recommended farming system, 1,000 acres is an economic size of holding; on similar soils but with a less favourable climate for crop growth and where mixed farming is recommended, 10,000 acres are considered necessary for an economic holding; in areas where the climate is too dry for cropping, up to 60,000 acres are required.

In those parts of Southern Rhodesia where climate is the primary influence on crop growth, the farming system will depend on the attainable levels of crop yield; Donovan discusses the parameters of climate as they affect crop yield.

(a) Available moisture

As a general rule in Southern Rhodesia, potential evapotranspiration (PE) increases as rainfall (P) decreases, and only in a small part of the country does P exceed PE. However, in considering the moisture available for crop growth, this picture needs to be modified in two important ways:

(i) About four-fifths of the rainfall is concentrated in a few months of the year (December to March inclusive), while less than half of the PE occurs in this period.

(ii) The actual evapotranspiration (AE) of a crop rises from about one-third PE at emergence to PE as it approaches maturity and then falls rapidly again to one-third PE.

These two factors make some form of cropping possible in areas where P never exceeds PE and is far below it in annual total; the type of cropping and the yield levels are better assessed from the relation between AE and P over the growing period. The amount of moisture available for crop growth depends only to a limited extent on soil factors. Under the same rainfall, different soils produce different moisture conditions. A comparison of the estimated moisture balance between the Basalt and the Kalahari sands shows that they differ mainly in the time that moisture is available and not in the amounts.

(b) Length of season

Although much of Southern Rhodesia has too low a total rainfall to make crop growing possible, there are so-called low-rainfall areas in which it is shortness of the summer rainfall season rather than inadequate moisture that limits crop growth and yield. Apart from irrigation, there is nothing that can be done where moisture is inadequate, but where it is shortness of season that limits crop yield, better adapted cultivars can be used. In much of Southern Rhodesia, temperatures are suitable for crop growth before and after the period of maximum availability of moisture, although temperature may be a limiting factor at the end of the season above 5,000 feet elevation.

Donovan accepts the conclusion that the critical level of moisture supply in differentiating between cash cropping areas and those areas in which cropping should be regarded as supporting livestock production is 20 inches effective rainfall, and that below 15 inches no cropping is to be recommended. A crop's productivity depends, however, more on whether it can reach the desired stage of development on a moisture consumption defined less by the crop than by the prevailing climatic conditions, particularly rainfall and temperature. In discussing the amount of moisture available it was said that soil factors play a relatively small part, but in considering length of season it is essential to take soil factors into account; a light soil may have a 20 to 25 per cent longer season than a heavy soil.

(c) Temperature

Due to the marginal and variable moisture conditions and the shortness of the season of adequate available moisture in much of Southern Rhodesia, temperature is seldom a limiting factor in crop production. However, cotton, rice, sesame and some other crops require higher temperatures for optimal yields than are found in most areas which have sufficient moisture, and it is this group of crops which has too long a growing season for the areas with suitable temperature conditions. Even for those crops commonly grown, temperatures of the growing season are not far above the lower limits. For sorghum, temperature has a considerable effect on yield levels. Temperatures are more likely to be found important in their effect on crop growth, particularly of perennials, in areas with long seasons of moisture supply.

Donovan concludes that estimates of available moisture or 'effective rainfall' for various areas are of limited value in assessing crop production potential. This can be estimated reliably only by matching the moisture use pattern (which is an inherent characteristic of the crop) with the favourable and operative portion of the growth period as defined by climate. The major features of growth period patterns in Southern Rhodesia are considered to be:

(i) the lack of moisture reserves in the soil at or before planting time and therefore the dependence on marginal moisture conditions early in the season,

(ii) the very rapid rise in favourable moisture conditions,

(iii) the very short period during which conditions remain favourable and the occasional occurrence of drought conditions within that period,

(iv) the very rapid decline of favourable moisture conditions and therefore dependence late in the season by crops on soil-moisture reserves built up in the current season, and

(v) the virtually complete absence of available moisture during the dry season.

Although the broad categories of farming systems may be defined from the ecology of the natural vegetation, the soil types and the effective rainfall, so that any particular area can be classified as most suited to cash cropping, mixed farming, semi-intensive or extensive ranching, the system of farming practised in any specific area will depend on the level of crop yield that is attainable. The factors involved in estimating crop yield potential are not those used in defining agro-ecological areas. Further, economic pressures and demands may have an equal or even a greater effect in the determination of a system of land-use than have the natural characteristics. The choice of crops or of types and standards of animal husbandry and therefore the farming system to be employed will depend as much on economic conditions and size of holding as on natural conditions.

This last statement applies even more to the Mediterranean and Near Eastern regions and to the monsoonal lands to the east, where ancient and rigid forms of land-use have recently come under review in the FAO/Unesco/WMO project mentioned above [39]. A theoretical assessment of the rainfall limit above which regular dry-land farming is possible shows this to lie between 200 and 300 mm of annual rainfall. The members of this project therefore suggest that the limit between semi-arid and arid conditions may be defined as falling along this borderline zone for dry-land farming. Presumably in Southern Rhodesia, Regions III and IV would on this basis be regarded as semi-arid, Region V as arid.

The region selected as the basic area of investigation was the semi-arid zones of the Near and Middle East, or more precisely the area between the Mediterranean Sea in the west and the border between Iran on the one side and Pakistan, Afghanistan and USSR on the other, in the east, with reservations regarding Israel, Turkey and Arabia. This limits the basic region to Jordan, Lebanon, Syria, Iraq and Iran (see Table C in draft report). In a zone of some homogeneity, the area for regular cultivation of a given crop can, according to Perrin de Brichambaut and Wallén, be determined by a study of precipitation and its variability, the precise limits being related to the type of soil and the topography. The variability of precipitation in the driest zones creates a great variety of ecological conditions reflected in greatly differing crops and in areas sown. In general, in Tunisia, Morocco and Syria, autumn rains have a particular bearing on the proportion of land sown every year, while rain falling during the period of rapid growth is a major factor in production. Because of the variability of rainfall it is not possible to use highly selected cultivars with very specific requirements.

Within the semi-arid zones as defined in this study, the cultivated land is left fallow at least one year in two. It has been shown in Canada and the Great Plains States of U S A that clear or worked fallow in conjunction with dry-farming methods not only favours the control of weeds and the increase of soil nitrogen, but also the storage of soil moisture from one year to the next. Research in the basic region has not yet been sufficient to indicate whether the bare fallow in that rainfall regime permits this conservation of moisture. On the contrary, idle or un-cultivated fallow which is generally grazed does not promote either weed control or moisture conservation; it provides a rest period for the soil and also stubble grazing without causing erosion and the leaching of nutrients. Following the main theme of this chapter, there is a great need for research in breaking this cereal/fallow system wherever possible for the cultivation of the short-term grazing and fodder crops which would contribute to the resting of the natural vegetation.

Although all the cultural practices used for 'dry farming' are not recommended in the basic region of the Interagency study, the great importance of surface tillage without turning the soil is recognised, to encourage water infiltration and to reduce evaporation. Deep cultivation combined with turning of the soil is seldom recommended where the rainfall is less than 350 to 400 mm, and is often considered detrimental to soil structure and to conservation. Every possible action should be taken to eliminate the bare fallow and to protect the soil by a plant cover or by stubble mulching.

Rough tillage practices for preparation of seedbeds and for seeding have to be carried out during the short period when soil moisture is favourable, and this has been greatly facilitated by the use of tractors. In fact this has caused the excessive encroachment of cultivation in arid zones, and has created a difficult situation for animal husbandry without substantially increasing overall cereal production. The Syrian government now forbids as far as possible the use of tractors in dry-land areas where the rainfall is below 250 mm.

The insufficient moisture available in semi-arid zones makes it essential to limit the density of the plant cover, either by light sowing in broad-spaced rows, or by limiting vegetative growth. For this reason, in the driest stations of Cali-fornia and Australia where the precipitation is less than 300 mm, nitrogenous fertilisers are not recommended. Their application is often uneconomic and causes excessive development of vegetative growth after the fallow. On the other hand, phosphates are always recommended in these countries, whatever the rain-fall. Generally speaking, the effectiveness of fertilisers and the economy of their use increase in proportion to the moisture and to the yield of the crops.

Cultivars of the semi-arid zones

It may be said that about 50 per cent of any increase in crop yields in semi-arid as in other zones will depend on agronomic measures, and some 10 per cent on the production of improved cultivars capable of responding to the improved cultural conditions so provided. A cultivar is the accepted modern name for a variety, strain or type, and refers in particular to registered or otherwise authenticated named varieties of agricultural and horticultural plants.

How does the plant breeder concerned with the crops of the semi-arid zones approach the problem of improving the existing crops of this area, and how does he set about introducing new genotypes or even new species? Ideally his breeding programme should be worked out and actually carried on in full collaboration with the ecoclimatologist and the plant physiologist, but this has rarely been possible due to limited resources. Empiricism must still remain the major approach, and the specialists involved are more likely to be the general crop agronomist and the crop improver rather than the cytogeneticist [41].

Drought resistance of actual or potential cultivars will be an important, in fact primary, consideration, but drought resistance is a very general term which may refer to several characters that give a plant its adaptation to a semi-arid environment. It is necessary to distinguish between plants that are drought-resisting and those that are less resistant but are drought-escaping. In the first instance, the resistance may be inherent in the metabolism and water regime of the plant, or related to the morphology of above-ground parts and root systems. Drought-escaping plants require a combination of short season of growth with maximum yield in the time available, before the growing season ends and drought conditions recur. Resistance to drought may not be required at a uniform level throughout the season; we have noted the requirement emphasised by the workers in Southern Rhodesia, ability to withstand moisture deficiency at the beginning or at the end of the growing season, or both.

The main crops of the semi-arid zones are annuals, combining the drought-resistant and drought-escaping characteristics. The typical cereals, *Pennisetum typhoides*, finger millet or bajra, the sorghums, wheat and barley develop rapidly in their respective habitats within the semi-arid zone while moisture is available and at the same time can resist periods of temporary drought after the susceptible seedling stage has been passed. Some grain legumes with a low water requirement come into the picture at a certain threshold value, can develop rapidly in the periods between droughts and make possible the first introduction of simple crop rotations. In the Near East, the grain legumes are more sensitive than the cereals to lack of humidity and high spring temperatures, while low winter and high spring temperatures set limits to the growth of cereals.

In common with plant breeders in general, the worker in the arid zones must of course be familiar with the breeding behaviour of his plant material, and also with the genetical history of the crop [42-45] and the distance from its primary and secondary centres of origin. When it comes to a question of plant exploration and introduction of wild species and primitive forms from these centres of origin, or of more advanced cultivars from other regions of cultivation, it may be argued that consideration should be given to the ecoclimatic comparability of the environments of collection and receipt.

It must in the first place be emphasised that, for plant exploration in underdeveloped countries and remote areas and for plant introduction in general, only regional climatology permits the necessary comparative delimitation of environments [46]. A regional climate may be defined as the climate of an area potentially suitable or already developed for a relatively restricted type of land-use with specific cropping systems and practices. This is a much more limited and precise definition than that usually adopted for the so-called agroclimatic regions and homoclimes. The cultivars in such a region will have several common characteristics of adaptation which may make them suitable for experimental introduction in other climatic regions where they are likely to meet comparable conditions.

Distinctions may be made on the following basis:

(a) The plant explorer who wishes to collect wild species and primitive forms generally goes to one or more of the primary and secondary centres, of which our region of study (Atlantic to Ganges) contains so many. These are often located in highly dissected topography, creating a mosaic of contrasting environments and thus geographical and genetic isolation in Vavilov's mountain-and-valley regions nearest the equator (eastern Turkey, southern USSR, Iran, Iraq, Afghanistan, the western part of the Himalayan range, and Ethiopia). In these areas, climatic studies on a very restricted geographical basis are neither feasible nor perhaps of great relevance.

(b) Land races are the local and characteristic types of crop plants which arise in relatively limited localities. Each race is a mixture of many related forms, and thus has the potentiality to become adapted to a range of conditions differing quite markedly from those in its place of origin. When introductions are to be made, general information on the regional climate of the area from which such races come should be sufficient.

(c) Comparisons between regional climates would appear to be promising for the introduction of improved but not highly bred cultivars and of ecotypes collected from the natural vegetation. They should make it possible to avoid certain errors. A good knowledge of the climate of the site and region from which a cultivar has come, and a field or laboratory study of the response of that variety to certain climatic extremes, should indicate where it is likely to be climatically adapted. Introduction nurseries are, however, still essential in the different climatic zones in the receiving country.

(d) The introduction of cultivars which have been highly bred for maximum yield under very specific conditions cannot be based exclusively on the comparability of regional climates. Only microclimatic investigations combined with soil and general phenological observations can indicate the areas of adaptation of these cultivars.

Current interest among plant breeders, particularly in western Europe, is in the valuable genotypical resource represented by the wild species and primitive cultivated forms of our region of study [41]. It would be desirable that greater interest in this resource were taken by the plant breeders and ecologists in the region itself, as this precious material is in great danger. The wild species are being progressively eliminated by the destruction of the natural vegetative cover to which we have already referred. The primitive cultivated forms are being eliminated by the advance into remote regions of bred forms, sometimes introduced from other semi-arid environments and possessing a much narrower range of genetic variability, and therefore adaptability to different environments, than the primitive forms from which they were originally selected.

Intelligent and forward-looking plans for land-use should cater for protected areas of natural vegetation. These would meet not only the requirements of the vegetation ecologist already mentioned, but would also provide under appropriate management for a gene bank of wild species which are still of great significance in the breeding of certain plants for semi-arid and other conditions. These protected areas would be linked to regional plant introduction stations to which collections of both wild species and primitive cultivated forms could be taken for study under conditions as similar as possible to their place of origin [41].

To improve the grazing and fodder resources of our region, it is probable that useful work can be done with the collection, selection, and to some extent hybridisation and breeding of the grasses and legumes in this region. At the Indian Agricultural Research Institute in New Delhi, B. D. Patil had demonstrated the wide range of variability which awaits the attention of the plant breeder in material collected from the wild stands of *Cenchrus ciliaris* and other drought-resistant grasses of western India and elsewhere [3]. Some of the new types that have been evolved have already been used successfully in reseeding projects on panchayat lands in Rajasthan. In Iran, H. Pabot has collected a wider range of species from their natural habitats, and considers that many could be multiplied and used to reseed degraded range land or to re-establish a plant cover on abandoned fallows.

But the selection and breeding of grass cultivars from wild species raise special problems. Among the important natural adaptations of wild species to their environment are a long season of seed maturity to spread the risk of exposure of seedlings to drought or other unfavourable conditions, and a seed shattering habit to ensure maximum fall of mature seed. In the transformation of a wild species into a cultivar, it is essential to remove both these characters as far as possible, to ensure a maximum seed yield at one specific time of harvest (early, medium or late according to agricultural requirements), and maximum retention of that seed in the harvested heads. It would seem to be necessary to reverse the usual process adopted by herbage plant breeders, first to select on the basis of these and other characters which will make the production of abundant, viable seed an economic proposition, and then to select or breed in the usual way to produce the desirable growth form, earliness or lateness, yield, leaf–stem ratio, and other characters of economic significance.

Plant introduction specialists tend to pay too much attention to the biological status of the introduced genotype prior to introduction and too little to the drastic evolutionary changes that are likely to occur under the impact of the new environment. Frankel [47] has considered this important aspect with particular reference to the invasion or introduction and subsequent evolution of plants in Australia and New Zealand (see also [48]).

As a consequence of the vast exchange of plants and animals between countries and continents following the extension of European influence beyond the borders of the Old World, migrants have, with, without, or against the will of man, established themselves in habitats differing to a greater or lesser degree from those which they had come. Over vast regions, exotic plants and animals have displaced or replaced indigenous ones, often in a remarkably short space of time. 'What has been the impact of this new environment on the genetic structure of the migrants and how far and in what manner has the process of re-establishment been aided by genetic adaptation?'

The scope for genetic adaptation in cultivated plants is not independent of the nature and degree of environmental adaptation. Glasshouse crops often receive in their new environment such a high degree of environmental adaptation that the scope and need for genetic adaptation are greatly reduced. At the other extreme, plants which establish themselves in native grasslands in competition with an adapted flora and without modification of the environment, except by the grazing animal or by light applications of fertilisers, are given full scope for adaptive genetic changes. Field crops lie between these two extremes. Weeds would presumably be likely to show as much or even more genetic adaptation than herbage plants.

Frankel selected as examples a range of plants differing in breeding system, nature of introduction into Australia and intensity of deliberate selection. All are annual or perennial field crops or pasture plants. Even in the six crop plants studied, a great diversity of evolutionary pattern can be seen. Perennial ryegrass and maize – the latter with some help from mass selection – have shown rapid and specific selective responses to different environments, resulting in ecological adaptation and genetic stabilisation. Wimmera ryegrass on the other hand has retained and possibly increased its adaptive polymorphism. Subterranean clover, although biologically conservative, brought with it or developed sufficient variation for a measure of ecological segregation into a range of types including among others some short-duration and therefore drought-escaping types suitable for extending the crop into more semi-arid areas. This range of variation, solely or mainly imported, would, according to Frankel, not suffice for a crop plant like wheat, which requires for agronomic fitness a greater appreciation and specialisation of genetic adaptation.

Genetic adjustment may improve adaptation in an established habitat; it may also extend the habitat range into, for example, more drought-affected environments. This possibility may require particular attention in species with limited genetic representation, which may be the case in deliberate or accidental introductions and especially in species with restricted recombination. Through the study and use of long-term changes in hybrid populations of self-fertilised crops in marginal zones, it may be possible to extend the ecological range of a species and also our knowledge of physiological and genetic barriers [49].

The plant breeder in semi-arid zones, where levels of production are low, technical staff scarce and of indifferent quality, and soil and climatic variations considerable, may therefore be encouraged by these examples of genetic adaptation and adjustment to be more adventurous in his choice of localities from which to introduce new genotypes. Major attention may continue to be given to the areas of comparable regional climates discussed above, but it will also be profitable to introduce from other less comparable areas and to expose the material to the adaptive influences in the area of introduction.

Trees in the desert

No student of the land in the broad sense as defined here can afford to be party to the unseemly rivalry, antagonism or mere disregard for the other's views and interests that exist at the present day between the different sectors concerned with the use of land. Admittedly, there is a high degree of competition between the land users for the limited space available, but the response to this situation should rather be maximum collaboration towards an integrated land-use policy, allowing for the optimal development and use of the three major land classes, forest land, grazing land and cultivated land. This statement applies with particular force to the arid and semi-arid lands, those lands that we have defined as the largest area of marginal land in the world.

The shrub and thorny savannah and *Acacia/Commiphora* woodlands in Africa south of the Sahara, for example, are essentially zones in which forestry and grazing must be closely integrated. The difficulty and indeed the main conflict arises because under the present system of management, grazing requires annual late burning, whereas such burning is a main factor inhibiting tree growth. Opponents of burning agree that the destruction of any organic matter is always undesirable. The forester burns early not by choice but in order to avoid more serious fires at the height of the dry season. The pastoralist would prefer not to burn if there was a practical alternative method of controlling encroaching scrub, eliminating ungrazed tussocks and removing unpalatable plants. Early burning at the beginning of the dry season damages the perennial grasses because food reserves have not returned to roots and crown. The dormant grass plant is not seriously damaged by burning although the new growth that appears after burning is produced at the expense of the reserves. The deliberate stimulation of growth by burning followed by grazing kills the most vigorous plants within two or three years. Tree lopping, particularly in seasons of fuel shortage, is a common practice in the sub-Saharan arid zones.

Similar remarks could be made with the necessary ecological variations for the so-called winter-rainfall environment of the Mediterranean and Near Eastern regions with their characteristic vegetation, and for the monsoonal environment and vegetation which begins again to the east, from the highest points of Baluchistan on into the Indian subcontinent.

In planning for Africa south of the Sahara, three different requirements need to be distinguished:

(i) Those areas to be used primarily for grazing on which the improvement or even the survival of woody vegetation is considered secondary, but for which basic criteria have by no means been established. Here burning could be regulated solely with grazing management in mind.

(ii) Areas where close integration between forestry and grazing is desirable. Here, early burning must be the practice since this is less harmful to woody growth than late burning.

(iii) Areas set aside specifically as forest reserves, with both productive and protective functions, either formed of natural species or plantations of fast-growing species. In these areas, burning must be prohibited, although it may be difficult to exclude accidental fires.

The silviculturist of the desert, if we can employ such an apparent contradiction in terms, has to accept the situation as it is. The members of the silvicultural section of the Central Arid Zone Research Institute in Jodhpur have to accept the lopping for fodder of *Prosopis spicigera* as an essential fact of land-use in an area of excessive pressure of human and animal populations. They cannot realistically propose the cessation of the practice, but rather need to find out how much lopping the trees can stand on a fixed rotation without being killed. They may at the same time hope that the work of the fodder agronomists and animal husbandmen may relieve the pressure on this upper storey source of feed and make better management of the trees possible.

There are a few special examples of commercial production on a large scale from trees of the semi-arid zone, such as the production of gum arabic from *Acacia senegal* and *A. seyal*. This industry in the Sudan provides an export commodity which is second only to cotton fibre and cottonseed in value. But, as we have seen, the main demand on the natural tree and shrub vegetation is for fuel wood. Everything must be done to provide this resource from special plantings in and around villages and oases, until such time as the raised standard of living permits a change to oil. Much has been written about the relative merits of various tree species as fodder, but these should be regarded rather as an emergency source of fodder. There is also the question of windbreaks. These demand careful consideration by the land planner, particularly in relation to microclimatic studies and the creation or regeneration of protective belts between the desert and the sub-humid zones.

Irrigated land plays an important role in relation to desert silviculture, since excellent tree crops may be grown along canals and irrigation ditches without reducing the yield of arable crops. In fact, the economic return per hectare from poplars in Syria is higher than that for any food, cash or fodder crop.

Related to the forest resource is the unique heritage which the magnificence and variety of wildlife represent in the arid and semi-arid zones of Africa. A clear and constructive declaration of policy and the establishment of machinery to implement it are needed in each country, so that this great asset can play its proper role in the economic, scientific and cultural development of the continent. It is essential to designate areas where wild animals are to be protected, and to carry out surveys on the economics of wildlife management as part of an overall land-use policy.

Thus a research programme concerned with silviculture and forest management in arid and semi-arid zones should cover:

(a) growth and development of nursery seedlings,
(b) forest establishment,
(c) artificial regeneration of trees and shrubs,
(d) forest management,
(e) forest protection against fire, and damage by game, rodents, termites, etc.,
(f) forest influences, including control of wind erosion, water yield of forested areas, etc.,
(g) miscellaneous uses for trees, shrubs and forest products (emergency fodder or top feed, indigenous products, use of alkaloids, essential oils and other substances for their medicinal properties).

Classification of land

In the first section of this chapter, we have defined land as a complex of all the factors on, above and below the earth's surface which affect or govern the degree and the way in which that land may be used by man for agricultural, pastoral and forest production. Throughout the chapter we have repeatedly spoken about different ways of classifying land and its component parts, as well as types of land-use, animal husbandry, cropping systems and individual crops. The possible confusion in the reader's mind is only equalled by the confusion which exists in the technical world regarding the relative merits and the degree of general application of various approaches to and techniques for the classification of land.

In Fig XV-25, an attempt has been made to provide a basis for discussion by bringing all these into one sequence, which may be applied in the classification of land and the analysis of its potential [50]. This is an idealistic framework, starting with the Australian technique of surveying the various components of 'land', and concluding with the evolution of an integrated land-use pattern. It would demand great time and expense to carry through from beginning to end. Specialists in the different approaches and disciplines that are involved may, however, consider whether their own place in the sequence has been correctly stated.

Throughout this chapter, the major emphasis has been placed on climate as a primary criterion in classification and on the significance of climatic or humidity gradients in considering land-use in the arid and semi-arid zones. This has not been introduced into the schematic sequence because it is assumed that it will always be in the mind of the land classifier and that it will be applied at all stages.

Under (C) it might be desirable to introduce a fourth class, 'Land for purposes of conservation', e.g. the 4 per cent of the area of East Africa already mentioned. It could be argued that in nature there is no such thing as waste land, and that a partially or completely unused area of natural vegetation may be filling a most important role in conserving soil and the water that is so precious in an arid environment, and in reducing the force of the factors of desiccation. Sometimes

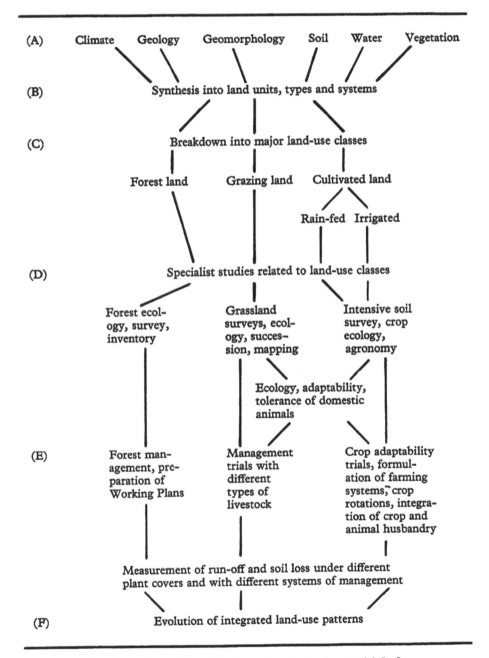

Fig xv-25. Six phases in land classification and analysis of potential [50]

it is desirable for land planners to look beyond the immediate economic return and to consider their resource on a wider canvas. The intensification of the use of certain more favourably situated land units or systems or parts thereof should make it possible to release other areas from the tremendous pressure under which they have suffered for so long.

Integrated land-use

In the final analysis, an attempt must be made to bring all these facts and considerations, past, present and future, into what we may call an ideal biological system of land-use, and then to consider to what extent this may be applied in the social and economic context in any area in which we are working. This appears to involve three different but nevertheless closely related types of integration:

(i) integration between different zones along a climatic or humidity gradient, providing for interchange and interdependence between them all,

(ii) integration of land in the three major land-use classes (forest, grazing and crop land) within a particular country to the extent that they are present, and

(iii) integration within individual farms or ownership units, again to the extent to which this is possible according to the sufficiency of acreage in the land-use classes concerned.

This ideal classification of land into the three major classes (C in Fig xv-25) and the planning of their use on a well-integrated basis are much easier to achieve in new lands than in old – in northern Australia, Nicaragua and parts of Brazil as compared with the lands of the ancient and modern civilisations of the Near and Middle East and the catchment area of the Indus. In the one case, it is a matter of finding an ecological equilibrium which will permit man and his animals,to use the land and its partially unused resources without reducing too far its capacity for conservation. In the second case, it is necessary to attempt to recover some new type of equilibrium to replace the one that was passed centuries or millennia ago during the evolution of civilisations that were ignorant of the principles of land ecology. This in short is the basic problem of the arid and semi-arid zones of the Old World, the recognition of ecological over-development [51] and the initiation of actions along several different lines, the cumulative effect of which shall be to bring man and his animals again into harmony with their environment.

'God himself that formed the earth and made it; he hath established it, he created it not in vain, he formed it to be inhabited' (Isaiah xlv. 18).

R. O. WHYTE

References

[1] BIROT, R. and J. DRESCH, *La Méditerrannée et le Moyen Orient*, Vol. 2, Paris, 1956.

[2] PAN-AMERICAN UNION, *Irrigation Civilizations. A Comparative Study*, Washington, 1955.

[3] WHYTE, R. O., 'The grassland and fodder resources of India' (2nd ed.), *Indian Council Agric. Research, Scientific Monograph*, No. 22, New Delhi, 1964, p. 553

[4] BOBEK, H., 'Vegetationsverwüstung und Bodenerschöpfung in Persien, und ihr Zusammenhang mit dem Niedergang alterer Zivilisationen', *Geogr. Inst., Univ. Wien* (Union Internat. Conserv. Nature, RT 7/Coll./4), 1958.

[5] MONOD, T., 'Parts respectives de l'homme et des phénomènes naturels dans la dégradation du paysage et le déclin des civilisations à travers le monde méditerranéen *lato sensu*, avec les déserts ou semi-déserts adjacents, au cours des derniers millénaires', *Colloque d'introduction, Union internationale pour la protection de la nature et de ses ressources, 6ᵉ Assemblée générale, 7ᵉ Réunion technique, Athènes-Delphi*, 1958.

[6] WHYTE, R. O., 'Evolution of land use in south-western Asia', *A History of Land Use in Arid Regions*, Unesco, Paris, 1961, pp. 57–118.

[7] WEST, O., 'The role of pasture in farming systems', *Proc. Fifth Ann. Conf. Professional Officers, Dept. Res. and Specialist Services, Fed. Min. Agric., Rhodesia/Nyasaland*, 1960, pp. 53–9.

[8] TAYLOR, B. W. and R. O. WHYTE, 'The role of vegetation studies in land classification', *Proc. Symp. Humid Tropics Vegetation, Indonesia*, Unesco, 1958, pp. 121–31.

[9] ADDISON, H., *Land, Water and Food. A Topical Commentary on the Past, Present and Future of Irrigation, Land Reclamation and the Food Supplies they Yield*, London, 1955.

[10] DROWER, M. S., 'Water-supply, irrigation and agriculture', in Singer *et al.*, *A History of Technology*, Vol. 1, Oxford, 1954, pp. 520–57.

[11] JACOBSON, T. and R. M. ADAMS, 'Salt and silt in ancient Mesopotamian agriculture', *Science*, Vol. 128, 1958, pp. 1251–8.

[12] EVENARI, M., L. SHANAN, N. H. TADMOR and Y. AHARONI, 'Ancient agriculture in the Negev', *Science*, Vol. 133, 1961, pp. 979–96.

[13] TADMOR, N. H., L. SHANAN and M. EVENARI, 'The ancient desert agriculture of the Negev. VI. The ratio of catchment to cultivated area', *Ktavim*, 10, 1960, pp. 193–206.

[14] SMITH, L. P., *Weather and Food*, FFHC Basic Study No. 1, World Meteorological Organisation, Geneva, 1962.

[15] VINCENT, V., R. G. THOMAS and R. R. STAPLES, *An Agricultural Survey of Southern Rhodesia*, Part 1, Agro-Ecological Survey, Salisbury, n.d.

[16] SWYNNERTON, R. J. M., 'African farming transformed', *Commonwealth Journ.*, Vol. 5, No. 3, 1962, pp. 137–40.

[17] EAST AFRICA ROYAL COMMISSION, *Report, 1953–5*, Cmd. 9475, London, 1955.

[18] COMMONWEALTH SCIENTIFIC AND INDUSTRIAL RESEARCH ORGANISA-
TION, *Soil and Pasture Research in South-Western Australia*, Melbourne,
1962.

[19] WHYTE, R. O., 'The significance of climatic change for natural vegetation and
agriculture', *Proc. Unesco Symp. on Climatic Change*, Rome, October 1961.

[20] REIFENBERG, A., *The Struggle Between the Desert and the Sown. Rise and
Fall in the Levant*, Jerusalem, 1955.

[21] RAINS, A. B., personal communication, 1962.

[22] PEREIRA, H. C., 'The land-use problem and the experimental environment.
The research project', *E. Afr. Agric. For. Journ.*, Vol. 27 (Spec. Issue),
1962, pp. 42–6.

[23] RUSSELL, E. W., Foreword, 'Hydrological effects of changes in land use in
some East African catchment areas', *E. Afr. Agric. For. Journ.*, Vol. 27
(Spec. Issue), 1962, pp. 1–2.

[24] STEWART, O. C., 'Fire as the first great force employed by Man', in W. L.
Thomas, *The Role of Man in Changing the Face of the Earth*, Chicago, 1956,
pp. 115–33.

[25] ROSSETTI, CH., personal communication, 1962.

[26] WEST, O., *Report of the Sub-Committee on Bush and Bush Encroachment*,
8th Meeting of SARCCUS, Mimeogr., 1961.

[27] EIG, A., 'Les éléments et les groupes phytogéographiques auxiliaires dans la
flore palestinienne I, II', *Fedde. Repert. Spec. Nov. Regni. Veg. Beih.*, No.
63, 1931–2, pp. 1–201.

[28] AWAD, H., 'Nomadism in the Arab lands of the Middle East', *The Problems
of the Arid Zone, Proc. Paris Symp. Unesco, 1960*, Arid Zone Research
XVIII, 1962, pp. 325–39.

[29] BARTH, F., 'Nomadism in the mountain and plateau areas of South-west
Asia', ibid., pp. 341–55.

[30] BREMAUD, O. and J. PAGOT, 'Grazing lands, nomadism and transhumance
in the Sahel', ibid., pp. 311–24.

[31] CAPOT-REY, R., 'The present state of nomadism in the Sahara', ibid., pp.
301–10.

[32] FOOD AND AGRICULTURE ORGANISATION OF THE UNITED NATIONS,
'Nomadic pastoralism as a method of land use', ibid., pp. 357–67.

[33] BRAIDWOOD, R. J., 'The earliest village communities of south-western Asia
reconsidered', *Atti VI Congr. Internaz. Sci. Preistor. Protoistor*, Vol. 1,
1962, pp. 115–26.

[34] ZEUNER, F. E., 'Domestication of animals', in Singer *et al.*, *A History of
Technology*, Oxford, 1954, pp. 327–52.

[35] REED, C. A., 'Animal domestication in the prehistoric Near East. The
origins and history of domestication are beginning to emerge from archaeological excavations', *Science*, Vol. 130, 1959, pp. 1629–39.

[36] KENYON, K. M., 'Jericho and its setting in Near Eastern history', *Antiquity*,
Vol. 30, 1956, pp. 184–95.

[37] TOYNBEE, A. J., *A Study of History* (2nd ed.), Oxford, 1935. 3 vols.

[38] DONOVAN, P. A., 'Crop ecology and farming systems', *Proc. Fifth Ann. Conf.
Professional Officers, Dept. Res. and Specialist Services, Fed. Min. Agric.*,
Rhodesia/Nyasaland, 1960, pp. 81–95.

[39] FOOD AND AGRICULTURE ORGANISATION OF THE UNITED NATIONS, *FAO/Unesco/WMO Interagency Project on Agroclimatology, Draft General Report on a Study of Agroclimatology in Semi-Arid and Arid Zones of the Near East*, Rome, 1962, Mimeogr.

[40] ANDERSON, R. and R. R. STAPLES, *An Agricultural Survey of Southern Rhodesia*, Part 2, The Agro-Economic Survey, Dept. Native Agriculture of Southern Rhodesia, Salisbury, n.d.

[41] WHYTE, R. O. and G. JULÉN, *Proc. FAO Techn. Meeting on Plant Exploration and Introduction, Rome, July 1961*, Genetica Agraria (Special Number), Vol. XVII, 1963.

[42] FRANKEL, O. H. and A. MUNDAY, 'The evolution of wheat', *The Evolution of Living Organisms*, Symp. Roy. Soc. Vict., Melbourne, Dec. *1959*, 1962, pp. 173–80.

[43] HELBAEK, H., 'Domestication of food plants in the Old World', *Science*, Vol. 130, 1959, p. 365.

[44] — 'Notes on the evolution and history of *Linum. Kuml*', *Aarhus*, October 1959, pp. 103–29.

[45] — 'Late Bronze Age and Byzantine crops at Beycesultan in Anatolia', *Anatol. Stud.*, 11, 1961, pp. 77–97.

[46] PERRIN DE BRICHAMBAUT, G., 'Contribution de la climatologie à la prospection et à l'introduction du matériel végétal', *Proc. FAO Techn. Meeting on Plant Exploration and Introduction, Rome, July 1961*, Genetica Agraria (Special Number), Vol. XVII, 1963, pp. 71–82.

[47] FRANKEL, O. H., 'Invasion and evolution of plants in Australia and New Zealand', *Caryologia*, 6 suppl., 1954, pp. 600–19.

[48] WHYTE, R. O., 'Plant exploration, collection and introduction', *FAO Agric. Study No. 41*, 1958.

[49] HARLAND, S. C., 'Genetics and the world's food', F. le Gros Clark and N. W. Pirie (Eds.), *Four Thousand Million Mouths*, London, 1951, pp. 93–108.

[50] WHYTE, R. O., 'Grazing resources and land use planning,' *J. Brit. Grassl. Soc.*, 1963.

[51] — 'International development of grazing and fodder resources. XI. India (sequel)', ibid., Vol. 17, 1962, pp. 287–93.

ADDITIONAL REFERENCES

AUDRY, P. and CH. ROSSETTI, *Observations sur les sols et la végétation en Mauritanie du SE et sur la bordure adjacente du Mali*, FAO, Rome, 1962.

STAMP, L. D., *A History of Land Use in Arid Regions*, Unesco, Paris, 1961 (Arid Zone Research XVII).

UNESCO, *Land Use in Semi-arid Mediterranean Climates*, Unesco IGU Symp. Iraklion, Greece, Unesco, Paris, 1964 (Arid Zone Research XXVI).

CHAPTER XVI

The Improvement of Animals through Introductions and Breeding

In all arid countries the animals kept by man provide food, fibre, hides and transport. Because of their close association with man these animals are generally regarded as being domesticated, and during the course of our present civilisation they have wrought vast changes in the arid parts of most countries [1]. These changes emphasise the need to remember the interactions between the environment and the animals we seek to improve through introductions and breeding.

In many arid areas, low levels of animal production per unit area of land result largely from inadequate forage, which in turn is partly due to the effects that grazing animals have had upon plant communities. Therefore in planning the improvement of domestic animals in the arid zone consideration must also be given to the extent to which low productivity is due to the environment that the animals and local management have created, as well as to the obvious disadvantages of aridity, extremes of temperatures, parasitism and disease, and any lack of critically conceived plans for animal improvement through introductions and breeding. Research into the improvement of animals is in progress in practically every country in the arid zone. Reference can be made in this paper to but few of these efforts, and those mentioned have been selected mainly to illustrate certain points. However, research is not enough; progress can be attained only if the knowledge that research has won can be inculcated into the thinking and methods used by the innumerable herdsmen who breed animals in the arid zone.

The use of animals in the arid zone

Domestic animals are kept primarily as a source of food, although their use for this purpose is restricted amongst some Hindus. The meat and the milk provided by horses, camels, donkeys, cattle, sheep and goats are consumed by humans in many arid countries, and all animals have been used in varying degrees as beasts of burden. However, some animals are required to provide special products. In

363

pastoral Kenya the Masai tribesmen add themselves to the burdens of parasites that their animals bear by drinking blood repeatedly drawn from live cattle [2]. In some other countries flesh is consumed after its removal from living animals, which are subsequently fattened for sale or slaughter.

It should not be assumed that the majority of the animal products from the arid zone are consumed by those who own and tend the animals. Many arid countries have a considerable export trade of meat, hair, wool, and of hides.

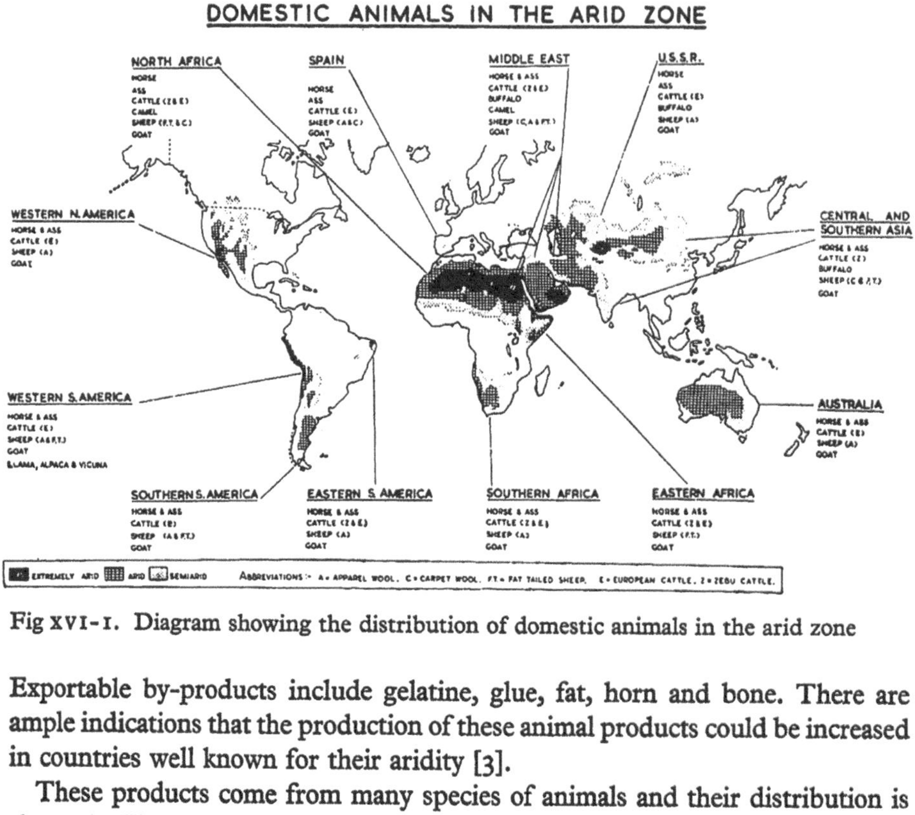

Fig XVI-I. Diagram showing the distribution of domestic animals in the arid zone

Exportable by-products include gelatine, glue, fat, horn and bone. There are ample indications that the production of these animal products could be increased in countries well known for their aridity [3].

These products come from many species of animals and their distribution is shown in Fig XVI-I. Camels can be divided conveniently into the two-humped or Bactrian, found mainly in northern Asia, and the single-humped or Arabian camel, which is common in northern Africa, the Middle East, Pakistan and India. Horses can be classified in broad terms as being light, i.e. for riding and for the drawing of light carriages, or as heavy, i.e. those used mainly for draught purposes. Cattle are usually classified according to their origin and conformation. Those with humped shoulders are of *Bos indicus*[1] origin and are usually referred to as being Zebu type; those without humps are of *Bos taurus*[1] origin and are commonly referred to as European cattle. Sheep are classified according to their wool as being either apparel or carpet wool producers, and according to their

---

[1] As these two types of cattle are completely interfertile, they should be classified as races of a single species.

conformation as being fat-tailed, fat-rumped or thin, long-tailed. Goats are classified according to their uses, i.e. for the production of meat, milk or of fibres, or a combination of these.

Principles of animal breeding

Many biological principles underlying the practice of animal breeding are now well established. The animal's inheritance (genotype) is controlled by genes, the frequency of which partly determines the number of individuals in an unselected population that express the various alternative forms (known as the phenotype). Some of these forms are more desirable than others, depending upon the place, and the purposes for which the animals are kept.

As the frequencies of the various genes that influence the expression of a particular character determine the genetic merit of any population of animals, i.e. their worth as parents of future generations, it is important to consider the ways in which gene frequency can be changed.

The two most important methods of altering gene frequency in domestic animals are the introduction of new genes (i.e. crossbreeding), and selection. Introducing animals from other population changess gene frequencies in the resulting crossbred populations, provided that gene frequencies differ between the two groups. This is the only way to introduce genes that are lacking, and the continued introduction of animals of the same breed (known commonly as 'grading up') is an important way of increasing the frequency of rare and/or desirable genes. By this process gene frequency can be greatly changed in a relatively short time. However, the rate of change of gene frequency may decrease as the recipient flock or herd becomes more like the donor flock or herd.

Selection within a population is the method most commonly used by the majority of animal breeders to change gene frequency. However, in the arid zone selection and improvement through introduction may be seriously hampered through the lack of control over breeding animals.

The amount of selection that can be practised in a breeding flock will depend upon the net reproduction rate, which is estimated by the number of young females that attain breeding age [4, 5].

Differences between animals in an unselected population provide the basis for selection. If the same character is measured for each animal in an unselected population and the results are arrayed to show the percentage of animals falling into each class interval, it will quickly become apparent that there is a normal distribution around the mean.

However, not all of the variance between the animals is due to inheritance [6]. Some characters are expressed only in a particular environment; for instance differences between animals in their ability to conserve body fluids may be neither obvious nor important in a temperate climate where adequate green feed and water are readily available, but its importance could be quickly manifest in a hot, arid environment where animals walk considerable distances to drink and to collect sufficient of the sparse forage in order to survive. Such interactions between inheritance and environment will be a hindrance to selection, especially when introductions of breeding stock are chosen from populations long established in an environment very different from that of the recipient flock or herd [7]. This mistake has been made repeatedly; ample evidence of it exists in Africa, the United States of America and in Australia.

That part of the differences between animals resulting from environmental effects and from the interaction between environment and inheritance is usually larger than that due collectively to the inherited factors alone. The variance due to genetic factors acting alone is known as the heritability of a character. Heritability is usually expressed as a decimal, or as a percentage, and it is conventional to refer to characters with a low heritability, a medium heritability or a high heritability [6].

The results likely to be achieved from selection depend upon the heritability of the character and the intensity and accuracy of selection. If selection is both accurate and intense the average merit for the character of the animals chosen as potential breeders will be considerably superior to the average merit of these two groups, known as the 'selection differential'. For instance, if five rams whose mean fleece weight is 7 kg are chosen from a large flock whose mean weight is 5 kg a selection differential of 2 kg (7 — 5 kg) is achieved. However, not all of this difference will be transmitted to their offspring; the approximate amount that will be passed on to the average performance of the unselected offspring can be estimated from the formula:

$$\frac{SD \female + SD \male}{2} \times H$$

where $SD \female$ = selection differential from the selection of females
$SD \male$ = selection differential from the selection of males
H = heritability of the character under consideration.

In most programmes of animal improvement by selective breeding, consideration has to be given to several characters at once, so that the intensity of selection and therefore the selection differential achieved for each character will be less than if only one character is considered. The position is further complicated by the occurrence of genotypic correlations between characters which might be either positive or negative. In the former case selection for one character leads automatically to progress in those that are genetically correlated. This situation can be exploited if a desirable character that is difficult to measure, or which does not become manifest until later in life, for example age at first calving, is positively correlated with one which is easy to measure, or one, such as growth rate, which is manifest early in life. In the latter situation selection for one character may also embody selection against another, and so in the event of both characters being desirable a compromise must be made.

Therefore the animal breeder is confronted with the need to decide upon the method of selection that will give the fastest rate of improvement for the range of characters that he considers desirable. There are three alternatives [6]. In the first method, known as 'tandem selection', there is direct selection for one character and then for another amongst the animals remaining after the first selection. The second is known as the method of 'independent culling', in which culling is first on the basis of one character and then on the basis of another so that the selected group comprises only animals which reach the required standard for each character. A third method is to combine all the required characters in a single index, allowing each character to be considered according to its relative importance, so finally choosing as parents those animals with the best total scores.

Selection based on heritability estimates of productive characters has certain limitations. For instance, selection can operate only on the variation that is already present in the population, and if this variation is small the rate of progress is severely limited. In these circumstances it would be preferable first to increase variation by crossbreeding, and then by selection to establish new populations of animals that have enhanced productivity or special adaptive functions.

If the heritabilities of the desired characters are low such different methods as progeny testing may have to be used in order to select prepotent sires. However, progeny testing is time-consuming and, because the sires are usually old before they are proven, it increases generation length and thereby tends to retard the rate of genetic progress.

Some physiological requirements of animals in the arid zone

Animal production is essentially the outcome of the interaction of a number of physiological processes [8], but the arid zone makes special demands upon animals. Some parts of the arid zone may be altered by irrigation, and by the planting of shade and shelter belts, but in the final analysis this is likely to affect comparatively small areas and proportionately few domestic animals. For the majority of the animal population the tremendous problems of scarce, poor-quality food, inadequate brackish water, merciless heat or cold, intense radiation, high winds, dust and persistent parasitism are likely to prevail.

In general terms the type of animal most likely to succeed in the arid zone needs to have an integument that will protect it from insolation, and from extremes of temperature. This protection may be achieved through a light-coloured or shiny coat of high reflectance, yet dense enough to provide insulation. Mechanisms for heat dissipation need to be well developed, and whilst the most important of these are based on evaporative loss of water to provide respiratory and/or cutaneous cooling, ability to conserve body fluids is an important physiological prerequisite of animals for the arid zone. Fluid conservation can be achieved largely by the concentration of urine by the kidneys, although reserves of fluid in the rumen and in the extracellular space may also make important contributions to the adaptation of animals to aridity. Reserve stores of fat such as occur in the humps of the camel, and of *Bos indicus* cattle, and in the fat tails

N

or rumps of some sheep, are useful to assist survival in areas where food supplies are uncertain.

The advantages or disadvantages of seasonal breeding have yet to be assessed in the arid zone. Many factors contribute to satisfactory reproduction rates; mating may be influenced by variations in day length, in feed, or in both; conception rates should preferably be high and can be affected adversely by high air temperatures [9]. The majority of the new-born animals should survive; survival depends partly upon the plane of nutrition enjoyed by the breeding females, and also upon the effects of local environmental factors on the new-born animal and for a period after birth [10] including, in some countries, animal predators. The extent to which unfavourable conditions can be avoided, and favourable ones can be made to coincide with the physiological requirements for reproduction and survival, vary with climatic conditions, the length of the gestation period and husbandry practices.

Animal behaviour is also an important factor in the adaptation of animals to the arid zone, for example an ability to graze in the bright sunshine where temperatures may be high, a capacity to eat and digest xerophytic plants, to gather feed without frequent recourse to water, and an ability to walk considerable distances [11]. These aspects have been dealt with in greater detail in Chapter IX, but the underlying principles apply equally to domestic animals, especially where introductions from other breeds, and from other climatic zones, are made in order 'to improve' the indigenous domestic livestock.

Animal breeding in the arid zone

A good deal of information is now available about many of the breeds of domesticated animals [12–16]. However, care is necessary not to place undue emphasis on breed. It is true that there are marked differences between breeds in their potential for certain types of production, for example the difference between dairy and beef breeds of cattle in their capacity to produce milk and meat. However, in all breeds developed by man considerably more attention has been paid to the fixing of characters which are more in the nature of 'trade marks', than to increasing the frequency of those that are positively related to productive or adaptive functions.

There is field evidence that wide physiological variations related to productivity [17] occur between sheep of the same breed. Some of the mechanisms that underlie these differences, for example those that influence reproduction by the ram [18] and the survival of new-born lambs, have been investigated. Evidence such as this suggests that physiological studies of this kind could well provide useful leads for selecting animals especially suited to the arid zone. The remarkable ability of the camel [19] and of the Merino sheep [20] to conserve body fluids, and thereby to retain nitrogenous waste that would normally be excreted in the urine, may also be physiological functions that assist these animals to adapt to shortages of food and of water, so characteristic of the arid environment.

Other useful information is available about the associations between the nature of the coat of cattle and their ability to withstand high air temperatures. The physiological relationships between these traits have not been established, but the association is sufficiently strong and important enough to warrant the consideration of coat type in the selection of cattle for hot, arid environments. Unfortunately no information about measured productive performance of draught animals and those used as beasts of burden seems to be readily available.

The basic biological strategy for the improvement of domestic animals in the arid zone can be advanced in a step-wise fashion as follows:

1. Define and rank the productive characters required in the livestock; the number of characters should be kept as few as possible.
2. Measure the productive potential of the indigenous breeds and those that have already been introduced. If further introductions are considered they should be sought first in proprioclimes [21].
3. Estimate the genetic situation that pertains to the productive characters in the chosen breeds. This procedure will need to provide estimates of the heritability of the defined productive characters, the relationships between them, and the interactions between genetically different animals and the environment under consideration.
4. Determine the influence of sex, age, seasonal and environmental conditions on productivity.
5. Formulate breeding plans based on the results obtained in the four previous steps, and involving decisions upon:
   (i) The type of management to be followed; i.e. the kind of product, whether the animals will be pure bred or crossbred, and whether the breeding group will be self-maintaining.
   (ii) The methods of selection.
   (iii) Mating systems.

The logical extension of this step-wise approach is to investigate the physiological basis of differences in production and to estimate the heritability of these traits. Unfortunately our knowledge is only just beginning to advance sufficiently for this stage to be exploited. However, the physiological requirements of animals in the arid zone must be kept in mind, especially as many are relatively simple to measure and there is already suggestive evidence of their value.

The improvement of cattle

SELECTION

Fortunately, estimates of the heritability of some of the productive characters in British breeds of beef cattle are now known [22] and these provide an opportunity for more rational selection, directed towards improving the productivity of existing breeds. The use of this information by stud breeders can ensure that genetic improvement will be passed on to those herds that consistently purchase bulls from studs that are progressing. Breeders of beef cattle are generally interested in a combination of growth characteristics, efficiency in converting forage to flesh, the quality of the flesh, and in regular reproduction. The characters that determine these qualities all appear to be highly heritable so that selection for them should be worth while. Characters that can be measured easily are weaning weight, and live-weight gain while grazing, and final feed-lot gain. These characters are important, especially as many cattle bred in arid parts find their way into feed lots, or to lush pastures for fattening before slaughter.

The heritability of most of these characters is sufficiently high to warrant the selection of sires on the basis of their own records for measured performance, and there is no need for delay in order to select sires on the basis of the measured performance of their offspring.

Extensive feeding tests are difficult and costly to conduct. However, some private owners of cattle herds in arid parts of the United States of America and of Australia are recording rate of gain by periodic weighing of animals that are grazing or being 'lot fed'. The combination of these objective measurements and the heritability estimates for the characters has been of great value as an aid to selection programmes. However, the amount of actual improvement is also influenced by the genetic correlations between such characters; in other words, it is necessary to know what effect selection for one trait has upon another. Unfortunately this kind of information is not yet available, although promising beginnings have been made to estimate them. For instance it has been shown that there is a positive genetic relationship between growth rates at different ages, so that selection for size or growth at one age will lead to some improvement at all ages. For growth comparisons to be valid, all animals must be subjected to the same environmental conditions.

There is also some evidence of negative genetic relationships between certain characters, for instance between the nursing ability of cows and the ability of their calves to gain weight [22]. This situation requires that selection be for both characters, since undue emphasis on one would decrease the other. Selection may also need to be practised for special characters. For example, in both the Union of South Africa and in Australia, Hereford cattle which have red coats but have white faces, are very prone to cancer of the third eyelid. Successful selection for a ring of brown-pigmented skin around the edges of the eyelids is reported to have overcome this problem. Similarly there is evidence that in tropical climates smooth-coated cattle have advantages in body-heat regulation over woolly-coated cattle [23], indicating that the selection of smooth-coated cattle is well worth while.

THE INTRODUCTION OF EXOTIC CATTLE

Numerous examples of the successful introduction of exotic cattle into arid regions can be cited. One of the most specialised examples comes from Israel. In the circumstances that prevailed under the Palestinian Mandate, the Syrian and Lebanese cattle produced little milk, although they provided power to draw carts, and were finally slaughtered for meat. Jewish settlers improved the level of nutrition and general husbandry of cattle in Israel and thereby doubled milk production to about 600–800 litres/year, but this low figure seemed to be the limit of the genetic potential. Crosses between the local cattle and animals imported from genetically different breeds in Syria and Lebanon led to increases in milk production to 1,000–2,000 litres/year. However, the establishment of

Fig XVI-2. Boran cow – a Zebu type well adapted to tropical conditions in semi-arid Africa. (*Photo Kenya Information Services*)

improved pastures under irrigation, combined with adequate measures for the control of disease, paved the way for the introduction of Friesian cattle whose geneticoproductive influence was rapidly extended by the use of artificial insemination, with the result that milk production has increased now to 6,000–8,000 litres/year. Artificial insemination is also being used in Iran to hasten the grading up of dairy cows to Brown Swiss bulls.

In Kenya, a stud of Sahiwal cattle has been established by the Department of Veterinary Services, and is being used by artificial insemination to grade up the local cattle owned by the African small-holders and farmers. In consequence substantial increases in milk production are being achieved. In both tropical Africa and tropical Australia, cattle of the British breeds introduced by early European settlers are now being crossed with cattle of the *Bos indicus* type in a deliberate attempt to improve adaptation to the tropical environment (see

Fig xvi-2). Effort is also being devoted to this subject by scientists in the USA, Africa and Australia. In Australia attempts are being made to identify those characters that might contribute to improved animal production. Breeding animals are being selected on the basis of:

(a) Ability to hold their body temperature within normal physiological ranges during exposure to high air temperatures. For example, animals able to maintain rectal temperatures at 39·1°C during the summer gained 100 kg in live weight more than those whose temperatures rose to 39·1°C.

(b) Suitable coat type, as it has been found that cattle with a short sleek coat gain weight faster and are more fecund than animals with a long rough coat.

Special attention must always be given to fecundity in any breeding programmes that include Zebu type cattle, as their reproduction rates tend to be lower than

Fig xvi-3. Beefmaster bull at Falfurius, Texas, USA. The animals in this herd are selected for docility as well as for such productive traits as high growth rates

those of British breeds. However, individual animals adapted to arid tropical conditions occur within some of the existing populations of British breeds of cattle [24]. Thus it is always advisable to examine the existing populations under reasonable conditions of nutrition and disease control before embarking upon an expensive, long-term project of introducing new breeds, or seeking animals that might be useful for grading up. This precaution is especially important in countries where serious infectious diseases are endemic, as there are many examples of these having destroyed the majority of animals in importations of exotic breeds. However, this difficulty can be largely overcome by importing deep-frozen semen from suitable sires in other countries.

THE DEVELOPMENT OF NEW BREEDS OF CATTLE

During this century notable progress has been achieved in the development of new breeds of cattle [13]. The Bonsmara breed which was developed at the Mara Experimental Station, South Africa, is now being tested under field conditions for adaptation to high temperature, tick-borne diseases and aridity. The Bonsmara contains $\frac{5}{8}$ Afrikander, $\frac{3}{16}$ Shorthorn and $\frac{3}{16}$ Hereford blood.

The Santa Gertrudis breed, developed on the King Ranch, Texas, USA, by intensive selection among Shorthorn × Zebu crossbreds, crossed back to Short-horns, is now established in North and South America, Australia, Africa and the Philippine Islands. The Beefmaster breed [25, 26] is of special interest as it has been established in less than fifty years by intensive selection for six characters, viz. disposition, fertility, weight, conformation, hardiness and milk production. No selection effort was dissipated on colour or similar distinguishing marks. The degree of success that has been achieved is clear from Fig XVI-3 (see Fig XVI-4).

The improvement of sheep

Broad-tailed sheep apparently originated in the Gobi Desert, and the fat-tailed variety migrated to Africa. Long-tailed sheep have their origin somewhere in Europe. Fat-rumped sheep are a variant of the fat-tailed breeds. Originally all wild sheep had a covering consisting of a stiff coarse hairy outer coat, and an extremely fine woolly undercoat. Many of the carpet wool breeds today resemble the general appearance of their ancestors, but considerable changes have been wrought in the apparel wool sheep.

MEASUREMENT AS AN AID TO SELECTION

The majority of the apparel wool-growing sheep in the arid zone are in Australia, the Union of Soviet Socialist Republics, South Africa, Argentine and the United States of America. These countries are favoured with a highly developed tech-nology, whilst in the USSR sheep and wool production is under government control. Considerable effort has been directed towards the improvement through breeding, of apparel wool-growing sheep, although different approaches have been used in each country.

The majority of sheep in Australia are Merinos that are privately owned and are part of a stable, as distinct from a nomadic, pastoral ecosystem. Few sheep are allowed access to seasonal ranges, as occurs on the intermountain plateaux of the United States of America [27].

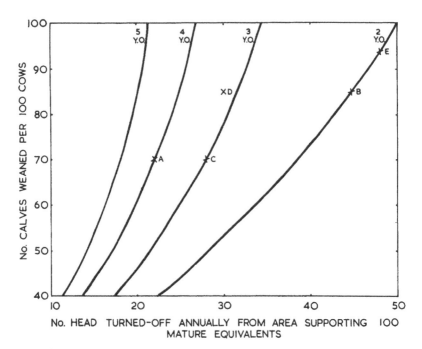

Fig XVI-4. The effect of changes in fertility and growth rate on the number of cattle that can be marketed each year. The turn-off is expressed, not precisely as a percentage of the total number of head run, but as the number turned off from an area capable of supporting 100 mature cows, or a proportionately larger number of cattle of mixed ages. For instance, in particular circumstances, such an area might carry 42 breeding cows with calves at foot, 29 yearlings, 28 two-year-old, and 26 three-year-old, a total of 125 head.

For illustrative purposes, the curves have been derived on the basis of the following arbitrary assumptions: (i) Cows have a mean active breeding life of 5 years, so 20 per cent of their number are replaced by heifers each year. (ii) The number of cows sold after allowing for mortality, and the value of these old cows, are such that half the market value of replacement heifers is recovered. (iii) The area needed to support a beast is proportional to its body weight. (iv) Mortality in calves after the first year is 5 per cent per annum.

These curves show, for instance, that the net turn-off would be doubled if a fairly typical herd (Point A), with a weaning percentage of 70 and a market age of 4 years, were improved to have a weaning percentage of 85 and a market age of 2 years (Point B). If *Bos indicus* bulls are put into a herd, an immediate increase in growth rate without change in fertility may be obtained (Point C). Crossbred cows may lift fertility but produce calves growing not quite as fast as first cross animals (Point D). Simple selection procedures without crossbreeding may permit fairly rapid movement from Point A to Point D, and more critical selection over a longer period will produce progressive upward movement.

Pasture improvement and other aspects of nutrition and management will materially affect these statistics. Independent of changes in these, the aim of genetic improvement is to modify the characteristics of cattle so that, in specific environments, the situation moves towards a goal such as Point E, from a starting point such as Point A. The available statistics indicate that in Texas, USA, herds of Beefmaster cattle are approaching points B and E. (*Figure by courtesy H. G. Turner, CSIRO, Cattle Research Laboratory, Rockhampton, Queensland*)

The Australian wool-growing industry supports numerous studs, which supply rams to the country's 110,000 independently owned flocks. The hierarchy [28] of the stud sheep flocks is situated predominantly in the arid zone and comprises a small group (i.e. 15) of 'parent' studs that have been closed to introductions for 50-100 years. Next in the hierarchy is a larger group of about 300 'daughter' studs. Each daughter stud consistently buys rams from one particular 'parent' stud or from another 'daughter' of the same 'parent' stud, so that genetic changes in the parent stud are transmitted quite rapidly, i.e. within 3–5 sheep generations, to its daughter studs. Below the daughter studs a large group of about 1,800 'general' studs exist. The general studs do not consistently obtain rams from any particular parent or daughter stud; they may buy sires from either or from one another. Practically all the rams used in the commercial wool-growing flocks originate in one or other of the stud flocks.

The net effect of this situation is that the parent studs are able to influence the genetic changes in virtually all Australian commercial wool-growing flocks. This breed structure permits any increases in production achieved by selection and breeding in the parent studs to be spread throughout the nation's flocks as a whole.

The breeding ewes in stud flocks are usually classified into three grades depending upon merit, according to standards set by the stud master [29]. Since stud rams are generally used for five to seven years, intensive selection can be exercised in choosing replacement rams to mate with the small group of top-grade ewes.

A tentative choice is made each year when the young rams are about eighteen months of age. About three times the number of rams that will be required are placed in a selected group that is reserved from sale. When these 'reserve' rams are about thirty months of age they are re-evaluated, and a final choice is made of those that will be used in the stud flock.

The heritabilities of the most important characters that contribute to the quantity and the quality of the wool produced by Australian Merinos have been estimated, and the measurement of variation in fleece weight, staple length, fibre diameter, percentage clean scoured yield and crimp frequency can be an important aid to the selection of high producing sheep [30]. As the heritabilities of the majority of these fleece characters for which selection should be practised is high, it follows that the repeatability of the order of ranking from year to year is also reasonably high. Therefore measurement of such fleece characters on the first adult fleece shorn at eighteen months of age are useful indices of future performance and can be used as an aid to selection.

There are certain disadvantages in the operation of the Australian stud sheep industry and its methods of selection, which extend generation lengths (i.e. the average age at which breeding sheep have offspring), and it may also favour late maturity. However, the adoption of objective measurements in the operational practices followed by the industry is becoming acceptable within its organisation and social structure [31]. Special attention may have to be given to such characters as skin wrinkling and folding, and excessive wool round the eyes, since both these characters are related to lowered fertility [32]. Similarly, the selection of lambs for fineness of birth coat, which is reflected in more uniform fleece wool in later life, may depress reproduction rates [33].

Fig XVI-5. Weighing fleeces from Merino sheep in an Australian shearing shed

THE IMPROVEMENT OF SHEEP IN THE USSR

Apparel wool sheep, which include sixteen breeds, dominate two of the Soviet Union's six principal sheep-farming zones. The arid parts of these zones extend eastward from the Caspian Sea through Kazakh SSR, Altai, Irkutsh Buryat-Mougol, and Chita, whose frontier is with Manchuria. Spectacular increases in the numbers and productivity of apparel wool sheep have been reported from the USSR [34] and these have been attributed largely to the widespread use of artificial insemination, and the development of new breeds by crossbreeding followed by intensive selection.

All the sheep in USSR are on either state or collective farms, where breeding and selection programmes are controlled. Leading Merino sires are reported to produce greasy fleeces exceeding 17 kg, and their body weights to exceed 100 kg. The average weight of fleeces from fine-woolled sheep on state farms in 1957 was 5 kg. However, the average fleece weight of ewes on the state farm at Stavropol is 7 kg, and although it is not easy to obtain a clear picture of the position, there seems little doubt that productivity per head of the Soviet flock has increased considerably during the last 20-30 years. As the industry is organised and has uniform control, adequate knowledge of genetics, good technology and the extensive use of artificial insemination, it seems reasonable to attribute an appropriate proportion of the increased productivity to genetic changes [35].

There is a long history of importation of western breeds of sheep into Russia, and there are numerous accounts of heavy mortality and low productivity of survivors. Many of the importations were mutton type sheep, and failures were partly due to inadequate nutrition. Nevertheless, considerable success was achieved with Merinos, the original importations of which were Württemburg and Précoce Merinos from Germany, Rambouillets from the USA and the remainder Australian. The Mazaev Merino was developed in the south-eastern Ukraine, and was later crossed with the German Merinos and Rambouillets to produce the modern Novocaucasian Merino, and these two Russian breeds are the parent stock of a large portion of the sheep population in the USSR today.

New breeds, whose names spring from the territories in which they were developed, are a feature of the sheep industry in modern Russia. Thus the Altai breed was evolved in the region of the Altai mountains when Rambouillet, Australian Merino and Caucasian Rambouillet rams were mated with the indigenous Siberian ewes. The result appears to be a fertile breed, well suited to the arid, hilly regions, and which produces up to 6·5 kg of wool. The rams weigh up to 110 kg and the ewes up to 65 kg.

The fat-rumped sheep indigenous to the Republic of Kazakhstan were improved by crossing with Précoce rams, and line breeding with Askanian Rambouillets, Altai and Grozny rams. This new breed, known as the Kazakh Finewool, is reported to have a dense fleece of 60–64's quality and 8–9 cm staple. Fleece weights average 4·5 kg.

The Kirgiz Finewool breed was developed in the mountainous republic of Kirgizia, south-east of Kazakhstan. The indigenous fat-rumped, coarse-woolled sheep were first improved by crossing with Novocaucasian and Siberian Merinos, and later with Württemburg Merino, Rambouillet and Précoce rams to form the new finewool breed, which produces 4·5 kg fine wool with a staple length of 7–8 cm.

Another breed was developed especially to use the extensive pastures at high altitudes in the mountains of Kazakhstan. Merino ewes were inseminated artificially with semen from wild Arkhar rams and the resulting breed, which was developed by careful selection, is reported to have produced wool of a 64–70's quality and 7–8 cm staple length, and the sheep were able to remain at pasture throughout the year, in a very severe environment.

THE IMPROVEMENT OF CARPET WOOL SHEEP

The carpet wool sheep are the oldest, but least improved of the domestic sheep, although there is a strong body of opinion in favour of effecting improvement. Average fleece weights from carpet wool sheep in USSR, China, the countries of the Middle East, India and Pakistan have shown little signs of increase in recent times, and complaints about the large proportion of parti-coloured fleeces are long standing.

There are many types of carpet wool sheep, which are found mainly in the different environments in which they were developed. The different types tend to take their names from their specific localities, whereas there are probably only a few genetically distinct breeds.

The fat-tailed Karakul sheep, indigenous to Bokhara, Samarkand and Persia, are of special interest. Some of the young Karakul lambs have a curly birth coat, known in the trade as 'Persian lamb'. The males are killed at three days of age, and their tightly curled black, grey or silver skins are highly valued for coats. 'Astrakhan' pelts are obtained from other types of inferior Karakul lambs and similar related breeds. Adult Karakul wool is coarse and long, and is used for carpet manufacture. Karakul rams have been imported into many countries, and have been particularly useful in South Africa when used on the fat-tailed Afrikander ewes, found in country that usually receives only 7 inches of rain a year. There is also considerable scope for the establishment of a 'Persian lamb' industry in other arid regions, in Somaliland, for example, by crossing Karakul rams on the indigenous ewes. The development of such a project would have a very advantageous effect upon the economy of the tribesmen in these regions.

A project started in Jordan in 1958 for the improvement of the Awassi sheep which predominate in the Middle East and in Africa, has already achieved considerable increases in both milk and lambs, and a substantial decline in the number of coloured and off-type fleeces. The Karadi is probably the dominant breed in Arabia, and its wool is in demand because it is mostly white. The Colchian, which originated in Asiatic Russia, is closely related to the Karadi and is important in China. The new-born Colchian lambs also have curly coats, like the Karakul. Generally, wool from Chinese sheep has a high proportion of kemp (i.e. medullated fibres), which dyes differently from non-medullated fibres. Attempts to improve Chinese sheep by crossbreeding have not been successful, due largely to inadequate nutrition [36].

The sheep of Pakistan and India are bred mainly for carpet wool, and for mutton, although they are sometimes used as beasts of burden. Some of the most sought-after carpet wools, which come from Afghanistan and Baluchistan, are exported through Karachi, although Pakistan is also concentrating on the production of carpet wools. In addition Pakistan has some useful breeds of fat-tailed sheep. Rajasthan is the principal wool-growing state of India. The indigenous sheep seem to be well adapted to the hot arid conditions and, although average fleece weights are not very high, the individual fibres combine softness with resilience.

The sheep of central and southern India are generally smaller than those of the north and north-west and they are hairy [14]. Outstanding improvements have been achieved within experimental flocks of the Deccan breed of sheep, which take their name from the plateau on which they occur. Selective breeding carried out at the Sheep Breeding Station, Poona, Bombay Province, has established greater uniformity in both the coarse fibres of the outer coat, and the fine fibres of the inner coat, while the quality and yield of both have been greatly improved.

It seems likely that Turkey, which is using fleece measurement and all other available technical methods, will provide a useful lead for the improvement of carpet and apparel wool sheep in the Middle East and Asia generally. There has

Fig XVI-6. Marwari ram – a carpet wool type of sheep indigenous to India

been little crossing of the indigenous sheep of Turkey, which are found in distinct geographical areas. The Karaman has a large fat tail, and produces coarse, coloured carpet wool, of 32–36's quality. One strain of this breed, the Akkaraman, is white and produces a superior quality white wool. The Daglic resembles the Akkaraman, and is also a carpet wool breed.

In the past, considerable effort has been expended on the selection of sheep with fat tails, as these are considered to be a delicacy by Moslems, whose religion precludes the consumption of pork. However, in the light of present knowledge, considerable improvements in wool production could be achieved by selection within indigenous breeds for desired fleece characters and fleece weight, and against coloured fibres. These steps can be taken with a minimum of technical assistance and without expense.

The improvement of goats

Goats are used for the production of meat, milk, wool and hair. They occur in all arid areas, where they make important contributions to human welfare. Despite the importance of milk goats there is comparatively little published information about efforts to improve milk production. In many arid areas, and especially in the Middle East and in Russia, the milk of the ewe is used for human consumption, and more effort seems to have been directed towards the

Fig XVI-7. The camel is an important beast of burden in many areas in the arid zone

improvement of the milking capacity of sheep than of goats. Even the more productive of the Russian breeds of goats like the Mingrelian and Romvit yield only about 600–800 litres milk/year.

The Saanen, which is an outstanding milk producer, has been imported into many countries and in Russia crossbred Saanens are reported to have produced 800–1,200 litres/year. However, these levels of production cannot be expected unless nutrition is also adequate [34].

The Angora is the outstanding fibre-producing goat. This breed is strongly established in Texas, USA, as well as in Turkey and the Union of South Africa. Angoras were imported into USSR in 1936, but succeeded only in Askaniya Nova in the Ukraine, and fleece weights of the recipient herds rose to 5-6 kg. In evaluating these importations Russian workers concluded that the introduced goats produced superior crossbreds. More recent reports mention the development of new breeds of goats in Russia, although fleece yields are not very different from those of the indigenous animals.

The improvement of camels

Although camels are important in most arid countries, little seems to have been done to effect their improvement through selection. Camels of average size and strength may conveniently carry between 120 and 140 kilos, although heavier loads are often placed on them. Good riding camels travel at not less than 6 to 7 miles per hour. Camel hair is used for making blankets and floor rugs, and for these purposes the hair of young animals is preferred.

A camel-breeding farm has been established conjointly by the Indian Council of Agricultural Research and the Government of Rajasthan at Bikaner. Selection is based on conformation, and especially that of the feet and limbs, and the animal's ability to travel long distances, i.e. 50 miles in the relatively short time of five hours. Bull camels bred on the farm are distributed free of charge to local breeders.

The strategy and tactics of animal improvement in the arid zone

There is an obvious need to improve the productivity of animals in the arid zone, but many factors combine to impede progress. Useful technical information is available, but incorporating it into the thinking and practices of those who keep livestock remains a difficult problem. The first essential is adequate control over the animal population, see Fig xvi-8. The work of the African Livestock Marketing Organisation in Kenya furnishes a useful example of how this essential control may be achieved. Kenya tribesmen rate their wealth by the number of cattle they own. This created a situation in which cattle numbers increased until, as was readily predictable on available ecological knowledge, disease, parasitism and/or drought supervened with disastrous effects. After studying the potential demands for cattle and the most practicable transport routes, the Department of Veterinary Services created the African Livestock Marketing Organisation and instituted quarterly sales at which it is obligatory for tribal owners to dispose of a portion of their cattle. Difficulties encountered in implementing this scheme included the absence of a personal motive amongst the Africans to sell their cattle, the arrangement of stable prices and a shortage of useful consumer goods for the tribesmen to purchase with the money from cattle sales. The two most important factors in securing the participation of the African tribesmen in the programme were:

(a) Confidence in the veterinary administration created through its disease control programme, which wisely aimed at limiting animal numbers to reduce undernutrition, as much as to control infectious disease and parasitism.

(b) The development of an understanding amongst the tribesmen that catastrophic falls in cattle populations were not so likely to occur if the seriousness of undernutrition was reduced by keeping total numbers of animals within reasonable limits.

Any programme aimed at the improvement of animals in arid zones through introductions and breeding must be based on an inventory of the resources of the region where animal production is to be improved. This inventory should

Fig XVI-8. Adequate control over animals is essential. Here an African farmer is shown formulating his plans for the control of his herd during discussions with a veterinary officer. (*Photo by courtesy Kenya Information Services*)

include a survey of soil types and their potential, a study of the climatic factors and an assessment of the hydrological and agronomic prospects of the region. The careful documentation of the social factors in terms of customs, beliefs, culture, education, the recognised needs of the communities and human resources available to assist, is equally important.

When these factors have been assessed a plan for improvement can be formulated. An example of this approach occurs in Algeria, where a carefully planned attempt is being made to secure popular support for the improvement of animal production [3]. The science of animal breeding has attained a high degree of sophistication in the arid parts of some countries. Those who live in countries that lack highly developed scientific services need not despair. Much could be achieved by adopting simple approaches consistently applied so as to bring a project to fruition. Progress can be achieved by selection of beef cattle for growth rate from birth to weaning, for growth from weaning to 6–12 months later, for calving rate, and in hot climates, for coat type. Sheep can readily be selected for fleece weight, fibre size, non-pigmented wool and high reproduction rates. Simple measurement and selection procedures for these characters may not give maximum rates of improvement, but they will give sufficient improvement to make worthwhile contributions within a reasonable period.

In many arid countries the rate of improvement would be greatly enhanced by the more widespread use of artificial insemination to 'grade up' local animal populations, with already available suitable animals.

It is almost certain that, in many arid countries, governmental agencies will have to provide superior animals and skilled services, and in many instances to control husbandry practices in order to improve animal production. For example, the milk production of goats in arid regions of Australia has been greatly improved by making it illegal for any individual to own an uncastrated male goat among the herds that grazed upon the extensive common grounds surrounding each township. The town councils continually provide sufficient pure Saanen male goats as sires for the entire herd, and in the course of fifteen years these were graded up to the Saanen breed. By contrast, the services of a superior jack donkey imported into Iran to improve the local breed were not sought because there were already more than sufficient uncastrated males competing for the available jennys!

Camels, the ponies and the sheep of Tibet, the Navajo sheep of the United States of America, the fat-tailed and the carpet wool sheep of the Middle East, India and Pakistan are all indigenous to the arid zones. For many generations these animals have been exposed, with little constructive interference by man, to natural selection that has demanded adaptations to an extremely harsh environment. Those who imagine that the problems of animal production in the arid zone can be easily solved by crossbreeding or grading up to exotic breeds, should heed the stern warning contained in the difficulties that have been encountered in the United States where sheep of numerous breeds were introduced 'to improve' the Navajo sheep [36].

In its modern concept the husbandry of animals in a pastoral environment consists of the strategy and tactics of management designed to ensure the welfare, the efficient reproduction and rearing, and the optimum productivity of animals consistent with the maintenance or improvement of their respective ecosystems. Thus it is more than likely that in the wider sphere of applied animal husbandry some measure of control similar to that exercised in the United States of America over the grazing of high-altitude ranges in the Rocky Mountains, will be necessary over the grazing of common grounds in many arid countries, in order to ensure that there is sufficient forage for the improved animals with their potential for greater productivity.

Improvements in animal production through introductions and breeding can be achieved in the arid zone only if the indigenous people realise that this is a convenient way of fulfilling their wants. Therefore the success of programmes of animal improvement will be influenced largely by the extent to which the benefits they provide can be fitted to the community's requirements, and upon the way in which the findings of research can be grafted into the practices of the pastoralists of the arid zone without undue disturbance to their social, aesthetic or religious practices.

G. R. MOULE

References

[1] ANON., *The Future of Arid Lands* (Ed. G. F. White), Amer. Assoc. Advanc. Sci., Washington, Publ. No. 43, 1960.

[2] FARSON, N., *Last Chance in Africa*, London, 1949.

[3] ANON., 'Perspectives d'amélioration des productions fourragères et animales', Tome I: Productions fourragères et parleurs. Tome II: Productions animales, *Centre Algérien d'expansion économique et sociale* (CAEES), June 1961.

[4] GRANGER, W., 'Selection of breeding ewes: dependence of practicable degree of selection of young ewes upon vital statistics', *Aust. Vet. Journ.*, Vol. 20, 1944, pp. 253–60.

[5] — and J. D. WALSH, 'Equations for determining the components of beef cattle populations with special reference to fluctuations in the number of breeders', *Aust. Vet. Journ.*, Vol. 36, 1960, pp. 398–402.

[6] LUSH, J. L., *Animal Breeding Plans*, Ames, Iowa, 1945.

[7] MCBRIDE, G., 'The environment and animal breeding problems', *Anim. Breed. Abstr.*, Vol. 26, 1958, pp. 349–58.

[8] MCMEEKAN, C. P., *Principles of Animal Production*, London, 1943.

[9] MOULE, G. R., 'Basic physiological and managerial considerations in the artificial insemination of sheep', *Proc. Conf. Art. Insem. Sheep*, Univ. NSW, 1961.

[10] — 'Observations on mortality amongst lambs in tropical Queensland', *Aust. Vet. Journ.*, Vol. 30, pp. 153–71.

[11] BONSMA, J. C., J. VAN MARLE and J. H. HOFMEYER, 'Climatological research on animal husbandry and its significance in the development of beef cattle production in colonial territories', *Emp. Journ. Exp. Agric.*, Vol. 21, No. 83, 1953, pp. 154–75.

[12] JOSHI, N. R. and R. W. PHILLIPS, 'Zebu cattle of India and Pakistan', *FAO Agric. Studies* No. 19, Rome, 1953.

[13] KELLEY, R. B., *Native and Adapted Cattle*, Sydney, 1959.

[14] LALL, H. K., 'Breeds of sheep in the Indian Union', *Indian Coun. Agric. Res.*, Bull. 75, 1956.

[15] MASON, I. L., *A World Dictionary of Breeds, Types and Varieties of Livestock*, CAB Publ., England, 1951.

[16] — and J. P. MAULE, 'The indigenous livestock of eastern South Africa', *Com. Bur. Anim. Breed. and Gen.*, Tech. Comm. 14, 1960.

[17] MOULE, G. R., 'The performance of sheep in semi-arid Queensland. Climatology and microclimatology', *Proc. Canberra Symposium*, 1956, pp. 235-242, Unesco, Paris, 1958.

[18] — and G. M. H. WAITES, 'Seminal degeneration in the ram and its relation to the temperature of the scrotum', *Journ. Reprod. and Fert.*, Vol. 5, 1963, pp. 433-46.

[19] SCHMIDT-NIELSEN, B. K., 'The camel: Facts and fables', *The Unesco Courier*, 8-9, 1955.

[20] MACFARLANE, W. V., R. J. MORRIS and B. HOWARD, 'Water economy of tropical Merino sheep', *Nature*, Vol. 178, 1956, pp. 304-5.

[21] WRIGHT, N. C., *The Ecology of Domesticated Animals: Progress in the Physiology of Farm Animals*, J. Hammond (Ed.), Vol. 1, London, 1954, pp. 191-251.

[22] ANON., 'Breeding better beef cattle through performance testing', *Agric. Res. Serv.* 22-62, US Dept. Agric., 1960.

[23] BONSMA, J. C., 'Breeding cattle for increased adaptability to tropical and subtropical environments', *Journ. Agric. Sci.*, Vol. 39, 1949, pp. 204-21.

[24] DOWLING, D. F., 'Heat tolerance in animals', *Aust. Vet. Journ.*, Vol. 32, 1956, pp. 246-9.

[25] ANON., 'Beefmasters – best plus best is the pattern for perfection', *The Cattleman*, May 1960.

[26] WILSON, R., 'Breeding cattle to please commercial cowmen', *The Cattleman*, May 1961.

[27] STEFFERUD, A. (Ed.), 'Grass', *Year Book of Agric., 1948*, US Dept. Agric., Washington, DC.

[28] SHORT, B. F. and H. B. CARTER, 'An analysis of the records of the registered Australian Merino flocks', *CSIRO Aust.*, Bull. No. 276, 1955.

[29] MOULE, G. R. and S. J. MILLER, 'The use of fleece measurement in the improvement of Merino flocks in Queensland', *Emp. Journ. Exp. Agric.*, Vol. 24, No. 93, 1956, pp. 37-51.

[30] TURNER, H. N., 'Measurement as an aid to selection in breeding sheep for wool production', *Anim. Breed. Abstr.*, Vol. 24, 1956, pp. 87-118.

[31] MOULE, G. R., 'Fleece measurement for Queensland Stud Masters', *Qld Agric. Journ.*, Vol. 81, 1955, pp. 351-5; Vol. 82, 1956, pp. 27-30, 103-8, 235-40, 275-9.

[32] DUN, R. B., 'Breeding Merino sheep for higher lamb production', *Wool Technol. and Sheep Breed.*, Vol. 8, 1961, pp. 9-13.

[33] SCHINCKEL, P. G., 'Inheritance of birth coats in a strain of Merino sheep', *Aust. Journ. Agric. Res.*, Vol. 6, 1955, pp. 595-607.

[34] RITCHIE, G. R., 'The use of western breeds of sheep and goats in the USSR', *Anim. Breed. Abstr.*, Vol. 27, 1959, pp. 375-87.

[35] BRUNSWICK, R. and J. M. MALECKY, 'Wool in Communist countries', *Wool Economic Research Dept.*, No. 1, Bur. of Agric. Econ., Canberra, Aust., 1960.

[36] PHILLIPS, R. W., 'Breeding livestock adapted to unfavourable environments', *FAO Agric. Studies*, No. 1, Washington, 1948.

CHAPTER XVII

The Individual in the Desert

Accent on the individual

With the passing of the pioneer and the aggregation of people into towns and cities, the majority of us are apt to forget the prime importance of the individual in the remaining thinly populated and geographically rugged parts of the world. Ironically enough, this forgetfulness is no less marked in those countries which were themselves on the pioneer fringe only a century ago, but today offer aid and advice to their less fortunate fellows. Even the author, with some experience in the arid regions of three continents, has difficulty, sitting comfortably in his air-conditioned office, in recalling the mixed feelings of irritation, achievement, expansiveness and frustration that attended his earlier excursions. It took Croce-Spinelli's graphic account [1] of the recent 'SOS Sahara' expedition to bring back the endless bumps, pestilential flies, sandy monotony and perpetual heat of arid zone travel.

He will try here to present, in so far as words permit, the essential problems that face and must be solved by the individual who seeks to come to terms with the open desert, away from the material and moral support of close and settled communities. Failure can be costly, and it is hoped that these words may do something to mitigate the risk. Space will not permit detailed scientific exposition, but reference should be made to the increasingly available publications [1-9] in which it can be found. The term 'desert' will be interpreted broadly to include the semi-arid regions, where conditions may be less dramatic but more people are affected and the economic consequences have greater impact.

Outstanding problems

It may be a truism to say that the overriding problem of the desert is lack of water, but any effective consideration must commence with this stark fact. The lack may be extreme: it is more than a local phenomenon and may extend for hundreds of miles; such water as is present may call for digging or even engineering development, and when found it may be far from potable.

387

Without replacement of water lost from the body, man will suffer dehydration, physiological disturbance and death. The more rapid the water loss, the quicker the inevitable end. Intense insolation by day, continuing heat by night, hard going on sandy or rocky terrain, desiccation by dry winds, and emotional stress all increase the rate of water loss up to 8 litres or more per day (Fig XVII-I).

By comparison with lack of water, shortage of food is unimportant to the desert castaway; death from thirst will far outstrip the progress of starvation.

DAILY WATER REQUIREMENTS FOR THREE LEVELS OF ACTIVITY

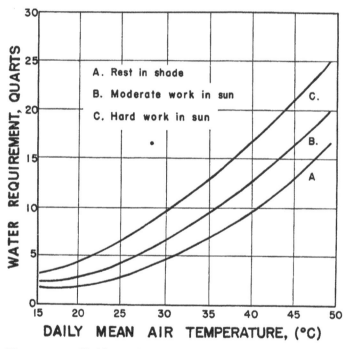

Fig XVII-I. Fluid requirements recommended by US Army Quartermaster Corps in relation to air temperature and activity for normal daily activity pattern. (1 quart = 1·1 litre)

But for one provided with a sufficiency of water, the scarcity of food may loom large. The nomad has learned over the centuries to meet this situation by adjusting his needs, activities and location to the desert's terms – the dates, herbs, goat milk and flesh, camel milk, crude legumes, etc., provided by oases or transient pasturages. Technological man wants to occupy predetermined locations on his own terms and relies upon supporting his existence there by importing necessities from elsewhere – in short, he subsidises the local economy or even replaces it. Such an investment can be justified only by a strong-felt need, such as the exploitation of mineral resources, or the occupation of a strategic position. Cut off from outside resources, untutored in the ways of the indigene, and fixed in his habits, the technological man would be in sorry case if his plans failed. Stripped of collective support and unskilled in working out his own salvation he becomes virtually helpless.

To those who know not how to woo her, the desert is an unrelenting mistress. Not contented with merely withholding essential water and food, she seems deliberately to make almost every action more difficult, more exhausting and more demanding. Soft sand, sharp declivities, rugged wadis, knife-edged stones, spiny vegetation and venomous animals wait to trap the unwary, rash or desperate. Heat shimmer and dust obscure the vision; every sandhill or stony depression looks like the last except to the initiated. And the mirage is her final supreme mockery! Demanding much and giving little, she may be overwhelmed by the force of modern technology; but woe betide any lesser endeavour by one who does not know how to win her good graces.

A typical summer day may be summarised as one of high temperatures (usually above skin temperature), low absolute as well as relative humidity, intense solar radiation, and high reflection and emission of heat from the terrain, with some offsetting opportunity for radiant heat loss to the clear sky [3, 5]. To cope with the large gain of heat from his surroundings, together with his own heat production, man has at his disposal only the evaporation of water from his skin and respiratory tract. Fortunately, man has a greater facility for providing water for evaporation from his skin than almost any other animal. A fairly active man in the open desert may need to evaporate one litre of sweat an hour to maintain his heat balance, but this is well within the capacity of a normal, healthy, acclimatised man, at least for a few hours.

Should the heat loss lag behind the heat gain, the body temperature will rise, and heat stroke will threaten. *Heat stroke* is a rare event in healthy, acclimatised men, as is evidenced by the fact that some 190 cases of heat stroke developed in United States troops while training at home bases in World War II [10], but practically none in overseas theatres. Heat stroke occurs fairly frequently during the summer months in the cities of the temperate zone [11, 12], but seldom in those of the subtropical or tropical zones. Disturbances such as dehydration, infection, circulatory failure or even severe emotion can, on the other hand, precipitate disaster by upsetting the heat balance.

While the risk of death for an acclimatised healthy man may be small, he does not escape unscathed from exposure to severe heat. In a complex system like that of the human body, a major demand such as adjustment to heat can hardly be met without involving various functions in secondary disturbances. Dilatation of the superficial blood vessels and sweating, which are major adaptive reactions, tend to bring about a crisis in the blood circulation by simultaneously increasing the capacity of the blood vessels and reducing the amount of blood available to fill them. The body makes some compensatory adjustments, such as closing down blood vessels to less important areas, reducing elimination of water through the kidneys, and mobilising fluid from the tissues. But these do not always meet the needs, and the blood flow, like the water supply in some towns on hot days, will be inadequate in some areas. The brain, like the hill-top house, is most likely to be deprived of a normal supply, and its inhabitants, being the most sensitive of all the body's population, will be quick to show effects. Lassitude, headache, nausea, dizziness, uneasiness, and finally fainting mark the development of typical *heat exhaustion*. Anyone who has been exposed to heat almost certainly has experienced some degree of heat exhaustion. It is, in effect, a 'fail-safe' mechanism: the sensations discourage activity and thus lessen heat production, and fainting restores the brain to a position of gravitational equality with the other tissues.

Various bodily functions may become involved in the secondary consequences of man's otherwise admirable heat regulatory effort. Loss of appetite, constipation, swelling of the feet, and even urinary stone may be traced back, at least in part, to that prime cause. *Heat cramps* constitute a third acute heat effect of clinical importance, and follow a marked reduction in the chloride content of the blood brought about by large losses of salt in the sweat [13]. This is likely to occur only in an unacclimatised person, doing hard work, in a very hot and dry environment. Dramatic though heat cramps may be, they are readily relieved by the administration of salt, and seldom occur in acclimatised persons. Dr Dill and the author both failed to induce heat cramps in themselves, in spite of subsisting (the word is appropriate) for several weeks on a virtually salt-free diet and working hard in quite hot environments. Dr McCance [14] has the dubious distinction of being, as far as I know, the only one to have succeeded in this effort; but perhaps he had the advantage of starting from the non-acclimatising environment of England.

The effect of ultra-violet light in producing sunburn is too well known to need emphasis here, but an appreciation of its role in provoking chronic skin reactions has been slow to develop. These chronic changes range from a simple increase of the horny layer (hyperkeratosis), through rodent ulcer to a markedly malignant squamous carcinoma. They are well described by Belisario [15], who discussed briefly the geographic distribution in Australia in relation to sunshine, citing the data in Table XVII–I.

TABLE XVII – I   *Incidence of dermatosis in four Australian cities in relation to hours of sunshine*

	Melbourne	Adelaide	Sydney	Perth
Latitude	37° 49′	34° 46′	33° 52′	31° 57′
Average mean hours of sunshine	5·6	7·0	6·8	7·8
Percentage of total dermatoses with precancerous or cancerous lesions	14	20	30	32

This information is strengthened by data reported by Auerbach [16] for the years 1947–8 in the United States, as expressed in Table XVII-2.

TABLE XVII-2   *Total white male and white female skin tumour incidence rate (per 100,000 population). Ten cities, United States, 1947–8. Age-adjusted rates*

City	North latitude in degrees	Incidence rate White male	White female
New Orleans	29·95	123·2	68·7
Dallas	32·78	164·4	95·6
Birmingham	33·35	182·2	132·3
Atlanta	33·75	110·0	82·9
San Francisco	37·77	97·1	68·0
Denver	39·73	86·3	60·6
Philadelphia	39·95	41·3	32·8
Pittsburgh	40·43	43·0	32·5
Chicago	41·87	25·3	22·4
Detroit	42·33	33·6	24·2

From this summary of symptoms, it will be seen that it is not sufficient merely to provide conditions in which man can manage to keep his body temperature within bounds. One must aim at something better, namely conditions which will avoid invoking the heat-regulating processes to the point at which they interfere seriously with normal bodily functions or the maintenance of efficient performance.

One last major physical factor – wind – provides a transition from the physiological to the psychological effects of desert environments. Local convectional winds are characteristically high during the hours of high insolation, and topographic winds may develop down mountain slopes after nightfall; but the best-known winds are those representing mass transfer movements sweeping across wide expanses, often carrying dust clouds before them, whose names – harmattan, shamal, khamsin – have long lent awe to adventure tales. Apart from their impeding, desiccating and abrasive effects, or perhaps because of all these, high and persistent winds seem to produce noticeable psychological responses in some persons. At ·this point of time it is hard to say how much the reputed effects have been reinforced by tradition, or to what extent the wind symbolises dissatisfactions and frustrations of multifarious origin. Climate has long been a convenient scapegoat for troubles which have their root in social conditions.

Confusion between necessary effect and the force of tradition clouds almost all discussions of man's psychological reactions to desert environments. The literature is abundant, but diffuse, nebulous and descriptive. Experimental psychology appears to have not yet advanced to the point that it can apply the usual procedures of analytical science to elucidation of the problems. As has been pointed out elsewhere [4], isolation is frequently cited as a major factor affecting the psychological attitude of many arid zone dwellers, with consequent effects upon nutritional habits, hygiene and housing practices, and these in turn influence physiological states and performance; but it is difficult to assemble concrete evidence for its operation or significance. Croce-Spinelli [1] illustrates the profound difference between the attitude of the desert Arab who lies calmly in his broken-down truck for days awaiting help, and that of the average European who frets at his inability to do anything, and may very well do something disastrous in his impatience. Isolation and austere conditions can make or mar character and personality traits, but the outcome surely turns as much upon the nature of the individual as upon the cultural group to which he belongs. The most important group factor would seem to be selection pressure. A European who fails to cope with the desert environment can usually go back whence he came; but for a nomad who fails there is little but extinction. As the desert dwellers become engulfed in the technological tide, inadequacies may appear in the reverse direction. According to newspaper accounts, the displaced residents of Tristan da Cunha want no part of modern progress and wish only to return to their island fastness. It is sincerely to be hoped that the studies of psychological factors in desert dwellers, both nomad and sojourning, initiated in the Sahara by PROHUZA [8, 9] can be continued and that the example will be followed elsewhere.

Protection from incidence of environmental heat

Everyone knows that shade is of prime importance in the open desert; but what not everyone knows is that there are good and bad ways of securing shade. A simple opaque layer, such as a tent fly, located at some distance from the person, is quite effective. Attempts to improve upon this may not only be ineffective, but actually reduce the value of the initial layer, unless they are designed with some care. The reason is quite simple, but not always appreciated. An opaque layer, in cutting off the sun, itself becomes heated. For effective protection this heat must be returned to the surroundings without involving the person beneath. A tent fly 2 feet or more above the person does this fairly well, but as one seeks to accomplish more things with the shelter, or to make it more durable, the chances of error increase.

The principles of effective shelter are fairly easily stated [3]:

(a) Shade to outer surface – as by trees, bank, another structure, screens, or (for walls) eaves and projections from walls.
(b) Reflective outer surface – as by light colour, or polished metal surface.
(c) Air movement over surface – secured by orientation to wind, slope, absence of wind barriers.
(d) Capacity insulation for permanent roof and wall structures exposed to sun – as by use of fairly thick layers of adobe, cement, stone or brick.
(e) Shade to openings – such as projections above windows, external blinds, shutters, screens.

(*f*) Ventilation of structural spaces – such as roof space, interval between layers of roof or wall construction.

(*g*) Controlled ventilation over undersurface of heated roof or wall – by appropriate arrangements of inlet and outlet vents, or properly directed forced ventilation to remove heat without mixing into room air.

(*h*) Minimisation of insolated surface – vertical slab structure running east-west, or tall structure of square section.

(*i*) Reduced temperature or reflectivity of surrounding terrain – as by grass or vegetation.

(*j*) Minimal introduction of heated air – as by closure of openings except as needed for item (*g*) above.

The application of these principles to specific situations will vary greatly with the circumstances, and may involve such additional considerations as availability of materials, cost, aesthetics, durability and compatibility with functional use of the structure. The range of possible answers for a temporary shelter such as a tent is limited – although it is surprising how many bad tent designs there are. The possibilities for a permanent, superior type house are legion; but it should not be difficult to arrive at something near optimal thermal results by a judicious consideration of the principles listed.

Clothing has been described as the logical end of a minimal housing programme; and, indeed, nomads with few physical possessions may be well advised to adopt it. In the humid tropics this process may be carried to its final extreme as in the adage that there are no clothes like no clothes. But in the desert, clothes, like shelter, have a definite protective role to play [3, 5] (see Figs XVII-3 and 4).

The first principle in the design of clothing for severely hot dry environments is that of extensive coverage, since every part of the body surface may be exposed to blasts of air much hotter than the skin, and many may come into contact with hot surfaces. At the same time, however, the coverage must permit transfer of evaporated moisture from beneath the clothing to the outer air. Loose fit and high permeability of the fabric to water vapour offer a compromise between these two functions – keeping heat out without locking water vapour in – especially if the normal openings at neck and limbs are adjustable to the needs of the moment. The loose fit is greatly helped if the garments hang freely from the shoulders, without constriction at the waist. The fabric needs to be sufficiently thick to drape well, without clinging to the skin, and to keep out hot air, but it should not be heavy. Natural fibres such as cotton and wool are more permeable to water vapour than most synthetic fabrics. Colour is not nearly as important as it is generally believed to be – in desert tests carried out by the US Army Quartermaster Corps, black uniforms were found to transmit very little more heat than white uniforms of the same cloth, even in full sunshine. (This apparent paradox is explicable by two facts: (*a*) white fibres, while reflecting a lot of light outwards, also reflect light inwards; and (*b*) over half of the radiation incident upon the clothing is in the infra-red band, in which colour is of much less significance.)

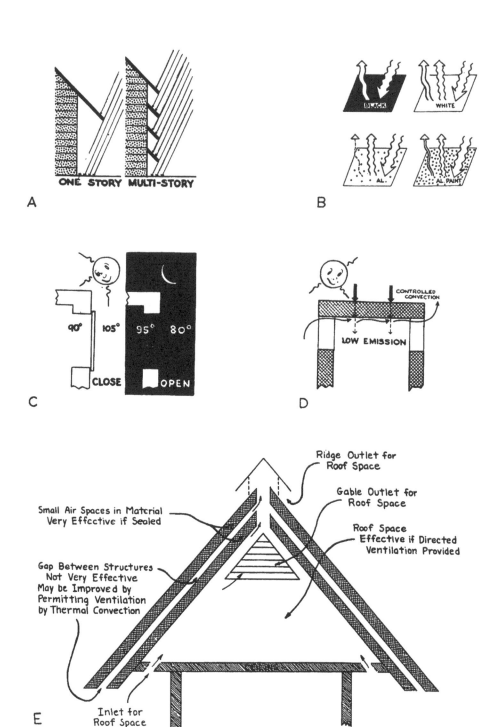

Fig XVII-2. Important principles of housing for hot dry environments

A.   Shade to wall from high-altitude sun by roof and wall projections.
B.   Effect of surface on absorption of solar heat. Black reflects none, but has good emission of long infra-red; white reflects visible and has equally good emission of long

The outer fabric may be reinforced over areas such as the knees, seat and elbows, which are likely to be pressed against hot surfaces.

Fig XVII-3. Central Australian aboriginal in a state of nature, ca. 1900

There are many occasions, even in the height of summer, when hot dry days are succeeded by cool nights. Clothing which is well designed for hot dry conditions – extensive covering, fairly insulative fabrics, controllable ventilation –

infra-red; polished aluminum reflects visible and some short infra-red but emits very little in long infra-red; aluminum paint reflects less but emits rather better in long infra-red.

C. Wall openings should be closed by day when outdoor air is hotter than indoor, but open by night when reverse gradient prevails.

D. Controlled ventilation over inner surface of heated structure very desirable. Controlled convection can be used to remove heat from inner surface of heated structure, without mixing into room space. Importance is increased and resultant protection is greater when the inner surface has a low emissivity for infra-red.

E. Conventional roof space is effective in protecting room beneath when designed so that convection, free and forced, removes trapped heat. (*From 'Physiological Objectives in Hot Weather Housing' published by U S Housing and Home Finance Agency, 1953*)

serves fairly well as protection against moderate cold, but an extra layer in the form of an additional garment may be desirable. A light, windproof fabric serves very well.

Fig. XVII-4. Heavily swathed Berber nomads dancing in a village deep in southern Algeria. (*Photo Ludwig Zöhrer*)

In the application of these principles to garments there is room for much individuality – and much error. Some examples [17] may be cited.

Headgear: The head, neck and shoulders receive the bulk of the incident direct solar radiation. The total load is not very great, and certainly presents no threat to the tissues of the region, but it may well be the last straw for a bodily heat balance which is already precarious. In any case, it is easily screened out. Unfortunately, western man is extraordinarily reluctant to adapt his ingrained notions of headgear to the physical realities (see Fig XVII-5).

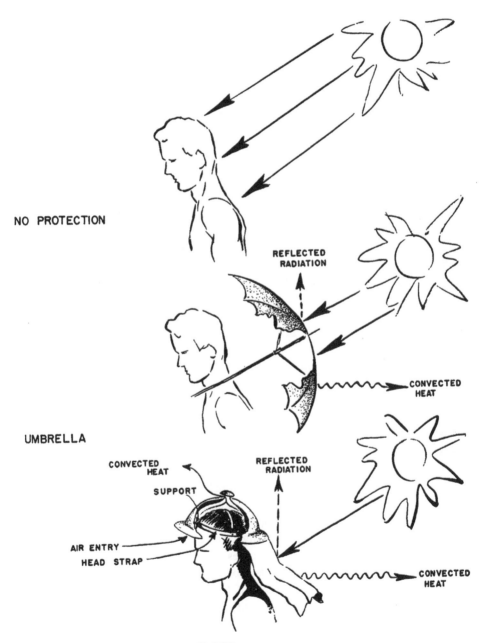

Fig XVII-5. Principles of shading for head and shoulders

By far the best screen is provided by the flowing Arab head-dress. A light-weight, light-coloured cloth, fixed to the crown, but falling freely over neck and shoulders, intercepts the solar radiation and returns the heat to the surrounding air, while simultaneously providing a means for screening the head from wind and dust. Military headgear popularised in stories of the Indian Mutiny and *Beau Geste* retained many of the good features, while conforming to army requirements, but they too, alas, have joined the psychologically unacceptable. I suspect that the dictates of civilisation have banned the large handkerchief that we tucked under the back of our hats as boys in the Australian bush. (Yet the same civilisation permits blue jeans and aloha shirts!)

Shade to the eyes is one thing provided by hats and caps that is not given by the Arab head-dress. The solar helmet or topee, when properly designed, can incorporate many of the advantages of the Arab type head-dress, and also permit airflow over the scalp between the hat proper and a headband separated from it by spaced supports. But the result is a highly stylised and cumbersome article, that falls or blows off with the greatest of ease during outdoor activity. It is useful for pre-empting the next seat in a crowded airport. A tight-fitting cap is probably the item of headgear least suited to hot climates, with the conventional felt hat not far behind.

One-piece main garment: There are certain theoretical advantages in having a single garment covering trunk and limbs, such as absence of constriction, ease of shoulder suspension, free bellows action, and freedom of bodily movement. Against these, however, must be set the real disadvantages of difficulty in removal, inability of discarding the top portion when free convection is desired, toilet inconveniences, and the necessity for renewing the whole garment when only a part is worn or damaged.

Upper garment: This should overlap the lower garment, preferably on the outside, with a drawstring or light elastic closure in the lower hem to give controllable closure. The neck and wrists should also have controllable closures. An extra thickness of material over the shoulders and upper back may assist insulation against radiation, and patches on the elbows may be desirable to give protection against hot objects.

Lower garment: Long trousers are probably better suited to a wider range of desert conditions than any other conventional type of lower garment, for both sexes. Suspension is best given from the shoulders, but if this is not acceptable, adjustable gathering straps at the hips may be used. The conventional belt is much less desirable. An extra insulating layer and wear-resisting layer may be incorporated in the seat and over the knees.

Footgear: The main consideration is protection against hot ground. Good sole insulation is required, as by materials of high thermal resistivity or by cleats which raise the sole from close contact with hard hot surfaces. Protection against mechanical and biological hazards is a relative matter, depending upon the extent of the hazards (rocks, cactus, snakes, etc.) to be expected, and the importance attributed to them by the wearer. Most people of western culture demand a high degree of protection, but it is evident that this desire is not shared by those who favour open sandals. One of the two extremes should be adopted – complete closure so that nothing gets in, or free opening so that what gets in can be shaken out. A compromise, in this case, is no solution. Other things being equal, one must admit that a dirty foot is to be preferred to a hot and wet foot. In the closed type of footgear, however, evaporation can to a certain extent be encouraged by the use of flexible, non-absorbent, open mesh inner soles and fairly thick absorbent socks that come above the shoe or boot line.

A fairly snug fit is to be preferred to an overly loose fit, in that it helps to counteract the tendency of fluid to accumulate in the tissues. Sandals and canvas shoes should be worn only by those who have acquired the necessary anatomical adaptation by long experience.

Underclothing: From a thermal point of view, underclothing is likely to have more disadvantages than advantages, unless the outer garments are very thin and translucent. Where underclothing is desirable for hygienic reasons, it should be thin and absorbent.

Additional methods of preserving the heat balance

In the previous section we discussed methods of protecting oneself against environmental heat. We need now to look at the other side of the coin, and consider rather more positive actions that may be taken to beat the heat. The first of these is positive only in the sense that it calls for decision – the decision to do things the easy way. There is a lot about habits developed in temperate cultures that is superfluous to the essential task, imposed for purposes of 'smartness', 'prestige' or 'elegance'. While it may be difficult to shake the social significance of these extra flourishes and self-imposed burdens, they are thermogenic, and at least call for careful scrutiny under circumstances where every little bit of unnecessary heat adds to the threat of breakdown. Those who have watched the adaptation of newcomers to hot environments must have remarked innumerable times on the way in which they learn to take things easy – not necessarily in the sense of being lazy or even of accomplishing less, but in the sense of omitting the unnecessary. One learns to walk loosely, slump in a chair, substitute deliberate for impulsive action, adopt a steady pace, and make all the other adjustments that bespeak increased physiological efficiency (ratio of useful output to energy expended) and reduced heat production. As with all other adjustments, of course, and especially with those learnt by experience, the pendulum can swing too far, and one needs to seek out the optimal balance between wasteful energy expenditure and sloth. (That much maligned creature, the sloth, has a lesson to impart – he can swing through the trees quicker than you can travel on foot; but when he is not running away from you, he just hangs.

The design for dispensing with unnecessary heat production should be extended beyond oneself to the work arrangements. By taking thought one may not add a cubit to one's stature, but one certainly can save oneself some work.

Efficiency experts have written much about arranging the work layout so that movements are reduced and output increased. We can apply the same principles in arranging whatever tasks have to be performed in the hot desert – erecting shelters, stacking stores, repairing equipment, inspecting operations. By a little prior planning, the materials or items to be used can be set close to the point of use and in suitable sequence, or the order of one's actions can be arranged so that backtracking is minimised. The result of such planning will be, not merely greater comfort, but lessened risk of breakdown, better work, and perhaps increased output.

It stands to reason that one should not add to the heat of the natural environment, but reason is not always heard. Both the design and use of cooking equipment could be greatly improved, the need for cooking reduced, the heat taken away from instead of over the man, and radiant shields provided where hot surfaces or open fires are unavoidable. It is amazing how often the inertia of habit triumphs over common sense. A critical appraisal of cooking practices in almost any desert encampment will reveal several ways in which the heat burden can be reduced, often with but little structural effort or cost.

Active cooling, rather than minimisation of heating, is certainly a positive matter. The fortunate few with access to air-conditioning have no further problems – except perhaps paying for capacity as large as they would like to have; but one can hardly speak of such folk as individuals in the desert – unless the machinery breaks down! For the rest, however, simpler but reasonably effective cooling is possible. Air movement by itself is unlikely to be effective, for reasons that have been discussed; but when air movement is combined with evaporation, considerable benefit can be obtained. The simple methods work as well as they did two thousand years or more ago. A burlap (hessian) cloth of fairly open mesh, kept damp by a water drip or periodic spraying, will produce a significant drop of temperature on the downwind side. The concurrent rise in humidity is of less significance, as long as an adequate air replacement is kept up. It is fatal to the scheme, however, if the resistance of the evaporative screen is such that air movement is substantially impeded. I have seen bamboo screens stuffed so full of twigs, that a dead calm prevailed in the room that they were supposed to cool. (The same calmness was not observable in the inhabitants.) Spraying the outside of a shelter can result in effective cooling, especially if the surface is water-absorbent. A layer of water on a flat roof can be used to produce the same effect, but its surface must be exposed to the breezes.

Good use can be made of infra-red radiation to the sky, especially at night. Zenith temperatures in the Arizona desert, for example, drop to 5°C when air temperatures are still at 23°C. The differential in temperature, operating over a wide expanse of sky, provides an important opportunity for a man to lose heat, especially if he is recumbent. As 'civilisation' overtakes people, there is a tendency for this experientially learned truth to be forgotten, or at least overridden by the status symbolism of being able to sleep in a house. 'Beneath the stars but above the snakes' is a good motto for desert sleeping.

Acclimatisation

The fact of acclimatisation is plain enough, otherwise we would all be like the visitors from headquarters who come to see desert studies but scarcely venture beyond the air-conditioned quarters or the swimming pool. (Lest my superiors feel insulted, let me say that when it comes to arctic conditions I react the same way.) Some of the major changes in bodily function that take place in the course of acclimatisation have been carefully studied; but it would be presumptuous to say that we know what acclimatisation is or how it is brought about. Hence any advice one might give on speeding up acclimatisation should be taken with reserve.

An important event in acclimatisation, as revealed by recent studies, is the release of a hormone, aldosterone, from the cortex of the adrenal gland. This occurs in response to the relative dehydration and inadequate circulation. Aldosterone acts upon the kidney, causing it to retain more water and salt, and so brings about some measure of compensation. This would explain, in part, the improvement in maintaining heat loss and the reduced pulse rate that occur over the first few days of exposure. At the same time, an important reduction occurs in the chloride content of the sweat, bringing the daily loss of this important substance into the range which can usually be handled by the daily diet. It is possible that the basal metabolic rate, that is the energy expenditure required for the basic activities of living cells at rest, may be reduced through the indirect action of heat on the thyroid gland. But a more important reduction of energy expenditure, and thus of undesired heat production, comes from learning to do the job one has to do more efficiently, without the unnecessary frills and flourishes I mentioned earlier.

These developments are interactive, one helping the other, and serve to explain much of the obvious improvement in one's adjustment to the situation. The periods of acute discomfort, lethargy, dizziness and even alarm get shorter and less frequent. Appetite begins to return. The overpowering thirst grows less. And then, somewhere in the second week, a phenomenon very similar to the runner's 'second wind' is felt. Suddenly one feels on top of the job, that one can come to terms with the conditions, that from here on the course is downhill. It is a delightful feeling when it comes, but one certainly has to put up with a lot of discomfort to arrive at it. There are, of course, extreme conditions to which even a healthy and vigorous person never would get accustomed; conditions in which each day would be worse than the last as breakdown accumulates. But these conditions are rare, and seldom last more than one or two days, unless aided and abetted by man. Those of us who spent a summer in Death Valley [17] admitted privately afterwards that, during the first few days, there were times when we thought that we had found the limit; but we got over that. From Spinelli's account [1] it seems that similar thoughts were not absent from the minds of some members of the SOS Sahara foray.

It is doubtful if one can speed up acclimatisation, but one can certainly interfere with it. The healthy, well-trained and well-motivated person is in a relatively good position to meet and adjust to the conditions. Fear or resentment towards the situation adds to the initial burden and obstructs adaptation by producing heat, increasing sweating, and inhibiting the process of learning how to get along. Retreat to cool quarters is necessary when the burden gets too heavy, but each retreat postpones the time of ultimate adjustment. As in training for athletic events, if one pushes too hard, injury and regression may occur; but if one does not push hard enough, the training will lag. Excesses of any kind take immediate toll and retard or even set back the acclimatisation process. The results of a 'night on the town' are liable to be evident for a few days thereafter.

Disease in the desert

The incidence of diseases is high in some desert communities and in irrigated lands, but, while there are conditions specifically connected with arid lands which adversely affect the individual, as discussed herein, for the most part those diseases which turn up frequently in desert areas are to be found equally elsewhere. Their appearance in deserts is dictated much more by accidents of biological distribution or by socio-economic conditions than by the climatic factors. By way of illustration, schistosomiasis occurs, but only in irrigated areas, and only where the appropriate snail abounds. There is probably more schistosomiasis in Puerto Rico than in most of the 'deserts' put together. Trachoma may appear to an Australian as a desert affliction, and dust may play some role; but a glance at a map of its distribution, such as that in Sidkey and Freyche's 'World distribution and prevalence of trachoma in recent years' [18] shows that there is virtually zero correlation. It is essentially the product of poor hygiene, linked in turn to socio-economic conditions.

There is in the USA one disease – coccidiomycosis, or San Joaquin Valley fever – which is limited to at least a semi-desert region; but it is far from certain that desert conditions have much to do with its spread.

So, although it is true that the medical aspects of desert development are of very great significance indeed to humanity, and that many complex and serious effects may result through the establishment of, say, a large irrigation project which will affect the physical environment and the incidence of disease vectors (mosquitoes, snails, etc.), the topic lies rather within the scope of general medical practice than of specifically arid zone matters.

Rules for survival

Croce-Spinelli and Lambert append a set of rules for survival to their illuminating 'paper-back', *SOS Sahara* [1]. I can think of no better conclusion than to follow their example.

Rule No. 1: Plan ahead!

So easy to say, so easy to do, so often forgotten! The seven soldiers who jumped in a truck and took off gaily to hunt gazelles, without water, without signals and without notice, exemplify only too well the casual attitude adopted by the inexperienced. The four deaths will not have been entirely in vain if they can be the means of driving the moral home to others. If you have to venture into waterless country, even for an hour:

> Make sure the transportation is in good order and fuelled;
> Take enough water for twenty-four hours;
> Tell someone just where you are going and when to expect you back;
> Take a prominent display or signalling device;
> Include a tarpaulin or other shading devices and sand mats;
> Include someone who knows the terrain.

Rule No. 2: Don't panic!

The more serious the problem, the more important is clear thinking. Work out what the best course of action is, and do not waste your resources on impulsive activity that does not contribute to the plan. Every bit of heat you generate by activity, and of sweat you lose, decreases your reserve against delayed rescue. The plan will vary with the circumstances, but obvious points for consideration are:

> Postpone heavy work until the cooler hours;
> Set up shade and stay in it, screening off wind if possible;
> Put out signals and signs to catch the eye of aircraft or vehicles at a distance (large objects set to throw shadows, fluttering or reflecting objects, smoke by day or fire by night);
> Ration the water as though you were going to be there for days (if you are found quickly it will not matter; if you are not, you will need it);
> Avoid unnecessary activity and excitement.

Rule No. 3: Stay put!

Vehicles are fairly easily seen from the air, especially if signs are displayed in the vicinity; individuals are extremely difficult to spot. All seven soldiers would have been saved in the Saharan episode if they had stayed with the truck. The truck was found first, the shade put up by the survivors second, and the bodies of the others last.

There may occasionally be an exception to this rule, as when you are stranded in scrub country where a small vehicle is hard to see, when no one knows that you are missing (there went Rule No. 1!), and you know for certain that help can be obtained within walking distance in a given direction. In that case, it is much better to walk by night, using the stars for guidance, than to expend one's limited resources fighting the hot day.

D. H. K. LEE

References

[1] CROCE-SPINELLI, M. and G. LAMBERT, *SOS Sahara*, Collection l'Actuel, Paris, 1961.

[2] LAMBERT, G. E., 'Modifications des temps de réaction et de l'électro-encéphalogramme au cours d'un séjour prolongé en zone aride', in *Environmental Physiology and Psychology in Arid Conditions*, Unesco Arid Zone Research XXIV, Paris, 1964, pp. 363-6.

[3] LEE, D. H. K., 'Proprioclimates of man and domestic animals', in *Climatology*, Unesco Arid Zone Research X, Paris, 1958, pp. 102-25.

[4] — 'Applications of human and animal physiology and ecology to the study of arid zone problems', in *The Problems of the Arid Zone*, Unesco Arid Zone Research XVIII, Paris, 1962, pp. 213-33.

[5] — 'Terrestrial animals in dry heat: man in the desert', *Handbook of Physiology*, Section 4, *Adaptation to the Environment*, Amer. Physiol. Soc., 1964, pp. 551-82.

[6] METZ, B. *et al.*, 'Métabolisme hydrominéral et bilan calorique du travailleur de force en zone Saharienne', PROHUZA, Paris, 1959.

[7] NEWBURGH, L. H., *Physiology of Heat Regulation and the Science of Clothing*, Philadelphia, 1949.

[8] PETIT, L., 'The attitude of the population and the problem of education in the Sahara', in *The Problems of the Arid Zone*, Unesco Arid Zone Research XVIII, Paris, 1962, pp. 459-70.

[9] PROHUZA (Centre d'Études et d'Informations des Problèmes Humains dans les Zones Arides), *Journées d'information médico-sociales Sahariennes*, Paris, 1960.

[10] MALAMUD, N., W. HAYMAKER and R. P. CUSTER, 'Heat stroke – a clinicopathologic study of 125 cases', *Mil. Surg.*, Vol. 99, 1946, pp. 397-449.

[11] KUTSCHENREUTER, P. H., 'Weather does affect mortality', *ASHRAE Journ.*, Vol. 2, 1960, pp. 39-43.

[12] SHATTUCK, G. D. and M. M. HILFERTY, 'Causes of death from heat in Massachusetts', *New Eng. Journ. Med.*, Vol. 209, 1933, pp. 319-29.

[13] TALBOTT, J. H., 'Heat cramps', *Medicine*, Vol. 14, 1935, pp. 323-76.

[14] MCCANCE, R. A., 'Experimental sodium chloride deficiency in man', *Proc. Roy. Soc. London*, Series B, Vol. 119, 1936, pp. 245-68.

[15] BELISARIO, J. C., *Cancer of the Skin*, London, 1959.

[16] AUERBACH, H., 'Geographic incidence of skin tumours', *Quart. Rpt. Biol. Med. Division*, Argonne Nat. Lab., 5-7, October 1955.

[17] LEE, D. H. K. *et al.*, 'Studies on clothing for hot environments. Death Valley, 1950, Part 1', *U.S. Army Quartermaster Corps, Environmental Protection Section*, Rpt. 178, 1951.

[18] *WHO Epidemiological Statistics Report*, 1949, Vol. 11, No. 11-12.

ADDITIONAL REFERENCE

UNESCO, *Environmental Physiology and Psychology in Arid Conditions*, Proc. *Lucknow Symp.*, Unesco, Paris, 1964 (Arid Zone Research XXIV).

CHAPTER XVIII

Social Life in the Arid Zones

The arid zones today are the meeting places of two types of cultures, traditional and modern. Builders of industries, explorers for fuels and metals, and transportation specialists have been active in the deserts and steppelands, where their contact with traditional cultures has brought about great changes.

The traditional cultures are basically of three varieties: hunters and gatherers of wild life; herdsmen; and irrigation farmers. Moreover, farming peoples of the arid zones have a long history of urban life, while pastoralists have been brought into even closer relations with the urban life of the farming peoples, and most recently have been developing a town and city life of their own. The hunters and gatherers have had no urban development other than the establishment of latecomers in their midst.

Despite the great range in variety of cultural types in the arid zones, is it not possible to draw general lines of ecological relations which they all share? This problem will be our main concern.

The hunters and gatherers are the simplest types in view of their technology, economic relations and demographic composition. Representatives of this category have been well described in the ethnographic literature, among them the Paiutes and Shoshones who inhabit the Great Basin of western North America, the Bushmen of south-west Africa, and the Australian aborigines of the dry interior of Australia. Data are needed to provide a basis for generalisations concerning their ecological adaptiveness, and that of pastoralists and farmers. Special reference to water economy will first be introduced for comparison with the other ecological types of arid zones cultures.

The hunters and gatherers of the arid zones

The Bushmen territory is perhaps the most arid of all those mentioned; it includes stretches of extreme desert as well as wetter, but still arid, parts. Residence in their present territory and their ecological relations are stable. The Bushmen culture includes many devices and arrangements that permit life in the desert, which few other societies have shown the capacity to match: these arrangements concern water and all other thirst-quenching devices. The Bushmen have no agriculture or animal husbandry; moisture is important to them for human sustenance in the first place, and then for the support of animals which they hunt or trap, and plant life, insects and grubs which they gather.

The characteristic form of Bushmen community organisation is the band, a loose grouping of families. Each band has its territory, and within the territory, reliable sources of water, whether as water-holes or water courses. Access to territorial boundaries and water sources are strictly controlled by the resident band. Dry seasons are winter (May–June) and summer (November–December); rainy seasons are January to April and September–October. Seasonality of rain and aridity control residence and movements, and hunting and gathering practices. In the dry seasons, bands move away from water-holes, so that prey which is then trapped, snared and mired, can come to water. Water-holes are poisoned with insect, plant, snake, scorpion and spider poisons in various parts of the Bushmen area. The animals are taken at the water-holes, slaughtered and eaten.

The settlement pattern is keyed to the adaptive ecological arrangements. Settlement is not uniform, but focused around water sources in the dry seasons. In the rainy seasons it is more thinly and uniformly spread out. Movement is seasonal, depending on the game cycle, which is closely related to water supply in the dry seasons. The Bushmen supplement the natural sources of moisture by water storage in ostrich egg shells and in the dried stomachs of large beasts. They have a device made of a hollow reed with a filter of grass, forming a sucking-tube to draw up groundwater during the seasons when lakes and water courses are dry [1, 2].[1]

The Paiutes and Shoshones of the Western Basin and Plateau of U S A occupied a generally more fertile and less arid country than the Bushmen. Much of the territory is watered the year round; true desert with seasonally dry water courses is present in pockets, principally over the southern part of the area, although the entire region is in fact arid. The traditional cultures have had to develop many techniques to deal with the problems of water shortage, including hunting and the gathering of wild plant products, of which pine nuts and cones were among the most important. Agriculture was sporadically practised by a few of the peoples of the Basin and Plateau.

The Western Shoshones had little or no agriculture; their subsistence source was mainly wild plant life, and game only to a minor extent, for game was less abundant in their territory than elsewhere in the region. For the greater part of the year, families travelled alone or in small groups, gathering over a wide area. They ordinarily ranged over 20 miles or more in each direction from the winter village. Their movements, although usually the same each year, were not always fixed, for seasonal variation in rainfall and plant growth frequently required that they alter their routine. Indeed, the rainfall affected the location of the winter

[1] See also Chapter x, p. 231.

village, which often had to be changed because of insufficient moisture. The storage of a surplus product of the hunt, or its failure, might cause the relocation of a winter village. The failure of the pine nut crop also caused bands and winter villages to break up and families to regroup [3, 4].

The hunters and gatherers of the arid zones had a complex set of social and cultural devices to meet problems of water scarcity. In other respects theirs is a simple life, but not with regard to water economy.

It is known that Bushmen covered a wider territory eight hundred to a thousand years ago than they do today; but while their desert focus is recent, their archaic habitat included arid parts of their present domain. During recent times, they have achieved a stable ecological adaptation to their present environment. The Arandas of central Australia, the Paiutes and the Shoshones have all attained a stable adaptive mode, as have the Bushmen, without the use of metals, pottery or the wheel.

They have adapted to their habitat with Stone Age techniques.

The pastoralists of interior Asia

The hunters and gatherers subsisting on the plant and animal life around them add nothing, but only detract from their food source. They are ecological parasites, possibly qualifying this relationship through exercise of restraint in the hunt if a natural source of food supply is diminishing. The farmers are in conjunctive symbiotic relation with their crops; and irrigation farming has intensified relations of symbiosis. That is, the farmer weeds, waters, ploughs, plants, fends off seed-eaters, forming an interdependent community with his crops while living on the harvest, permanently occupying his land.

The pastoralist has an amalgam of both parasitic and symbiotic relations with the surrounding environment. The herdsmen fend off predators; they provide windbreaks in the lee of which herds may find winter forage; intervene in calving and foaling; at times they have been known to isolate in quarantine diseased elements of the herds. They subsist on the herding products, wool, felt, cloth, milk, meat, hides, pelts, and on trade of stock and herd products. But both man and herds depend on natural supplies of grasses and water, which they do not replenish. Therefore, herdsmen are both parasitic and symbiotic, even more closely symbiotic with the herds than farmers with their grain. Herdsmen have developed deep empathy with the herds, while the imprint of domestication and of human contact is a vital factor in herd psychology and herd management.

The pastoralists of North Africa, south-west Asia and interior Asia have as their habitat the great, interconnected desert system that extends from the Atlantic to North China. The pastoralists are, with few exceptions, tent-dwelling nomads moving in a fixed annual cycle from winter camp to summer pasture. They have no permanent residence, but have highly predictable movements.

The cycle of annual movement of herds of domesticated animals, constituting nomadic pastoralism, has been traced back by the same pattern among wild genera of the same animals, by a Japanese scientist, Imanishi; that is wild sheep, wild horses, etc., also move in annual cycles [5, 6]. Nomadic pastoralism, however, is a cultural practice, exclusively the ecological pattern developed by human beings. The herdsmen, in exploitation of nomadic capabilities of the genera under domestication, have caused them to adapt their movements to each other and to the human communities which have controlled them. This cultural feature is a necessary part of traditional nomadic pastoralism. Differences in seasonal movements of wild genera are great; horses, sheep, goats, camels, cattle, all have different rates of movement, terrain and feed preferences. Pastoral practices of nomads and their herds have evolved out of the overcoming of these differences and ordering movements in a common pattern.

The patterns of movement of pastoral nomads and of hunting and gathering wanderers stand in contrast to each other. The relations to terrain, grass, water and weather provide the basis for the relatively fixed cycle of movement of pastoralists which we have called nomadism. The composition of the domestic herds is stable, comprising, in the interior of Asia, cattle, horses, sheep, goats and camels. The numerical proportion of the various stock genera within the herd is likewise a stable one, with some variation among the stock genera in the various sub-regions. Thus, in Russian Turkestan (Turkmenia, Uzbekistan) the ass, *Equus asinus*, is added to the list of stock species; in Mongolia and Kazakhstan the traditional pastoralism included few or no asses; the Mongols, nomadising at a higher altitude than the Kazakhs, have the yak among their herd species. The yak, *Bos grunniens*, belongs to the same genus as domestic cattle, *Bos taurus*; it is a cold-adapted bovine.

The degree of control over the environment is increased with domestication compared with hunting and gathering, as Kashkarov and Korovin [7] have pointed out.[1] That the movement of pastoralists is more predictable is a corollary to this. It should not be inferred, however, that the wandering of hunters is without plan or pattern. Nevertheless, the wandering within the hunting or gathering territory of the band is less determinate by the band than is nomadism of the pastoralists. The animals of the hunt are not subject to cultural contacts. Human hunters have control over their movements only in so far as they accumulate the nature lore concerning the movements of their prey. Steward [4] has indicated that the fixed point in the movement of the Shoshone gatherers, the winter camp, may be shifted through climatic necessity. This is less likely to happen among nomadic pastoralists. Thus, in establishing a typology of movement of peoples in the arid zone, a distinction may be proposed between *wander-*

[1] Kashkarov and Korovin have considered that man adapts to aridity with difficulty, the desert providing no natural product capable of sustaining human life, and that only with domestication could man dwell in the desert. In the territory of their investigation Kashkarov and Korovin could have encountered no people who resembled the Bushmen or the Shoshones in general technology, economy and ecological adaptation, for there are no cultures at present nor in historical reports in the arid zones of the interior of Asia with a primary subsistence base in hunting and gathering of wild plants and plant products.

ing of hunters and gatherers, and *nomadism* of pastoralists, with the characteristics indicated.

Transhumance is a pattern of movement of herds (and herdsmen) from the plains to summer pastures in the mountains. It differs from nomadism in a number of features: the transhumant herdsman has a fixed place of residence in a farming village, and a house to which he returns; the transhumant pastoralist, moreover, is not a member of an independent economy, but is a part of a larger whole, the economic complex of the whole countryside (e.g. Spain, Italy, Greece, western USA, south-eastern Australia). In contrast, the Kazakhs, Turkmens, Mongols and other nomadic pastoralists of Asia formed economic unities in the historic past, and Mongols do so at present. *Migration* is unlike any of the patterns of movement discussed above in so far as it involves a total displacement of habitat and ecology. There is no basis to consider on ethnological grounds the characterisation of nomadism as 'habitual or intermittent movement' as J. L. Myres has done. He further contrasts nomadism in this sense with transhumance which he calls 'seasonal alternation of pastures' [8]. In fact, both pastoralism, nomadism and transhumance possess the property which he attributes to transhumance alone. It is in other matters than those of herd–land relations that the differences between nomadism and transhumance lie: in the different residence patterns and in the relations to the greater cultural whole.

The mastery of domestication permits a greater population density than is found among hunters and gatherers of the arid zones. The amount of food produced per unit of ground is greater under conditions of domestication than those of hunting and gathering, considering the natural features of regions in question to be more or less the same. The density of human populations where animal domestication has been introduced is greater by a factor of ten than that of hunters and gatherers. The size of ethnic units forming pastoral societies is many times greater yet than those of hunters and gatherers. Complexity of social organisations and durability of their internal structures increases correlatively with size and density of settlement.

Irrigation farmers

While it is true that irrigation farming is practised to some extent in humid zones, nevertheless the most ancient irrigation practices were laid down in the arid zones of the Middle East, from the Nile to Mesopotamia, and irrigation farmers virtually monopolise agriculture in the arid zones, although not entirely. There are limited amounts of agriculture in piedmont areas of mountain ranges which take advantage of run-offs resulting from snow melts and thaws, and some farming is practised in the winter camps of nomads and in the vicinity of lakes and river deltas.

In irrigated land cultivation, high returns in concentrated areas make possible high sedentary population densities, and the irrigation farming populations usually form ethnic and national unities around the rivers which support them –

the Nile, Tigris, Euphrates, Indus, today; the Zerafshan and Amu-Darya until the twentieth century. The populations around these rivers, while densely settled, have been limited in size by water availability, crop areas and crop yields by traditional techniques.

The farmers practising irrigation have developed societies with far greater complexity of internal structure than the nomads. Social classes, estates and castes,[1] institutions of government, law, economy, religion, have been more intensively developed and have lasted longer than those of the nomads. The time depth of the structures and institutions of the irrigation farmers is much deeper than those of the humid temperate zone or humid tropics farmers, and probably of the greatest antiquity in the evolution of mankind. This is not only true of the Old World irrigation farmers, but with appropriate reduction of time scales it is also true of the irrigation farmers of the arid zones of Mexico and Peru.

Irrigation requires the co-ordinated, controlled and supervised effort of large numbers of people, as explained in Chapter XII. This sort of enterprise has taken place in societies differentiated into estates and castes, divisions which have proved to have great durability. On the other hand, some moderate irrigation networks covering hundreds of square kilometres were developed by traditional clan societies of Turks in central Asia, that is, without 'despotic' social controls.

Urban centres have been established from ancient times in the irrigation agriculture domains, many of them with populations in the hundreds of thousands. Functions of the direction of the irrigation undertakings, the government of empires and provinces, tribute collection, functions of cultural centres, depot and entrepot points, trade, handicraft manufacture, art and artisanry, scholarship, gardening have flourished in these cities.

The populous cities have had two kinds of water problems: those shared with arid zone peoples in general, and the intensified problem of providing water in large amounts for populations concentrated in a small space behind walls (see Chapter X). These cities have developed gardens for economic, artistic, cultic, sanitary and sumptuary purposes, thus multiplying the water problem.

The great cities of Shiraz and Isfahan in Persia, of Samarkand and Bukhara in Turkestan, have been garden cities of this sort. Within their walls, the populations have built courts and courtyards, each with its surrounding walls, divided into streets and quarters. In the centres of the cities are found open squares with magnificent architecture of palaces, bazaars and mosques. Surrounding the city centres with lavish display space were the inhabited quarters of the common folk, densely settled, poorly furnished. The gardens of the cities grew flowers and foods. The city with canals and underground water galleries was, in this sense, a settlement more closely packed than the surrounding cultivated countryside, performing the same functions of cultivation as well as others. Within the cities the population density lessened radially towards the walls, falling off beyond. The settlement pattern here is one of population focus with quantitatively

[1] But see Chapter XIII for a discussion of artisan castes among nomads.

decreasing density to the limits of cultivation; in the focal centre, qualitatively different functions were developed.

Settlement and movement patterns of the arid zones

The cultures whose traditional habitat is the arid zones comprise a range, from simple to advanced, as measured by the criterion of relative complexity of social organisation and of technology. These cultures, hunting and gathering, pastoral, and irrigation farming, have little in common on a superficial view, save a concern with water as reflected in their various technological adaptations. Nevertheless, certain common features in settlement and related cultural patterns may be noted:

1. Cultures are concentrated in watered places, the nearby vacant interstices having a sparse population (if any at all) of different cultural type. The Arabian peninsula has two zones, one known as *Arabia Felix*, and the other as *Arabia Deserta*, otherwise Rub al-Khali or the Empty Quarter, which has less than 100 mm of rainfall annually. Iran and Afghanistan are comprised of a series of landlocked deserts, the Dasht-i-Lut, Dasht-i-Kavir, the Baluchi and Seistan basins. They are cut off from water-bearing atmospheric circulations by encircling mountain ranges which on their seaward faces are fertile and cultivated. The Sahara, which has the greatest expanse and lowest precipitation rate of any area on earth, is pierced by the Nile and surrounded by tropical Africa and the somewhat less humid Mediterranean zones to south and north. The main settlement of these parts of North Africa and south-west Asia lies along the Atlantic and Mediterranean coasts of Morocco–Algeria–Tunisia; in the Nile valley; along the Red Sea and Persian Gulf coasts of Arabia; and in loops encircling the desert basins of the Iranian plateau [9]. All of these territories contain empty spaces, and the word 'desert' referred to these parts of the world, which to the ancient Romans, measuring Arabia and the Sahara against their own more humid Mediterranean environment, appeared empty of people, deserted.

2. The so-called empty quarters are, of course, not utterly deserted, even in the most severe deserts. They are seasonally occupied by hunters and gatherers of wild life, and by pastoralists, when the desert flowers. But the settlement of the desert and most arid zones of all types is thin and impermanent, as is the case with the Bushmen, the Australian aborigines and the Mongolian nomads.

In Mongolia, domestic herds of sheep, goats, camels, horses move out on to the semi-desert of the Gobi as part of the annual round. The Gobi is seasonally occupied, but empty during part of the year. Camels and goats among these livestock have a better arid adaptation than horses, cattle or sheep.

3. Human travel, transport of goods and communication of messages take place across the desert, which, like the sea, at once joins and divides the human settlements. The Sahara, Arabia Deserta, and the Kara-Kum desert of Turkmenia are criss-crossed by caravan routes which afford means of travel, transport and communication just as do the Nile, the Tigris, the Indus, the Red Sea or the

Persian Gulf. The development of urban staging points has been discussed in Chapter X, and the relationship of nomadic and sedentary crafts to the trade of caravans is treated in Chapter XIII.

Caravan routes must follow the shortest distance between two points. The descriptions of North African caravan routes by Coon [10] and others show that in this region there are several routes between Lake Chad and the Niger river north to the Mediterranean coast. These routes are linked in turn by several east–west, north–east and north–west routes, with a great nodal point at In Salah, in central Algeria.

The caravan routes of the Sahara are coupled with the nomadic activity of the Tuareg of the central Sahara. Their winter camp is in the Ahaggar, which they occupy in January–February. The Tuareg are a pastoral group, combining nomadic and transhumant features in the tending of their flocks of sheep and goats, and herds of camels, horses, asses. They move north–east to the oasis of In Salah during late February and March, where they exchange dates for manu-factures of Sudanese provenance. During April and May they return to the Ahaggar, where they remain with their herds during the early summer. In late summer, August and September, they move north to Amadror for salt. In September–October they bring their herds and salt south to Talak for pasture, then proceed further south. They spend the end of the year to December in various pasturages and at places close to the border of Nigeria, exchanging salt for grain and manufactured goods, returning thereafter to their winter camp in the Ahaggar [10, 11].

These caravan activities are not simple. They combine four major occupations: pastoralism, date-picking, salt-mining and trading. Partly the stock are raised for subsistence purposes, partly for caravan and trade purposes. The exchange, and hence the caravans, are under Tuareg control. Arab, Turkish, Turkmen and Mongol caravans were and sometimes still are in the service of a merchant who hires the caravan to transport his goods. The caravan leaders will then leave their herds in the charge of their families and proceed only with the beasts of burden, which are camels in the main.

Less well known is the regime of the Taklamakan Desert in the Kashgarian region of north-western China. This region is a wholly enclosed basin, sur-rounded by the Kun-Lun range to the south and the Tien-Shans to the north.

Part of the basin is known as the Tarim Desert, containing the Tarim river (Yarkend Darya), which rises in the Kun-Luns and flows across the northern part of the basin, and is the main water course of Kashgaria. The rivers of the Kun-Luns flowing north into the Taklamakan from the south all have beds which reach the Tarim river in the north, although these beds are rarely filled. The most extensive are the Khotan Darya, Deriya Darya and Niyan Darya, of which the Khotan Darya (Darya is a Persian word for river) is the most signi-ficant, reaching the Tarim river when there is sufficient mountain run-off – a rare occurrence. Although these rivers shift their beds, are unreliable, seasonally dry and generally do not reach their destinations, the limits of their beds, they nevertheless provide the sole caravan routes across the Tarim on the ground. The only alternative is to skirt the desert [12]. North–south travel across the Taklamakan follows the beds of the Khotan and Kerye rivers. A third route in the eastern part of the Tarim basin is related to the old bed of the Tyertye river. Unlike the Sahara, the Tarim does not enjoy a network of routes, only those leading north–south. It is the distribution of water sources which determines the location of caravan routes from oasis to oasis, for the water lore and hydro-technology is comparable in both regions, the Sahara and Sinkiang.

Water rights in the cultures of the humid temperate zones

Settlement in the humid temperate zones is continuous depending on terrain factors. There is an assumption of confidence in availability of water by the inhabitants. Cities are the foci of dense population settlement, but rarely, save in mountainous areas, is one out of sight of populated places in the temperate latitudes, especially those surrounding the North Atlantic. Transportation routes compete for space with other land-uses; in the humid northern temperate zone there are seasonal occupancies of permanent settlement at resorts, but not seasonal settlements. Finally, there are no significant areas of zero population density.

Countries bordering the North Atlantic have a characteristic cultural pattern concerning water. This cultural pattern is to be seen in Britain, France, Germany, Denmark, Norway, Sweden, Poland and the Baltic countries generally, including north Russia and Finland. The Low Countries have a problem of fresh water control; therefore they too must be added to the list. Eastern parts of Canada and the United States are extensions of this pattern.

These countries have few cultural arrangements concerning water. Neither their laws and customs nor technology have any concern with water save its treatment, storage and use. Conservancy has no traditional base here, although lately it has become a problem of which a few thinkers and activists have become aware. The problem of access to water has no historical consciousness in these cultures.

The common law of England, as it has been extended to eastern parts of the United States and parts of Canada, has had little concern with water, nor with land-users' rights and other users' rights governing water. A rule of historical antiquity affirms that the user may treat water courses on his land as a common enemy and fend them off as he pleases. In so doing, naturally, he may do injury to the rights, uses and property of his neighbour; but in any case he may disregard those rights [13]. Rules of the common law may on occasion consider the matter of protection of land along a water course from such damage. Three conclusions may be inferred from the concern with water in common law: (1) There is no systematic treatment; in fact, water rules are contradictory, some dealing with protection from damage, others denying protection. (2) There is little or no community concern with water use; this matter is left to the individual as one who seeks redress before the law for an injury which he holds to be an actionable wrong. It should be noted in this connection that the onus is upon the one who seeks redress. (3) Rules are negative and do not concern use of water, but rather damage from use by others. An exception to this is the rule concerning navigable streams. Maintenance of navigability of streams has priority over other uses.

The assumption in all these cultures is that there will be enough water for everyone – barring drought – and that plethora will not require common action. The riparian right is only a part of the water right of common law. The code law of France and the countries which have received it or have been under its influence have not changed the underlying assumptions materially.

Arid zones – principles of human ecology

In contrast to the cultures of the temperate zones of the North Atlantic seaboard, the cultures of the arid zones have a traditional concern with water, its sources, storage and flow. The great water projects of the Nile, Indus, Tigris, Euphrates, Zerafshan, western Iran and the simple reed tubes of the Bushmen have common intents. Many skills have been developed by these peoples for the survival and flowering of their cultures under conditions of aridity. However diverse the range of these cultures may be, they have this feature in common – their adaptation to aridity and their mastery of water shortage. Among all these peoples, water is neither an enemy nor taken for granted. It is always treated as something precious, whether the culture has a means of measuring value, as among the commercial cultures of the Nile or the Zerafshan, or not, as among the Bushmen and the Paiutes. From this point of view we may draw the cultures of the arid zones together and consider them in contrast to the water-rich cultures.

There are yet other cultures with water management problems and related cultural arrangements. The Chinese, the people of India, and the peoples of south-east Asia have a joint concern with dearth and plethora of water. They have devised means of irrigation and a vast hydrotechnology to meet these problems. The farmers of the central Mexican plateau and valley, the Maya of Yucatan and the Pre-Columbian farmers of Peru likewise solved water problems with a highly developed technology. Some of these peoples, such as the Incas of Peru and north-west India, had problems concerning aridity, but most of these great farming peoples had problems rather of maldistribution in location or season, of space and time concerning the needed water.

The constant problem of water is recognised by the inhabitants of the arid zone. A next step in ecological theory as applied to humanity is to define the arid zones. These are zones in which the undersupply of water is normally a recognised problem in the cultures which occupy them. This may serve as an operational definition of arid zones, inasmuch as the standing problem of aridity is known to the unsophisticated inhabitant as it is to the scientist who investigates it: the cultures of the arid zones are those which have traditionally means of coping with the problems of water shortage.

These cultures have in common not a geographic milieu, but a mode of treating with that milieu. The geographic environments constituting the arid lands are diverse, ranging from very dry desert to steppeland and prairie. They are discontinuous and scattered in various parts of the earth. Therefore, we cannot think of one milieu. But there is also another sense in which we must consider the unity of the man-milieu problem. We, mankind, do not interact with nature directly, but through the screens of our cultures. The Bushmen and the Paiutes, the Shoshones and the Arandas of central Australia cannot be seen simply as creatures of their environment. They use water in the form of rain, snow, sleet, dew, surface water, underground water, fresh water, brackish water, flowing water. Water is not an enemy. Egypt, wrote Herodotus, is the gift of the River. In contrast to this view of water stands that of the Anglo-Saxon common law. The Chinese usage regarding water is ambivalent and stands in contrast with both. Water courses in China are to be fended off; the Yellow River, China's sorrow, is to be contained, constrained from doing harm to man. And yet the rivers are necessary for life. Of this the Chinese are more conscious than are Europeans and Americans.

Cultural area and cultural zone

The concept of cultures peculiar to the arid zones is a development of the classical problem of cultural areas. In anthropology, A. L. Kroeber [14] and others have written about the relation between cultural and natural areas. A culture is generally a group of people with a common territory, way of life, language, and some identifying symbol, such as a name of the group. A cultural area is by the same token a group of cultures which may or may not be of common origin, but which are contiguous and share some identifiable features. Because of geographic contiguity, environmental features may be shared by these cultures; adaptive features may be shared by borrowing or diffusion. The environment, as Kroeber [14] has put it, does not produce a culture, but stabilises it; the causes of cultural phenomena are other cultural phenomena.

Communication between cultures in a geographic area, such as an arid region, is aided by the fact that they have common concerns. Even though Berbers and Arabs speak different languages, they can come to an understanding and mutually exchange techniques and conceptions; Shoshones and Paiutes may do the same, or Bushmen and Hottentots. Thus we may here infer for purposes of cultural analysis various cultural areas of the arid zones – the North African, the South West African, the western American Basin, the central Australian, the west coast of South America, and so on. But these various cultural areas are not in contact with each other in so far as traditions are concerned.

In what sense can we speak of a single culture of the arid zones? We have carried our generality from a single culture to a cultural area. Studies of cultural areas of the arid zones involving not one culture but several have been published by Feilberg [15], J. Steward [4] and myself [16]. We have posed the problem of cultural features shared by all cultures occupying arid natural environments. They cannot be conceived as forming a single cultural area, still less as forming a unified culture, but they may be conceived as representing certain zonal problems. These problems exist not only in the minds of the outside observer, but also in the minds of the peoples who live in the arid zones. They concern water and hydrotechnology above all. We have limited our concern in the main thus far to the material aspects, the tangible things and visible practices of the cultures of the arid zones. It would follow that the next step should be to con-sider the social organisational and the spiritual features of these cultures, both in so far as they are shared, and as they differ. This, however, is insufficiently known for primitive cultures, and may, in fact, no longer be possible of achieve-ment because it is apparent that everywhere in the arid lands profound changes are taking place. These changes result from the introduction, and often the very sudden introduction, of modern industry and new means of transport, as discussed in Chapter XIV. Such changes affect peoples with different cultural backgrounds. Some, such as the Bushmen or the aborigines of central Australia, have an excellent and long-term adaptation to their habitats. However, they have few skills which are directly applicable to the machine age. They must be trained for specific jobs and they must be generally educated. These joint undertakings must be introduced by experts coming to them from outside their cultures.

Others have been settled in their habitats for shorter terms with a close participation in the industrial civilisation. The nineteenth-century migrants to Australia, California, and more recently Israel, brought a small amount of industry with them. They have little difficulty in further economic development of their regions, and may be counted on to initiate change in a desired direction.

Between these two extremes are the peoples of North Africa, south-west and central Asia, and Mongolia, who have made modest advances, have overcome some inner problems, but can develop further only with capital and labour skills introduced from abroad.

The impact of industrialisation advances and related changes upon these peoples is different in every case. Some peoples of the arid zones are part of the industrial civilisation, and hence development of their land is, at least in part, their own problem. Others can accept outside help in limited ways to be mean-ingful. Or to put the same thought another way, programmes of capital invest-ment should accompany training programmes up to the highest levels. Thus, the development of further skills beyond the particular programme can be generated within the country. This is a problem of economic development which concerns not only certain lands of the arid zones, but others as well.

Peoples of the arid zones, just as peoples elsewhere, can accept what they are culturally prepared to accept from outside. Industrialisation can be developed anywhere, but it requires general education and training in specific skills; and industrialisation can only be introduced where both the general and the specific education is introduced throughout the country. The alternative to this is a kind of capsule urbanism and industrialism which was characteristic for many years in the Abadan–Khorramshahr complex at the head of the Persian Gulf. Here, because of the limited nature of the capital development, the petroleum industry had at one time little impact on or relation to the rest of the economy of Iran.

We are viewing today a change in the rate of impact of one culture on another, or acculturation in general. The rapidly accelerating rate of flow of ideas and people, especially experts, and of commodities, is heightening the development of the resources of the countries of the arid zones. Capsulated industrialisations and limited impacts are being increasingly a matter of past history. Changes are being effected on a broad scale and in ever-decreasing periods of time. We must now turn our attention to the processes whereby these accelerated rates of change are being brought about.

Political change

Kashkarov and Korovin [8] have written that change in social life came about with the domestication of livestock. The stock was and is a means of transforming the meagre desert vegetation into meat and milk. We should interpret this thought figuratively rather than literally, and seek in the means of successful adaptation in desert life a whole series of cultural transformations. These transformations have each led people who mastered them to an ever greater control of the desert environment.

The first of these transformations was domestication of plants and of livestock. Cultivation coupled with irrigation and other forms of hydrotechnology made possible the settling of large numbers of people in small areas of desert – or former desert now tranformed. The early states and empires in the deserts surrounding the Nile, Tigris, Euphrates and Indus introduced means of organising the work force to build the irrigation works. They were means for disseminating information, innovations. A nation, state or empire is a large-scale communications system through which information travels with rapidity and efficiency.

The peoples of the arid zones today are organised with only a few exceptions into self-governing states. Some of these states are of older origin, some are recent. The advantage of a nation-state organisation is that leaders can seek out advice and aid, and promulgate these throughout their countries. Educational systems and technical programmes, both practical and general or scientific, can be developed under these conditions. Technological innovations can be broadcast throughout the nation. An excellent example of the use of a national institution for promulgation of change is the farm extension programme of the United States. In this programme and through a nation-wide network of country agents, farm production, marketing and management practices have been introduced quickly and thoroughly disseminated. The nations of the arid zones have organised similar institutions in a number of instances. This communications situation may be contrasted with the situation in which tribal divisions exist; we usually observe difficult communications between or among the tribes. Contrast this communications situation also with a colonial government: in a colony, social distance is usually great between colonists and indigenous peoples, making communication difficult.

New political institutions aid the process of flow of technological advances by relating whole nations and whole scientific communities to each other. The new political developments as well as older achievements may be expected therefore to accelerate cultural change in most parts of the arid zones.

There are certain difficulties, however, involving the nation as the unit of cultural change. National decisions to follow a certain practice are usually weighty ones; national commitments are difficult to change once they are made. There is no easy remedy for problems raised by means of national commitments; only careful study, an experimental attitude and readiness to change even the weightiest of decisions can overcome the momentum forces generated by national or nation-wide engagements. Nevertheless, these difficulties must be faced because the nation today is usually co-terminous with the cultural whole and profound cultural change generally involves the entire nation.

The change in many instances involves a shift from tribe to nation and from small community to great city. The tribe or the small community is a comforting place to be in in one sense. The individual lives among people who know him; relatives and friends are around to care for him if he or his family becomes ill, indigent or aged. The traditional family, tribe, community is a system of psychological social security.

The formation of nations, development of industries and migration to the cities has transformed the traditional psychological and social security systems. New modes of psychological and social security have to be developed in such cases.

One encounters a polarisation of opinion in regard to social change. There are some who say that the tribesman has a way of life which has too many advantages to be lost by modernisation. These are the romanticists, who consider the noble savage superior and civilisation deleterious. There are at the opposite pole realists who say that change is inevitable and that those who oppose it will be overrun. Scientific procedure can take neither view into account.

As scientists we can recognise resources problems, needs of inhabitants who occupy territories where the resources are found, world needs of resources, and local needs created by the meeting of world needs. Petroleum extracted from the Middle East and North Africa has met a resources need, but has created, or increased, local water needs. The local population has a cultural problem in meeting technical and economic changes introduced from the outside, and, just as comparative regional studies based on climate, soils and the like permits use to be made of accumulated experience in the natural sciences, so, in the social sciences, local populations may derive benefits from experience gained in other regions in solving similar problems. All such experience may be said to be added to the general reservoir of knowledge pertaining to human needs and human factors.

Technological solutions of problems regarding water, land and their resources may be expected in the future as in the past. As White has written on an aspect of our problem, 'There is needed a new synthesis of indigenous patterns with modern technology to meet the new demands of urban life in arid lands' [17]. This expression needs no emendation, but only expansion to embrace not only urban life but all of human life in the arid zones, rural as well as urban, and their various sub-categories.

LAWRENCE KRADER

References

[1] FROURIE, L. in S. and P. OTTENBERG, Cultures and Societies of Africa, New York, 1960.

[2] FORDE, C. D., Habitat, Economy and Society (7th ed.), London, 1949.

[3] STEWARD, J., Basin–Plateau Aboriginal Socio-political Groups, Bureau of American Ethnology, Bull. 126, 1938.

[4] STRONG, W. D., Aboriginal Society in Southern California, University of California, Publications in American Archaeology and Ethnology, Vol. 26, 1929.

[5] IMANISHI, K., 'Nomadism, an ecological interpretation', Silver Jubilee Volume, Zinbun-Kagaku-Kenkyusyo, Kyoto University, 1954.

[6] KRADER, L., 'Ecology of central Asian pastoralism', Southwestern Journal of Anthropology, Vol. 11, No. 4, 1955, pp. 301–26.

[7] KACHKAROV, D. N. and E. P. KOROVINE (Kashkarov and Korovin), La vie dans les déserts, Ed. Th. Monod., 1942.

[8] MYRES, J. L., 'Nomadism', Journal of the Royal Anthropological Institute, Vol. 71, 1941, pp. 19–42.

[9] COON, C. S., Caravan, New York, 1951.

[10] LHOTE, H., Les Touaregs du Hoggar, 1944.

[11] NICOLAISEN, J., 'Ecology and culture of the Pastoral Tuareg', National-museets skifter, Etnologisk Raekke, IX, Copenhagen, 1963.

[12] ARMAND, D. L. et al. (Ed.), Zarubezhnaya Asiya ('Asia beyond the borders'), Moscow, 1956.

[13] BUSBY, C. E., A. STEFFERUD et al., Water, US Dept. Agriculture, 1955.

[14] KROEBER, A. L., Cultural and Natural Regions of Native North America, University of California, Publications in American Archaeology and Ethnology, Vol. 38, 1939.

[15] FEILBERG, C. G., La tente noire, Nationalmuseets skrifter, Etnografisk Raekke, II, København, 1944.

[16] KRADER, L., 'Ecology of nomadic pastoralism', International Social Science Journal, Vol. 11, No. 4, 1959, pp. 499–510.

[17] White, G. Science and the Future of Arid Lands, Unesco, 1960.

ADDITIONAL REFERENCES

BEN-DAVID, J., *Agricultural Planning and Village Community in Israel*, Unesco, Paris, 1964 (Arid Zone Research XXIII).

UNESCO, *Nomades et nomadisme au Sahara* (in French only), Unesco, Paris, 1963 (Arid Zone Research XIX).

CHAPTER XIX

Deserts as Producing Regions Today

The preceding chapters have shown that the array of adjustments which man is making to arid zone conditions today is highly diverse and is changing rapidly. While the basic complexes of rock, soil, water and vegetation are altering slightly in some places, such as in managed grazing lands, or are deteriorating tragically, as in the irrigated flood plain of the Lower Indus, their productive use is fluctuating more widely. Increasingly, decisions are made by public agencies to expand cultivation in one sector – as in the dry lands of the USSR – or to curb it in another place – as in overgrazed mountain slopes of Baluchistan and the United States. These are large-scale efforts. But the growth of technology available to farmers, the opening up of monetary economies in isolated or semi-subsistence communities, and the enforcement of internal security have also made it possible for individual farmers, grazers, miners and recreation seekers to press into new areas and to shift their means of livelihood quickly. Many parts of both the interior and the margins of the arid zone are in a state of flux. The traditionally restless state of the nomad is more than matched by the surging margins and shifting present-day occupation of the interior of the zone.

The principal conditions of arid land environment and use which are involved in these changes have been described. Obviously, a great governmental investment in new irrigation as well as an individual's choice to push his tractor into dry soils is complex, taking account of many of these conditions. Each one is aimed at gaining in productivity either in monetary terms or in broader social benefits.

Understanding of how these changes are taking place every day and what conditions may be significant in future in rearranging the zone's productive use may be advanced by examining the spatial distribution of some of the major elements in the process of decision-making.

As suggested in Chapter II, man's decision to manage arid land is an intricate process in which he assesses the resources with which he has to deal, applies some part of the technology and organisation at his command, winnows out the measures that seem inefficient in reaching his aims, confronts certain of the

421

effects of his action on other areas, and readjusts his institutions to serve the desired uses. These five aspects of decision-making are closely related to each other in the course of arriving at a final choice of resource use. For example, a farmer on the alluvial fan of a semi-arid mountain slope may have the choice of planting any one of several dry-farming crops or of putting down a well for irrigation. His decision will rest in part upon his appraisal of the resources available: what crops he thinks the soils will permit him to cultivate successfully without irrigation and what crops with irrigation, and whether or not he perceives the availability of water underground. He also will be influenced by the kind of technology he can employ; a new cheap pump may change his view of raising water to the surface, or a new strain of drought-resistant wheat may increase his interest in dry farming. He will make some kind of estimate of the economic returns in prospect: this requires his assessment of the costs and benefits of each likely system of farming and the returns and risks he would expect. He will take account, if only by ignoring them, of the effects of his work on other users of the available water and soil. All of this will take place in the system of education, custom, economic exchange, taxation and property control of his particular community that restrains certain actions and encourages others. Some farmers consciously review a few aspects of their choice, others act entirely according to the custom of their group, and still others make careful calculations of the probable consequences.

When governments make the decisions – as they do in launching new irrigation programmes or regulating the grazing on open grass lands – the process of choice may be more evident, but the same elements are present: the resources must be estimated, however inaccurately; a certain stage of technology must be applied; the probable benefits and costs must be compared; the possible effects upon other areas must be weighed; and the social guides must be accepted or modified. Never are these considerations completely rational, comprehensive or accurate. Some unpleasant results may be ignored for political reasons, as when a new water development would benefit one group of farmers more than another; a whole sector of resource use, such as drainage of irrigated soils, may be overlooked; and even careful surveys, as involved in appraisal of average stream flow, may turn out to be far from the condition in nature. Oftentimes, the very action that government takes to improve resource use has the opposite effect, as when a levy on land yield to support an agricultural programme encourages the cultivator to progressively destroy his soil to meet the tax bill.

Each of these five elements in the process of decision will be reviewed as it affects productive use in representative regions. No attempt will be made to discuss all parts of the arid zone. Rather, the character of decision-making will be demonstrated against the background of areas that vary greatly in the conditions affecting decision.

Estimating the resources

One of the difficult aspects of developing arid land resources to their optimal productivity is the lack of basic information as to their extent and quality. It has been shown in the chapters on soils, water and plants that most of the data as to these resources are spotty and inadequate. Even with respect to the fundamental distributions of precipitation, there are large areas for which the coverage of rain gauges is so sparse that an estimate of annual precipitation is with the best of extrapolation methods little more than guesswork. This is strikingly the case in the Arabian Peninsula, where in 1958 there were records of only sixteen rainfall measuring stations in an expanse of one million square miles. Moreover, only a few of the stations have been in operation long enough to permit confident generalisations as to variability and short-term trends for those places.

The coverage of surveys and data-collection networks is generally insufficient to support solid estimates of resources for major sectors of the zone, and they commonly have accumulated in a sort of overlapping patchwork. A geological survey for metallic minerals is comprehensive in a dry sector of Mexico where stream flow measurements are missing, and reconnaissance soil investigations penetrate valleys in Iran where rainfall has never been measured. Extensive sectors have been blanketed with aerial photographs for purposes of petroleum exploration and these potentialities have been canvassed without attempting to appraise the character of the vegetative cover. It is a rare area where geology, landform, climate, soils and vegetation have all been assessed, and it is rarer yet to find a place where this has been done under unified direction. Australia provides the outstanding venture in this direction, and West Pakistan offers an experimental example of one valley where this was accomplished.

Necessary to most resource assessment is the base topographic map or the aerial photographs which may be used for landscape interpretation. In recent years aerial photography has become the common foundation for topographic mapping, but it also is employed for direct resource survey, as for soil mapping in Pakistan or for plotting of vegetation associations in Iran, where the base maps are not prepared. Without either, most other resource appraisal is severely handicapped.

From Figs XIX-1 and 2 it is evident that certain sectors are especially favoured and others grossly deficient. Virtually all of North and South American, South African, and Indian subcontinent arid lands have been covered by aerial photography. The greater part of the Sahara and adjoining lands, except for Libya and other scattered areas, have been flown. Australia lacks photography for two parts of the dry interior, and the coverage of the Arabian Peninsula and of other sectors of south-west Asia is incomplete.

The greater part of south-west Asia, except for the Arabian Peninsula, of the Indian subcontinent, of the north-western Sahara and of coastal arid lands in South America has been covered by large-scale topographic maps. Arid lands in

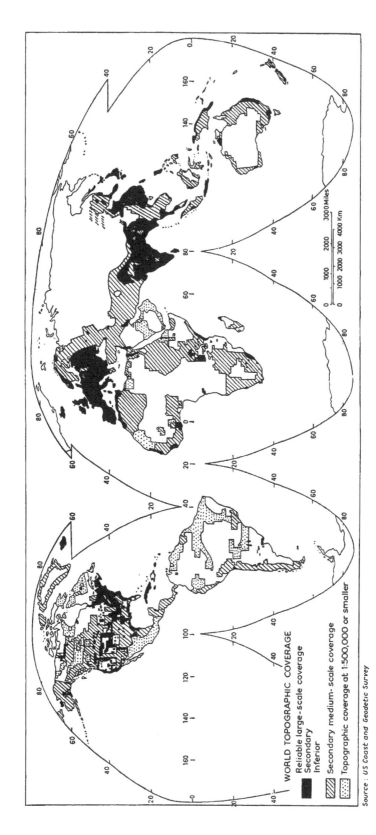

WORLD TOPOGRAPHIC COVERAGE

Reliable large-scale coverage
Secondary
Inferior

Secondary medium-scale coverage

Topographic coverage at 1:500,000 or smaller

Source : US Coast and Geodetic Survey

Fig XIX-I. Extent of topographic coverage of the world, 1957

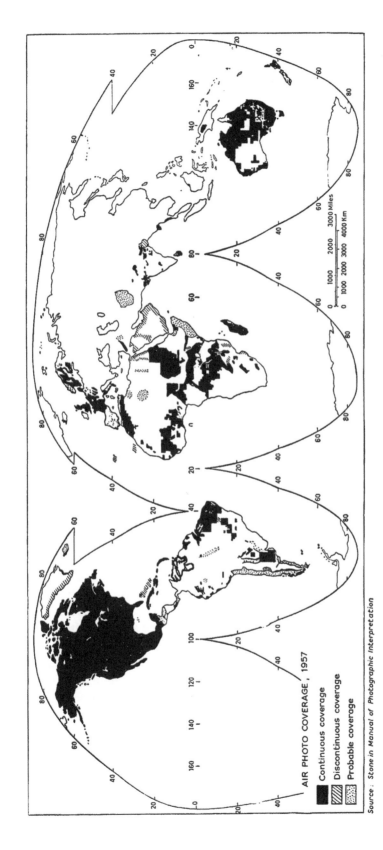

AIR PHOTO COVERAGE, 1957

■ Continuous coverage
▨ Discontinuous coverage
⣿ Probable coverage

Source : Stone in Manual of Photographic Interpretation

Fig xix-2. Extent of air photographic coverage of the world, 1957

North America, in Arabia, the middle and eastern Sahara and Australia are provided with maps only in spotty fashion.

Much of the survey work is based upon assumptions as to productivity which themselves are subject to change with new technology or economic conditions, and which will be altered as scientific understanding of land and water processes advances. Until very recently the recreational uses of desert lands did not figure in appraisals of future utility, but they now are taking on importance wherever cities are growing near deserts, and they are a dominant factor in distribution of new land-uses for both vacationers and year-round residence in southern California and Arizona. It often is convenient to assume that cost–price relationships will remain virtually unchanged when assessing capability of land below a proposed irrigation ditch, yet a major shift in the markets for cotton and sugar as reflected in prices would alter overnight the view of what is a feasible investment in providing water to grow those crops. Likewise, judgements of the availability of water for pump irrigation depend only in part upon knowledge of the size of water-bearing formations, and may be altered by assumptions as to the kinds of pumps and energy which will be used and by their costs, so that when new pump designs or energy sources are found an aquifer which once was considered useless takes on positive value.

Scientific surveys provide a generally accurate description of what natural resources are at hand, but they do not necessarily reflect the perception which the people of the area have of the same resources. For example, the Arab grazers of the steppes of northern Algeria and Tunisia for many centuries did not recognise as useful the supplies of fresh water that were to be found in shallow aquifers which had been tapped by Romans. When these old wells were renovated and new ones dug the opportunities for livestock production without going long distances to water were enhanced. On the other hand, local inhabitants may recognise subtle differences that are hidden from scientists lacking their experience. The Bindibu tribesmen in a sector of central Western Australia which early explorers considered almost waterless are able to show scientists scores of water resources, each with a distinguishing name and location, that otherwise would go unobserved [1, p. 274]. If people fail to perceive what the scientist sees it does not figure in their decisions, and their image of the resources available to them may be quite different than the image held by another cultural group.

Some of the variable assumptions in a resource appraisal are indicated by the agro-ecological survey of Southern Rhodesia described in Chapter xv. This careful assessment of land-use capability was based upon a soil classification, using distribution of vegetation as an indicator along with a measure of effective precipitation to group together lands having similar characteristics. To do so required judgement as to the methods of management which would not cause soil deterioration and as to the farming systems which would yield satisfactory net income to the farmer. It was assumed that European farmers would operate the farms, and thus the characteristics which might be important to successful management of small plots by Africans were not appraised directly. It was necessary to assume that the technology of seeds, fertilisers and tools available to the farmer would not change and that the markets for tobacco, maize and livestock would hold up. No attempt was made to estimate productivity under irrigation. The discovery of a new method of suppressing evaporation, or of new drought-resistant plants or a reorientation towards African farming, would alter the classification.

The full complement of scientific disciplines involved in an estimate of arid land productivity is assembled by the Central Scientific and Industrial Research Organisation of Australia in its investigations of the dry interior [2]. A team of experts drawing upon experience in geology, geography, climatology, botany and soils examines an area in reconnaissance fashion, using aerial photographs. Upon the basis of the observed associations and of more general relationships they divide the land into classes having homogeneous combinations which seem significant to potential human use. Because of the large areas to be covered and the small economic return which may be expected from such lands, this type of appraisal in the arid zone must be carried out rapidly and at low cost per acre if it is to be practicable.

A somewhat similar effort has been made for one small valley high in the plateau of Baluchistan. In the Isplingli area near Quetta the Pakistani government, with the advice of Unesco scientists, conducted an integrated assessment of resources and their possible management for a primarily livestock economy. Bringing together specialists from several scientific fields to work together at the same time is a troublesome task. Usually it is easier for each to work separately at their own pace, but they do so at the cost of opportunities to make comprehensive resource estimates.

In the United States the Soil Conservation Service prepares for individual farmers 'farm plans' in which an aerial photograph is used as a base for mapping land-use and land-use capability. This provides the framework for defining the precise location and type of conservation practices such as reseeding, water development and erosion control that should be undertaken within a scheme for optimal farm returns. The farmer and the soils and conservation experts work this out together so that as far as practicable they agree on the estimate of the resource and on ways of using it.

Technology

A striking feature of man's use of arid lands is the gap between the technology that is known and the technology that is applied in everyday husbandry of soil, plants and water. A livestock farmer runs excessive flocks of sheep on pastures with poor returns where a systematic use of grazing management, seeding and fertiliser would pay in increased yields. An irrigator applies water when not needed or neglects to do so when moisture deficiency might first be noted. A manufacturing plant withdraws three times as much water as required to process a product with water conservation devices that are available to him. The size of this gap varies according to the culture, place and time, but only in a few areas, such as in some parts of Israel, the USSR and the USA, is it small.

Great gains in productivity could be achieved if the skills and knowledge reported in Chapters XII, XV and XVI were to be widely accepted and practised. Larger physical returns could be obtained from each unit of water withdrawn and from each acre of land under cultivation.

Yet, in most parts of the arid zone the new understanding of technology resulting from scientific research is moving ahead faster than the ability of the people to apply it, with the result that as the opportunities for resource use increase the spread between the most advanced and least advanced farmers widens. A major challenge to improvement in productivity emerges in efforts to gain acceptance for the advanced methods. It is apparent on every hand that cultures and social groups within cultures differ in their response to new technology, some holding firmly to old ways, others making only modest concessions to the new, and still others adapting their methods rapidly to the promise of greater or more stable returns. This is shown most dramatically where Bushmen of the Kalahari confront a European equipped with machinery and technical skills. They resist changing their gentle ways, and although he may be unable to practise their highly sensitive methods of finding water and game he has other techniques which permit him to make a living on the margins. Acceptance of new technology may, however, come rapidly, as in the north Saharan areas affected by petroleum exploitation, where nomads in a generation shift to motor transport and skilled labour with accompanying dislocation in social organisation and values.

In many sectors of the arid zone the role of science in providing the basis for new technology has changed radically in recent years. For a long time the findings from research laboratories have given the facts and relationships upon which improved technical processes could be designed. As indicated in Chapters XV and V, research on genetics has offered the means of breeding drought-resistant plants, and laboratory studies of the movement of water in rock have made it possible to develop effective methods of recharging some aquifers or of determining the safe yield from a new well. Such adaptation of scientific results to practical situations may be expected to expand as science itself grows, much of the work having unpredicted value for arid situations. Increasingly, however, scientific research and technical development is being aimed at finding solutions for specific problems of resource use, and enters the array of capital expenditures, resource appraisals and social regulations as a planned means of managing natural resources.

Perhaps the most intensive input of scientific and technical skill in solving questions of arid land-use in the world today is in Israel. In its universities and research stations systematic studies are being made of water use in plants, the processes of sediment movement in arid stream systems, human physiology in high temperatures, methods of converting and storing solar energy, and a host of other problems which affect the productivity of its dry areas. Research is regarded as a tool of resource management, and as a means of changing the limits imposed by scarce water, soil and vegetation under present technologies.

If, for example, the rate at which water is lost from a soil–plant surface by evapotranspiration were to be reduced by as much as a third by some cheap means, the consequences for agricultural use of the upper Negev and of semi-arid lands would be momentous for, in effect, the available moisture would have been raised by that amount.

An instance of concerted application of modern scientific knowledge and methods is in the Lower Indus plain in West Pakistan. Faced with great and mounting loss of agricultural land through waterlogging and salt accumulation, notwithstanding modest public attempts to curb these abuses, the government of Pakistan in 1961 enlisted co-operation from scientists in the United States to work out feasible ways of stopping the wastage promptly and of rehabilitating the damaged land. Botanists, agronomists, chemists, engineers, econometricians, political scientists and various others joined in assessing the situation and in charting economic solutions. The procedure was not to take existing knowledge of soil–water relations as the basis for planning wider measures, but to identify points at which new knowledge would permit a more effective type of measure.

The inquiry into food production capacity in India in 1958–9 points out the massive opportunities for and obstacles to adoption of available technology in the dry north-western and central plateau areas of that country [3]. Beginning with estimates of population growth and demand which showed that the need for food grains would mount rapidly, a team of agricultural experts concluded that while India might reasonably expect to meet those needs by using proven agricultural practices and organisation, the nation would fall short of its goal by as much as 28 million tons by 1965–6 if it continued at the then-current rate of improvement. Steps considered necessary to reach the desired level of production included price stabilisation, a programme of small-scale public works, nine-fold increase in chemical fertilisers, intensification of irrigation and drainage, improvements in land tenure and holding patterns, and reduction in numbers of cattle. Here, in a country with large population and both arid and humid lands, several lessons for arid zone technology emerged. It is clear that intensified use of arid lands will be a part of the solution, that the technology is available to meet basic food needs, but that the nation will fail unless it can manage to apply the technology on a wide scale and in combinations that recognise the interaction among different measures.

Economic efficiency

Quite aside from questions of physical efficiency in using soil without destroying it or in withdrawing water with minimum loss, there is the persistent question of the economic efficiency of the many uses of arid land resources. If all cultivators were free to choose and if there were no social obstacles to resource use they might be expected to select those uses yielding greatest net marginal returns, and to develop trade among each other which encouraged each to enjoy his comparative advantage by producing whatever would have the greatest relative economic efficiency. Clearly, there is not complete freedom of choice: people have different perceptions of their environment, different commands of technology, and different access to markets and other producers. They do not always work out the most efficient systems of resource use. Much of the movement in the zone is in response to efforts to gain larger economic opportunities, as when new irrigated lands are opened in the Punjab, or to retreat from highly inefficient ones, as when farmers migrate from the parched fields of the Nordeste in Brazil following a serious drought.

One indication of the range of economic production is given in Fig XIX-3 which shows the estimated gross national product *per capita* as of about 1955 for each country. These estimates are very rough in many cases where the national economic statistics are sparse or inaccurate, but they do indicate some broad comparisons. There will have been increases since then and, in general, the gap between low-income and high-income countries will have widened. Among the countries lying wholly within the arid zone, the GNP per capita is everywhere less than US $125, considerably below the mean of $200 for the world's population. Countries lying partly within the arid zone display the world range, including nations such as India and Nigeria with means in the neighbourhood of $200, and such as Australia and the United States with averages of $1,200 and $2,300 respectively. Venezuela and the USSR are the other two partly arid nations with *per capita* GNP greatly above the world mean. The former depends upon petroleum revenues for much of its income, as do Iran, Iraq and the states of the Persian Gulf. Without petroleum the level of product in the wholly arid nations would be much lower than it is today, and their current use of this mineral income to invest in more lasting improvements will profoundly affect their productivity once the oil is exhausted.

To a large degree the economic returns from new occupation of the arid zone is affected by government judgement as to its consequences for the nation as a whole. Because of the sparseness, concentration and unreliability of rainfall, most expansion of cultivated land must depend either upon irrigation or upon improvement of dry-farming techniques. With a few possible exceptions the new irrigation works will require strong public support in the form of storage and diversion works or of extensive well drilling. The smaller, more easily developed irrigation projects have been realised and those remaining have larger scale and higher requirements for public planning and financing. In the High Plains of west Texas and eastern Colorado it was practicable for individual farmers beginning in the 1940's and 50's to open up irrigation over a large area by drilling relatively shallow wells for themselves, but generally the depths and costs are such as to require public support. A public decision as to the likely contribution of the new development to the national welfare therefore is involved in agricultural expansion.

Because of the long distances between population centres, much of the transport development requires public initiative, and under prevailing policies the exploitation of minerals is under government supervision. In these circumstances

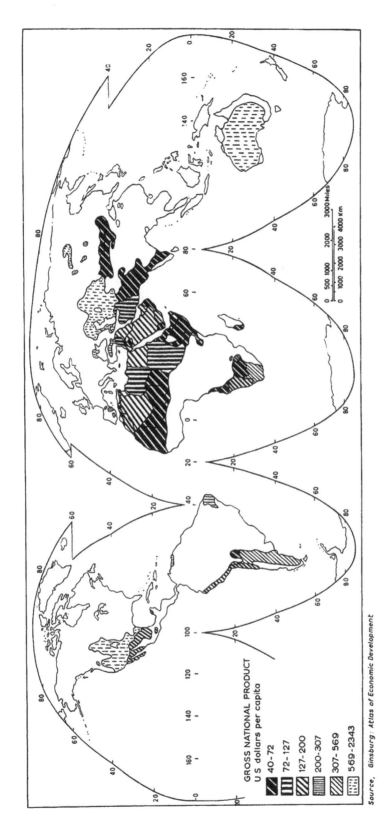

GROSS NATIONAL PRODUCT
US dollars per capita

- 40-72
- 72-127
- 127-200
- 200-307
- 307-569
- 569-2343

Source, *Ginsburg: Atlas of Economic Development*

Fig xix-3. Gross National Product in relation to population, in US dollars *per capita*, 1955

the welfare criteria and capacity of government action may be expected to steer the course of production shifts. A widespread and often accepted aim for public support of a dam, or well drilling, or fertiliser manufacture is to increase net national income. However, it may be extremely difficult to identify and measure the streams of benefits and costs which are expected to flow from efforts such as irrigation or range improvement. The accurate estimate of engineering costs with all of the uncertainties of weather, foundations and materials is itself difficult enough, and the estimate of what will be net returns from farm operations both on the farm and in other areas is extremely complex. The risk of occurrence of droughts, floods, hail and insect infestations, and the uncertainty of their coming, raises special problems of evaluation.

All too often in the past there has been a tendency for public bodies to underestimate the costs of new irrigation and to overestimate the benefits. Where the farmers were expected to repay the construction cost, as with the projects built under the Reclamation Act of 1902 in the United States, many of them failed to earn enough income to permit such payments, and special dispensations were required. Where no direct payments were required, as with irrigators in the Tigris–Euphrates, the net returns to the national budget have sometimes failed to reach the expected levels because the drains upon the national income were unexpectedly large. For example, the social overhead of providing schools, roads and health services sometimes had not been calculated in advance, and the expense of installing needed drainage works for irrigated fields had been neglected.

In low-income countries with severe financial constraints on their expenditures, as in India under its five-year plans, or in Peru where investment demands far exceed available savings, the wise choice of new investments for the limited capital is extremely important and the choice involves selecting among different scales and types of works in the arid zone, as well as selecting between alternatives in arid and more humid lands. For example, in India there are opportunities to expand production by different methods in both arid and humid lands, and these far exceed the available funds. Small dams should be compared with large ones, and both should be compared with fertiliser, seed and machinery programmes if the returns are to be maximised.

There may be other strong aims to government investment beyond increasing net income. Where certain areas or classes have become highly disadvantaged it may be public policy to redistribute income so that their living levels are raised, even though part of the cost must be carried by taxpayers in another part of the country. Thus, the government of the United States has made heavy expenditures for construction of additional irrigation facilities to aid in resettlement of American Indians. The Helmand Valley scheme in Afghanistan had as a major purpose the sedentarisation of nomadic groups, and one of the benefits in view in addition to increase in crop production was the political stabilisation of that roving segment of the population. The IBRD mission to Libya in 1959 placed heavy emphasis upon investment in 'measures to improve the lot of the small farmer and pastoralist', not only to meet needs from prospective population growth but to provide opportunities for workers who otherwise, going from agriculture into towns and finding unemployment, would foster social discontent and unrest [4, p. 329].

Few countries have made more judicious use of improvement of arid lands as a part of a national programme of investment assisted by petroleum revenues than has Mexico. In recent years the more rapid agricultural development has taken place in the northern, arid lands, where the opportunities for irrigation, for expansion of cultivation and for mechanisation in a sparsely settled area were promising. The total agricultural investment in 1939–50, accounting for one-fifth of national investment, was about evenly divided between public and private, with from 70 to 95 per cent of the expenditures for irrigation. This includes large projects such as the Papaloapan multiple-purpose enterprise, and several score smaller projects. During this time total agricultural production was growing at more than 4 per cent per year, but it is estimated that only about 20–30 per cent of the increase can be ascribed to irrigation, and much of this was due to the smaller projects which began to yield more promptly than the large ones where there was a lag between major works and farm output [5, pp. 22–5]. Improvement in varieties of maize has been a principal factor in expanding the nation's agricultural production.

Among the nations of the Mediterranean basin where improved management of grazing lands is essential both to holding the soil and vegetation resources and to raising national income, irrigation is expected to play a key role in advancing the economic change which is planned. This is illustrated by the programme for Iraq, Syria and Turkey prepared as part of the FAO co-operative programme. In each country during recent years the amount of arable land has increased by more than 40 per cent at the expense of ploughing up perennial pasture [6]. Livestock production *per capita* has decreased while crop production has increased and lands have deteriorated. To meet prospective needs of a growing population apparently will require reduction in destructive cultivation of dry lands, increase in production of food and forage on irrigation lands, and building up the livestock industry on grassland which will not be grazed beyond its carrying capacity.

Spatial linkages

As scientific knowledge of soil–water–vegetation processes enlarges, and as the economic organisation of human life in arid lands becomes more complex, the linkages between action in one place and that in another become more important to maintaining the zone's productivity. So long as only small parts of the normal stream flow were withdrawn from the great exogenous streams or as work of isolated nomads or oasis dwellers was only weakly connected with other areas, the repercussions of resource use upon the production of another area could readily be ignored, although sometimes, to be sure, with disastrous results. Some of these effects now are seen more clearly. The interrelations of weather condition, vegetative cover, soil, sediment movement, stream flow and groundwater in a drainage basin are now recognised as an intricate system which often requires management as a unit.

The division and management of water in the massive exogenous streams illustrates how future growth and the protection of present development will be dependent in part upon water storage and land-use in distant reaches of the same basin. The Indus, Tigris–Euphrates, Nile, Rio Grande and Colorado are among the principal exogenous streams, and each faces or has sought to resolve a pressing question of water allocation in which the value of a physical volume is related to where and how it is used. Agreements as to allocation of water are essential to large-scale development as the uses approach the volume of dependable supply in dry seasons. Without the early Nile Water agreement the timing of diversions for the Gezira scheme in the Sudan could not be scheduled so as to avoid injury to Egyptian irrigators dependent upon flood flows, and without a treaty between the United States and Mexico there would be no assurance that Mexican farmers near the mouth of the Colorado would not be injured by small or overly salty flows released by American farmers upsteam. Arrangements for the Colorado were linked with those for the Rio Grande, for although the streams have no physical connection, the Mexicans control the major flows downstream in the Rio Grande and the Americans in the Colorado.

Mention already has been made of the problem of integrating the development of an arid sector of a nation with total national development. This demands a solid understanding of financial and economic as well as geographic relationships within the nation.

The complexity of physical and human relationships within one arid basin is exemplified by the Middle Rio Grande basin in New Mexico. This is a plateau area through which the Rio Grande flows in a valley trough with strips of irrigated land flanked by broad reaches of dissected pastures. Dry-season water flow is used to the utmost, but is augmented by storage and is allocated by a set of agreements among New Mexico, the upper state of Colorado, the lower state of Texas, and between Mexico and the United States. Return flow becomes so highly charged with salts that it is unusable at points downstream. Sediment accumulation in the valley impedes drainage and builds up the stream bed so that local levees for flood control are threatened, water tables are raised, and the growth of water-wasting salt cedar is fostered in the old channels. To control silt inflows some reservoirs have been built in the lower reaches of tributaries, and soil conservation has been attempted through pasture management, reseeding and structures Thus, the welfare of farmers and city dwellers along the main stream is affected by the activities of livestock operators and foresters upstream, whose markets in turn are influenced by downstream activity. In the long run, the web of relationships must be viewed as a whole if a dam here, a pasture scheme there, or a government water treaty are to contribute to lasting development of the society and its economy. Any such efforts are handicapped by doubt as to some of the relationships. It is not known, for example, how much of the accelerated arroyo cutting in upland areas is caused by overgrazing and how much by changes in intensity of rainfall, and it still is not clear whether or not the river in its middle course is reaching an equilibrium state. Meanwhile, public decisions on remedial action must be taken.

Somewhat different but no less complex conditions prevail in the Helmand Valley in Afghanistan. Here the government established in 1946-9 a programme to deal with the development of the river for irrigation, flood control and hydro-electric power. Among the principal beneficiaries were to be nomads who for generations have moved their herds across the valley on their way from high summer pastures in central Afghanistan to winter pastures in lower areas of Pakistan, and tenant farmers who sought more land under their own control. Construction was initiated on diversion works, canals, storage works and the roads that must accompany such an effort. It became evident that readjustment of patterns of nomadic livelihood was extremely difficult and involved economic movements far beyond the project area. Tenants were unprepared to manage water and soil in the new conditions, and productivity of the few lands opened up was soon threatened by waterlogging and salt accumulation. Regulation of stream flow led to shifts in farming practice along the valley outside the project. To assure net increases in national production after bearing the costs of the new works would require measures, beyond the engineering works, aimed at affecting the whole organisation and practice of agriculture not only in the irrigated lands but in adjacent lands [7].

Social guides to resource management

The way in which governments guide resource management within their territories has a profound effect upon the outcome, and this is strikingly the case in arid lands, where public investment is essential to most new development and where the spatial impacts may be widespread and severe. Governments may select from a wide range of social guides according to their abilities, the desired effects, and their criteria as to what are desirable methods. The methods include direct ownership of resources, public investment, taxation, regulation of resource use, information activities and research. A promising research inquiry on drought resistance and plant genetics may be neglected because suitably trained scientists are not provided; a police method of curbing nomadic grazing of denuded lands may be rejected because it violates standards of justice; a scheme for raising live-stock production may violate basic religious belief. No measure for changing productivity of a green valley or a brown expanse of steppe land can be evaluated in practical terms unless account is taken of the social setting in which it is launched.

Closely linked with the question of effective combinations of social guides to resource management is the issue of what areas will be delimited for planning and administrative purposes. This typically is a question of whether resource use will be guided by agencies dealing with one resource, such as water, or one sector of the economy, such as agriculture, or by regional agencies having responsibility for most activities within a unified area. The idea of regional resource administration has great appeal because of the apparent unity of soil, water and vegetation within an area, yet it has its drawbacks, the chief ones being the difficulty of relating regional to national policies, and of correlating a regional agency with national bureaux.

An outstanding example of comprehensive guidance of use of soil and water in an arid situation is the Gezira irrigation scheme in the Sudan. In that area of more than 400,000 developed hectares in the almost flat, gently sloping plain above the junction of the Blue Nile and White Nile, the government of the Sudan, first working through a mixed corporation, directs the farming operations of long-term tenants who produce the cotton which gives the country its major foreign exchange [8]. Farmers are allocated plots of about 10 acres in which they cultivate cotton, beans, millet and fallow in proportions and at times specified by the agency. They use specified seeds, water and fertilisers in specified amounts, market their products through designated channels, and keep 40 per cent of the returns. The agency provides an elaborate system of agricultural advice and technical assistance to supplement normal government administrative and educational services. Farmers share in returns according to their productivity within the rigid schedule given them. Perhaps no other large irrigation scheme reaches as uniformly high efficiency in water use (see Figs XIX-2 and 3).

In contrast to the centralised and planned development of the Gezira is the state water plan of California, where larger investment is being committed to water regulation with much less regulation of the resulting economic activity. In a programme having an ultimate goal of regulating about 77 million acre-feet of water annually at a capital cost of over $10 billion, the state is beginning to store water in the more humid, 'water surplus' areas of the Sacramento Valley in the north, and to transport it by an intricate system of canals, aqueducts and exchanges to the parched San Joaquin Valley and to inland and coastal sections of the state further south [9]. Increased flows of water are expected to make possible an expansion of irrigated land by as much as 100 per cent, to support urban and industrial growth, to generate additional hydroelectric power, to supplement water recreation facilities and to partially recharge some aquifers now in danger of exhaustion. The plan no doubt will change markedly over the years. The role of government is to store, transport and sell water to farmers and urban centres, but it does not exercise control over how the water and land will be used.

A third example of comprehensive regional development is presented by the Iranian programme for its south-western province of Khuzistan. Using revenues from international petroleum exploitation in the area, the government established a special agency to regulate flow on the Dez river and its tributaries, to generate electricity and to promote agriculture and industry using the increased supplies of water and energy. This has involved the construction of a storage dam with canals and power plant, and the organisation of studies, experiments and demonstrations looking to the transformation of the agriculture of the area. Pilot operations in sugar cane culture, in mechanisation, in use of irrigation water and in many other aspects of rural life were undertaken as part of a regional scheme in which the government made the initial investments in capital works, research and education. These were not launched without difficulties with the established government agencies dealing with other parts of the nation.

GILBERT F. WHITE

References

[1] THOMSON, D. F., 'The Bindibu Expedition: exploration among the desert aborigines of Western Australia: III', Geogr. Journ., Vol. 128, 1962, pp. 262–275.

[2] PERRY, R. A. et al., General Report on Lands of the Alice Springs Area, Northern Territory, 1956–7, CSIRO, Melbourne, 1962.

[3] AGRICULTURAL PRODUCTION TEAM, Report on India's Food Crisis and Steps to Meet It, New Delhi, Ministries of Food and Agriculture and of Community Development and Co-operation, 1959.

[4] IBDR, Economic Development of Libya, Baltimore, 1960.

[5] COMBINED MEXICAN WORKING PARTY, Economic Development of Mexico, Baltimore, 1953.

[6] FAO, Mediterranean Development Project, op. cit.

[7] MICHEL, A. A., The Kabul, Kunduz and Helmand Valleys and the National Economy of Afghanistan: A Study of Regional Resources and the Comparative Advantages of Development, Washington, National Research Council, 1959.

[8] GAITSKELL, A., The Gezira: A Story of Development in the Sudan, London, 1959.

[9] CALIFORNIA, STATE OF, The California State Water Plan, Sacramento, State Department of Water Resources, 1957.

CHAPTER XX

Research and the Future of Arid Lands

Amid the many changes which today affect national political ideologies and the boundaries of states on the world map, it is truly remarkable that all the nation-states have accepted one philosophy of technological development, by which man is permitted and encouraged to stand sufficiently aloof from his environment to regard it as subject to his will within the limits imposed by the physical and biological facts and laws of nature and by economic limitations. Thus is envisaged the control or modification of natural processes by the direct intervention of man, to his material benefit. This has indeed been enormous despite local and individual failures; but whereas in ancient civilisations the right to perform or to authorise acts of this kind was the prerogative of rulers or priests to whom some attributes of divinity were attached, today the responsibility is assumed by a large and growing number of individuals. Generally, but not always, action is taken with the advice, approval or direction of scientists, engineers, technicians and administrators. The planned actions are based on the results of research which itself is directed and co-ordinated and is, in large measure, made available by demonstration and publication throughout the world. Some of the political, social and economic influences and limitations that affect the adoption of scientific and technological advances within the national sphere are discussed in Chapter xix, and accordingly in this chapter we shall concern ourselves purely with the research itself.

Basic surveys

In earlier chapters many references have already been made to research in specific fields, but it should be said that in each field there exists first the necessity for basic surveys to obtain data concerning the space–time distribution on earth of the fundamental elements of the topic under investigation, whether it be air temperatures, rock types, soil types, plants or animals, or the works of man. Accordingly, all sound scientific planning is founded upon such basic surveys, which are undertaken by government instrumentalities, universities, foundations, and by interested individuals, with the support, where appropriate, of the United Nations and its specialised agencies. This fact-gathering itself often involves research as well as routine investigations, and may, as is perhaps most obvious for meteorology, seismology and oceanography, involve international collaboration both for specific projects and for the collation of continually accruing data from all parts of the world. The notion of international collaboration is, however, now well established for all scientific disciplines. The accumulation of basic data being necessary in every field of scientific endeavour including those that relate to the arid zone, all that need be stressed is the need to continue and expand such work which, while often unspectacular, is nevertheless essential for planned development.

Causes and origins

If we are to influence the natural course of events or modify the existing condition of the earth or its inhabitants, we must be fully cognisant of the causes and origins of the existing situations, and we must also understand the interrelationships, often highly complex, between the many factors of the environment (e.g. climate, rocks and soil), and between the living things and their environment (e.g. soil-water–plant) and again between the living things themselves (e.g. plant–animal–man). Much of this understanding of causes and interrelationships arises in inquiring minds from considerations of the basic facts of distribution, as with plant or animal ecology, but must also be studied experimentally in order to achieve exact knowledge. Yet again, with a deep knowledge of causes, we may deliberately set out to modify the existing condition of things, as in engineering works for irrigation, in the fertilising and draining of soils, in plant and animal breeding, or the like. The scope for research is indeed conceptually unlimited, but is in practice determined by the availability of trained and competent research workers and by the available facilities for them to pursue their research.

The fields for research relative to the arid zone are categorised below. For each item the works of man are included within the item so that under Hydrology are included dams, canals, drains and other works. By Artefacts (6) is meant the purely cultural works of man.

1. Insolation and the planet earth, with its atmosphere – *climate and weather, meteorology* and *hydrometeorology*.
2. The Hydrosphere – *hydrology*.
3. The Lithosphere – *minerals and rocks*.
4. Interaction of the above items, together with the Biosphere – *soils*.
5. The Biosphere – (i) *plant life*, (ii) *animal life* including man as an animal.
6. Human Artefacts – dwellings, clothing, vehicles, roads, etc.

With the understanding that investigation and research will proceed on a broad front in all these fields as outlined above, the remarks that follow will refer to special aspects, to which particular attention has been paid because a major advance would hold out prospects of great benefits to the arid lands.

Insolation and solar power

The energy of the sun's radiation is strongly incident on the earth's surface in the arid zone, and is indeed the major determinant of aridity and of adverse effects on living things in the arid zone. Attempts to inhibit these adverse effects, or to put insolation to some useful purpose, have long been made. In agriculture, a mulch of vegetable matter or stable refuse, or a cover of stones laid on the ground around plants, have been used to conserve soil moisture, and today the efficacy of these and other ground covers such as asphaltum or sheets of black polythene is being tested. Although successful in reducing evaporation, asphalt and polythene do increase the soil temperature, which may adversely affect certain crops. Reference has been made in Chapter XVII to protective measures in regard to clothing and shelter for human beings.[1] Indeed there are no activities at the earth's surface in the arid zone for which insolation can be neglected, but aside from attempts to inhibit its effects, the more positive approach of putting solar energy to direct use is a challenge to technology. Although varying throughout the day and from season to season, the total energy incident on the vast desert areas is obviously very great. Measured on one square metre at right angles to the sun's rays on a clear day in the tropics it may attain 1 kilowatt, and in India it averages 0·30 kw over a year including both day and night. Although considerable in relation to the energy balance of living organisms, this incident solar heat is not intense from a technological point of view, and for this reason sunshine from a large area must be gathered and concentrated in order to attain high temperatures. The lens used as a burning glass has long been known to children and travellers, and the same principles of geometrical optics underlie use of reflectors in the form of parabolic mirrors, which are the concentrators used in solar furnaces and ovens. A reflector 30 feet in diameter, such as that at the Mont Louis Solar Energy Laboratory in the French Pyrenees, can produce temperatures of 3,000°C in a furnace at its focus, at a cost 25 per cent lower than electric arc methods and with no danger of contamination, which might affect the research that is done there on refractory materials. A flat mirror 40 feet across follows the passage of the sun across the sky and reflects rays to the parabolic mirror (Fig XX-1).

Solar cookers, which have reflectors about 3–4 feet in diameter and which themselves are adjustable to follow the sun, can boil water in 20–30 minutes. They have many disadvantages which are obvious, especially in a peasant economy, but more fundamentally they demonstrate clearly in their inability to be used near sunrise or after sunset, one of the inherent difficulties in the use of solar and also of wind-power – the variability of the source. This may not, however, be serious for operations that can conveniently cease during the night, especially as it has been found that, during the day, diffuse light when the sky is moderately overcast largely compensates for the loss of direct sunlight.

[1] Reduction of evaporation from free water surfaces such as reservoirs is discussed in Chapter III.

In USSR plants have been established for steam generation and used to operate canneries, for water distillation (including desalination of sea water), for refrigeration and for heating laboratory buildings. From one plant 75,000 tons of distilled water and 12,000 tons of ice are produced annually. Small semi-portable plants for saline water conversion are already in use in Israel,

Fig xx-1. Fortress walls built by Vauban, Louis XIV's great military engineer, provide a seventeenth-century setting for a spectacular modern research device. In the old fort at Mont-Louis, in the Pyrenees, flat mirror (right) follows the passage of the sun and reflects rays to parabolic mirror (left) so curved that it concentrates all rays into the orifice furnace (centre). (*Photo Laboratoire d'Energie solaire de Mont-Louis, France*)

Africa and Australia, and larger units for supplying stock and domestic water are in use experimentally (Fig xx-2).[1]

The production of electricity using solar-powered boilers to energise dynamos is subject not only to the variation of insolation through a 24-hour period but also to the inefficiency of power generation by steam. Water may be pumped by a solar engine to a high-level reservoir whence it flows back at night to operate a turbine and dynamo. A mine in New Mexico has been lighted for some years using this method.

More recently, electric batteries have been developed which yield a current of 2·5 amperes per square metre at 0·3 volt when heated by the sun's rays. This direct conversion of solar heat to electricity has an efficiency of 8–11 per cent, but the cells, which contain silicon plates of very high purity, are currently much

[1] See also Desalination, pp. 448–50.

too expensive for commercial or domestic purposes, and find their main use in earth satellites.

The amount of energy that falls on an area of 100 square metres of flat surface, an area about as large as that of the roof of a small house, is of the same order of magnitude as that for air-conditioning (both cooling and heating) such a house, or providing it with a hot-water service.[1] Quite successful use is already

Fig XX-2. 4,500 sq. ft. solar still as developed by the Division of Mechanical Engineering, CSIRO, installed at Muresk Agricultural College, Northam, West Australia. (*Photo CSIRO*)

made of solar energy for this latter purpose, but so far, systems of true air-conditioning have not been successful as against standard types commercially available. There are, however, several types of solar hot-water installations available in Japan, and more or less experimental units are being successfully operated in Australia, USSR, Israel and Africa.

An indirect application of solar power is the marine thermal plant, whose action depends on the temperature difference between the cold water deep beneath the surface (conveniently near land) and the hot water in shallow coastal lagoons. An experimental plant has been established near Dakar (West Africa), an additional feature of which is that it produces fresh water from the sea.

A more promising installation is, however, the solar 'tank', a shallow lake in which the bottom waters are heated by placing black stones over the bed to absorb and re-radiate the solar energy, while the upper waters remain cool. With a suitable arrangement of salinities, inlets and outflows, the bottom waters

[1] In hot climates a smaller collecting area will serve for these purposes.

can be drawn off, and a continuous supply of hot water of possible value in certain industrial processes may be obtained. Further research is in progress, especially in Israel.

Doubtless future research will discover more efficient and useful ways of using solar energy, some less direct and perhaps employing the life processes of plants or micro-organisms, which have the ability to cause chemical reactions to take place in the presence of sunlight. The anaerobic bacteria which reduce ferric iron to the ferrous condition, or those which reduce the sulphate radicle, may be mentioned.

Wind power

Used for centuries in some countries, wind power is strangely neglected in others. The power of wind is proportional to the cube of its speed, and of course the wind speed varies greatly from time to time. There are, however, especially in the arid zone, regions where steady and rather strong winds blow for long periods, and if some means of storing the energy output, or the product of it

Fig xx-3. Windmill and water tank, Northern Territory, Australia. (*Photo E. S. Hills*)

(e.g. pumped water), is available, wind power can be, and indeed is, very important in remote areas to which electricity cannot be supplied by a grid, or where fuel costs for local generation of electricity are high. The two most common uses are large slow-revolving windmills to pump water from underground to a tank which is placed on a high stand, and small fast-revolving mills designed to generate electric power for lighting. Wind-operated lighting-sets charge storage batteries to ensure continuity of supply. Large windmills (Fig xx-3) are expensive, but if properly attended require little mechanical maintenance. In the Australian scene they have been very widely used, but are being progressively replaced by diesel power plants, which are more reliable. In India, wind surveys to establish the best areas for using windmills have shown that certain parts are quite suitable for wind-power development (see Fig xx-4), and if the local

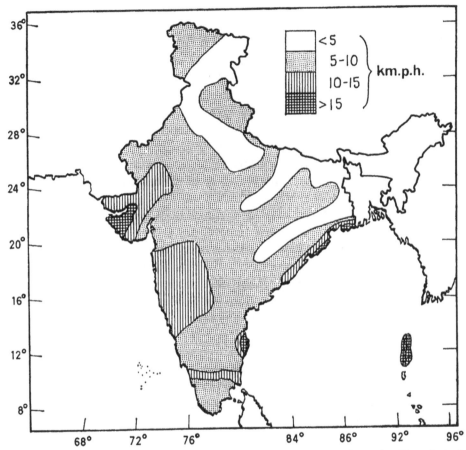

Fig xx-4. Annual mean wind speeds in kilometres per hour. (*From the climatological tables of the India Metrological Department*)

inhabitants are not well trained technically windmills are likely to be more successful than diesel or petrol-driven generators.

Apart from serving the individual or small local group, it is planned that wind-driven electric generators might be linked to an electric grid system using thermal or hydro-power, in such a way as to reduce the use of the conventional prime mover and thus save fuel costs or conserve water in the reservoirs. The saving of coal or oil may, in the coming years, become a matter of considerable interest as these fossil fuels are not inexhaustible, they will become more expensive with the years, and they are also of greater potential use as raw materials for the petro-chemical industries than as fuels. There are, too, so many difficulties attaching to the wide use of atomic energy that research on wind power is believed to be fully justified in a country such as India or other arid lands, most of which are low in energy resources and have many isolated groups of inhabitants to whom it is inconvenient to supply grid electricity [1].

Climate and weather

The planetary influences that determine the climates of the land areas of the globe have been treated in Chapter III. It is the aim of the meteorologist not only to be fully informed as to existing regimes but also to provide forecasts of local weather. On a short-term basis of one or two days in advance this is part of the duties of all meteorological services, but there would be enormous advantage to all but the most arid regions where it is always dry, if long-range forecasting could be given with reasonable accuracy some months or a year in advance. To be able to do so would require much more knowledge of events in the upper atmosphere than is presently available to meteorologists, and perhaps more fundamental knowledge of the interactions of air masses than is current. Nevertheless both empirical methods in which predictions are based on cycles or patterns of weather recorded over past years and projected to the future, and methods based on the fundamental facts of atmospheric circulation, are continually being tested, unfortunately without unequivocally correct results.

Suggestions to modify climates have so far been limited to secondary factors such as the creation of mountain ranges, the shifting of ocean currents, or the melting of the Arctic sea ice, all of which are perhaps conceivable using atomic energy either as blasts or as long-enduring thermal emission. The side effects of such actions are, however, incalculable, and any such acts must, in a rational world, be long delayed.

Of more immediate interest are experiments to induce or to increase rainfall ('rain-making'), and to reduce evaporation from free water surfaces in reservoirs. As mentioned in Chapter III, the results of rain-making trials are tantalisingly uncertain, and in any case the methods in current use, or likely to be developed, are of more use in regions where rain normally falls than under truly arid conditions, since clouds in a condition to precipitate, given some stimulus, are required. Regions of seasonal aridity such as the monsoon lands may benefit when the wet monsoon is delayed unduly, or some region in especial need is passed over by the rain clouds without precipitation occurring.

Microclimates and local influences

Leaving aside major or radical effects, all research relating to arid and especially to semi-arid regions points to the importance of small-scale deviations towards ameliorated conditions which affect living organisms. The shaded location, the aspect giving less direct impingement of solar radiation, the local surface feature that directs run-off to favoured spots – such and similar small-scale features having some microclimatic or other physical advantages are all-important in this zone. We try to induce these by building windbreaks (Fig xx-5), planting protective rows of trees or shrubs, and in many other ways inducing local amelioration of the severities or deficiencies of nature. The ancient and now revived schemes for impounding the waters of small streams to provide minor local storages ('water harvesting' is the term commonly applied in Australia for such

Fig xx-5. Palmyra, Syria, an oasis town and orchards with wind-breaks.
(*Photo by courtesy Unesco*)

works) is an example of minor manipulation which, despite its small scale, may have notable benefits for small groups.

In north Egypt, the cultivators of Roman times built broad embankments, known as *kharms*, of loose rock and soil, enclosing an area of a few acres. The rain falling on the banks runs off, to increase the water supply to the enclosed area apparently sufficiently to improve the cropping, although most have now been abandoned.

It is, indeed, fundamental that even were fresh water available freely in the deserts all the details of local topography, soils, subsoils, climate and weather, plant types suitable to the region, and other scientific details would be called into question in its utilisation. This is especially true since we know from sad experience that water ill-applied may in fact be deleterious to the extent that it can cause land to become waterlogged or salinised and useless. Thus it is that continuing studies of all the scientific aspects of arid lands are essential, whatever major break-through in research may take place in the future. Indeed, since water can never be free, economic considerations alone will require it to be applied where optimum results may be achieved, which leads us back again to surveys and researches on the infinitely variable arid environment.

Desalination of water

Undoubtedly the most significant benefit that could accrue to the world in relation to water would derive from the discovery of a cheap process for producing fresh from saline or mineralised waters, including of course the sea. With the certainty that world populations will double in the next 40–50 years, even lands that today are well-supplied with water will find, as is indeed the case today, that their use of water will need to be restricted, and the price of water as a commodity will rise. Accordingly, very large sums are currently being spent on research into demineralisation, and several large plants have been constructed to serve areas having special needs, such as oil refineries and growing cities where natural or piped supplies are inadequate. The cost of desalinised water is currently too high to permit its economic use for irrigation.

The oldest and still most widely used method of purifying saline water is by distillation. It has long been used on shipboard, and for industrial and military purposes as well as in the laboratory. A wide range of devices is, however, being tested, and are listed in a Unesco publication by E. D. Howe [2] as follows:

The many possible schemes for demineralizing water may be classified as thermal, mechanical, electrical, and chemical. Thermal and mechanical schemes include those involving phase changes of the product water and have energy requirements which are independent of the initial salinity of the water being processed. Chemical and electrical methods are sensitive to the initial salinity of the water being processed and require appreciably less energy for the purification of brackish water than for sea-water. It should be noted that at present the thermal and mechanical methods are cheapest for desalting sea-water while chemical and electrical methods are the cheapest for brackish water. However, since the thermal and mechanical methods are so very far from achieving the ultimate minimum power or fuel consumptions, substantial improvement in fuel or power costs could render them attractive for saline water conversion.

The great number of possible schemes for demineralizing water is indicated by the tabulations included in the annual reports of the United States Department of the Interior [3–5]. These tabulations show over twenty possible schemes or phenomena which could conceivably be used for demineralizing. Several of these have been investigated, found to be unpromising, and have been abandoned. Those remaining in the active list correspond quite well with the processes being investigated throughout the world, and are indicated below:

1. Processes using thermal energy, involving a phase change and requiring energy quantities essentially independent of the initial salinity of the water:
   Multi-effect distillation

Multi-flash distillation
Vacuum flash distillation
Supercritical distillation
Solar distillation
Solvent extraction (in which water is extracted by the solvent)

2. Processes using mechanical energy, involving a phase change and requiring energy quantities essentially independent of the initial salinity of the water:
   Vapour compression distillation
   Freezing separation
   Reversed osmosis

3. Processes using electrical energy, not involving a phase change and requiring energy quantities dependent on the initial salinity of the water:
   Electrolysis
   Electrodialysis
   Osmionic separation

4. Processes using chemicals and requiring them in quantities dependent on the initial salinity of the water:
   Precipitation
   Ion exchange
   Solvent extraction (in which the unwanted salts are extracted by the solvent)

Technical details about these processes are given in the same article by Howe [2].[1] He concludes as follows:

The period since 1956 has witnessed a very considerable increase in the application of demineralizing processes to the water needs of cities in the arid zones of the world. It has also seen a great acceleration in the pertinent research programmes in many countries.

Distillation continues to be the chief method used in commercial plants. Through the initiative of private manufacturers, the application of the multi-flash principle has resulted in the increase of water production per set of equipment from 120,000 gpd in 1953 to 1,200,000 gpd in 1960. Also, the GOR has been increased from 4·6 to 12 in the same period of time. It is likely that additional increases in size and economy will continue to make distillation an important method for obtaining fresh water from the sea.

If distillation is to be used for the very large capacities which will be needed in the future, some source of heat other than fossil fuels must be developed. The most logical source would seem to be the nuclear reactor. Studies by Brice et al. have dealt with combinations of certain reactors with specific distiller plants and have given results sufficiently encouraging that one of the USDI demonstration plants will employ this combination. The use of underground nuclear explosions was studied by Fatt [7] and found to be unattractive at the present very immature state of knowledge in this field. Geothermal sources have been studied under USDI contracts [8, 9], and have been found to be of very limited scope and hence un-important to the general developments.

Electrodialysis has the great virtue that its power requirement is sensitive to the initial salt content of the water. It has so far been firmly established that electrodialysis is thoroughly competitive with distillation in the saline water field. Development of the process and equipment should continue for some time to come and will undoubtedly bring the costs of demineralised sea water close to those attainable with distillation.

The use of freezing for separation is now in the pilot-plant stage, with one plant of 15,000 gpd in operation and a second of 35,000 gpd under design. The potentialities of the direct freezing process seem to be established as being very economical of power. Costs of equipment can be estimated firmly only after operation of these first two pilot plants has been studied.

The several studies being conducted on ion exchange and solvent extraction have not reached final conclusions as yet. Their attractiveness from the view-point of sensitivity to the initial salt concentration makes it imperative that their potentialities be fully explored, even though the first studies are not encouraging.

A scientist of the University of Samarkand has developed a portable device which solves the problem of obtaining drinking water in the desert. Weighing only 10 kg and resembling a suitcase, the equipment can distil 800 grammes of salt water per hour. Moreover, it can be used to cook a meal for two or three persons thanks to a special anodised aluminium reflector which concentrates the sun's rays on the helio-cooker [10].

Underground water and minerals

Research on the earth's crust, and investigations connected with the exploitation of mineral resources, oil, gas and underground water, are moderately well organised in most countries through geological, geophysical and hydrological surveys, and related technical services carried out by governmental agencies, research organisations and companies. Geological surveys afford basic data for all such exploitation as well as for soil studies and site development for large structures and works in which soil and rock mechanics are involved. Mapping on a large scale is required to afford a broad knowledge of regional geology, and detailed mapping is called for in connection with particular projects. Geophysical and geochemical techniques are widely used in conjunction with geological methods, but the drill, which obtains samples of rock from depths ranging down to 15,000 or even 20,000 feet, is still required to provide definite information about these hidden sub-surface rocks.

All these methods of investigating the earth's crust are subject to continual improvements. Geophysical methods have been developed particularly in the past thirty years, with a succession of new techniques finding rapid use and re-finement. In the same period, too, the use of aerial photographs has well-nigh revolutionised geological mapping, especially in arid and semi-arid regions where rock structure is often wonderfully displayed in photographs (see Chapter IV). Geochemistry, although it has long been practised, has also experienced such rapid advances of recent years in analytical techniques and fundamental understanding as to have also suffered a revolution in principles and in applicability. The rate of scientific advance in these disciplines has been so great that new major breaks-through in the near future are perhaps not to be expected. Nevertheless the great wealth that may be revealed by their application makes it imperative that adequate sums should be expended on them, and these sums are not small. Furthermore, surveys by geological, geophysical and geochemical methods call for highly trained and competent scientists. Each operation is indeed a research activity calling for sensitive interpretation rather than routine recording of data. Accordingly, many schemes for international aid to underdeveloped countries, in the arid zone and elsewhere, involve assistance in such surveys through the provision of teams of scientists who not only carry out the surveys, but also provide training for those nationals who work with them.

Although none of the methods of research listed below is peculiar to the arid zone, as all are of general applicability, in several instances there are special factors operating in arid regions, such as the widespread presence of saline waters which influence electrical conductivity of rocks, which call for very experienced and knowledgeable interpretation. It is also true that each area of the earth is unique, and any geological or geophysical study must therefore be made in the spirit of research rather than of routine data-recording.

Geophysical methods

The methods of geophysical survey for geological purposes may be listed as follows: (a) magnetic methods; (b) gravity methods; (c) electrical methods; (d) seismic methods; (e) radiometric methods. It will be understood from this list that geophysical surveys record physical data concerning the earth. In this regard they are, equally with topographical and geological surveys, of permanent value as well as of immediate application to the investigation in hand. Geophysics is rarely of great value unless geological data are also available [11].

MAGNETIC METHODS

These have been used since the seventeenth century. In much modern work, either the horizontal or the vertical intensity of the earth's magnetic field is measured with magnetometers over a grid of spots in the field, and the results are contoured to join points of equal intensity. The results are related to the normal local intensity of the earth's field, which is affected by the nearness to the instrument of bodies of more highly or less highly magnetic rocks. Not only are highly magnetic rocks, such as bodies of magnetic iron ore, revealed, but by correlation with local geology a great deal can be inferred about the more normal kinds of rock, for example granites, schists, shales, limestones and salt. The use of air-borne magnetometers not only speeds up surveys, but reveals a broader picture of the earth's magnetic field, in which many smaller anomalies are obscured but major structures more clearly revealed. Electromagnetic induction methods may also be carried out from aircraft. They are of particular value in the search for metallic ore-bodies.

GRAVITY METHODS

Almost all current gravitational geophysical surveys are made with sensitive, portable and rapid-reading gravimeters, the development of which is a triumph for the instrument-makers. The field of gravity at any place is affected not only by the main body of the earth, at great depths, but also by the rocks of the outer crust. Dense, heavy rocks cause a local increase in the acceleration due to gravity (g); light rocks such as certain salt deposits cause a local decrease. A number of readings of g are made, and corrected for various effects such as altitude and local terrain. The figures are then contoured, generally in 'milligals', and the resultant gravity pattern may be interpreted in terms of geology. The method is widely used in oilfield exploration and the examination of coal basins.

ELECTRICAL METHODS

There is a variety of geophysical methods using electrical properties of the rocks. They fall into two main groups: (a) methods measuring natural potential differences in the crust; (b) methods measuring induced or applied currents. Natural earth currents are formed in various ways. Sulphide ore-bodies generate currents chiefly as a result of chemical changes during weathering; bodies of underground water percolating through rocks also generate currents. Potential profiles may be measured at right-angles to the strike of a known structure or ore-body, and running further profiles beyond the known extent of the lode may reveal its further continuation. Again, equipotential lines may be surveyed. These appear like contours, and the 'highs' and 'lows' may be related to structures or to lodes.

When a current is applied to the ground through electrodes driven into the soil, the rocks are, as it were, 'probed' by the current passing between the electrodes, and as these are shifted from point to point, some indication of the different strata present may be obtained. The wider the spacing of electrodes used, the deeper the current passes and so the deeper are the structures penetrated. Equipotential lines may be plotted, or again, the resistivity of the rocks through which the current has passed.

Electrical methods are often used in surveys for underground water, as the electrical properties of rocks are modified according to their water content.

SEISMIC METHODS

These may be compared with sonic depth sounding in the sea. A charge of dynamite exploded in a shallow hole drilled in the ground sends elastic waves through the rocks. The waves are either reflected back to the surface from hard rocks, or are bent like light waves (refracted) on passing from one rock type to another. The waves on reaching the surface are recorded on small seismographs, called 'geophones'. According to which principle is used, we have 'refraction shooting' or 'reflection shooting', and within each group there are several possible arrangements of the instruments as in 'fan shooting', 'profile shooting' and the like. Seismic methods give the most accurate and detailed representation of structure, and are widely used in oilfield surveys.

RADIOMETRIC METHODS

The use of Geiger counters and scintillometers to detect radioactive minerals and their weathering products in soils and surface rocks is a geophysical method of prospecting that has yielded important results of recent years. Airborne scintillometers have enabled the reconnaissance of large areas to be made, followed, on the ground, by more intensive examination of likely prospects. There are many tricky points: indications of ore have proved to be worthless; in other instances ore has failed to be located although it is present, but the portable Geiger counter has directly led to many important finds of uranium minerals.

GEOPHYSICAL WELL-LOGGING

Various methods of logging the physical characteristics of strata intersected in bores have been developed, and bore-hole correlation by these means is now virtually a standard technique. The results are quickly obtained and of great use in geological work, especially in oilfields.

Geophysical surveys are very costly, and not to be lightly undertaken. They may be resorted to at the request of geologists, when normal geological methods of exploration have 'stalled' for lack of new data; or, again, they may be the first exploratory method tried, as in extensive aero-magnetic surveys. These reveal anomalies which are further investigated by prospectors and geologists. Radiometric surveys are also used in this latter way. But it must be stressed that, in the ultimate analysis, geology and drilling must be brought to bear in the field.

GEOCHEMICAL METHODS

The analysis of rocks, ores and naturally occurring water has long been carried out as a matter of standard procedure in relation to surveys. Advances in chemical and spectro-chemical analytical methods have, however, not only shortened the time required for each analysis but have given such refinement that 'trace' components may now also be studied with precision and speed. Apart from direct applications in soil chemistry of interest in agriculture and pastoral pursuits, these new methods have been of considerable value in the search for minerals. In the weathering of ore-bodies, metallic elements are distributed as a 'halo' about the deposit, due both to the spreading of grains shed from the ore, and to the solution of the elements in soil waters, with subsequent precipitation in soil and weathered rocks. Systematic sampling and analysis of the soils reveals the location of the hidden ore-body [12].

The chemistry of natural waters, especially of sub-surface waters, is of interest particularly in relation to their utilisation in irrigation, or as drinking water for animals and man. These latter aspects have been mentioned in Chapters V and XII, but much interest has also been shown in the use of saline waters for irrigation [13, 14]. Special problems arise where deleterious elements, such as fluorine in excess of a certain value, may be present in addition to more common salts. Excess fluorine in animal metabolism may cause diseases of the teeth and bones, known as fluorosis.

Soil science (pedology)

The scientific study of soils belongs almost entirely to the past fifty years. Soil is indeed a remarkably complex material. Based on and derived from the lithosphere, it is penetrated by the atmosphere and by the hydrosphere and the biosphere. It contains organic matter derived from dead and from living plants and animals; it contains water, both free and linked with clay minerals; it also contains atmospheric gases and gases of organic origin, and mineral salts in solution, derived from the mineral grains of the soil and from the rain and from drainage waters passing any locality. The basis of the soil is afforded by mineral grains of sand, silt or clay, and among or attached loosely to these the other constituents lie. This natural chemical and biochemical admixture, the soil, is penetrated in a most intimate way by the roots, rootlets and root hairs of plants, churned and dug by burrowing animals of all sizes, and by man. Fertilised and tended, it has since the dawn of civilised life been recognised as Mother Earth, from which all life on land derives its nourishment. Its complexity is such that research on it must, of necessity, be reduced to relatively simple terms by studying one or other aspect either in artificially compounded mixtures, or, in the field, by controlled experiments to test some additive such as fertiliser or some way of handling the soil, say by ploughing, irrigating, draining, or in other ways acting on it. Research on soils is, also, closely linked with that on plants, especially as regards plant–water relations and plant nutrition, and in the same general field may be mentioned hydroponics – the growing of plants in water containing plant nutrients, without soil. To treat with research on soil would require several large volumes, but some general principles and outstanding results may be mentioned here.

FIELD STUDIES – SOIL TYPES AND GROUPS

In using his soil the farmer or grazier should be aware of the areas and location of soils of different kinds on his farm. He is, moreover, likely to be interested in a great many properties of the soil, which could be used as a basis for soil mapping. Texture, depth and zonation in depth, porosity, permeability and moisture retention, mineral constituents and chemical composition including acidity or alkalinity[1] are the most important, as they affect tilling, drainage, soil-moisture retention, soil fertility and other properties of direct significance in farm practices. While it would be possible to represent each such property on a separate map, this would yield too complex a picture for ready use, and accordingly it is usual to map soil types, each of which has some uniformity of properties which distinguishes it from others. A soil type is, ideally, a soil mass which possesses essentially the same physical and chemical characteristics throughout its areal extension. It may have originated from essentially similar parental rock or alluvium and have undergone similar pedogenic effects; or it may, according to some pedologists, have been derived from different materials, and by a very long process of pedogenesis have become essentially uniform irrespective of the parental material from which any portion was derived. Notions as to origin are significant in soil mapping since observations cannot be made at all points over the whole area and in depth. Interpolation between points where bores are made and analyses carried out is required, and the soil surveyor is influenced in this by his ideas as to the space-distribution of different parent materials, essentially a geological matter, and also by his observations on geomorphology. Soil types are, short of very detailed mapping, to some extent subjective and conceptual, and to some extent generalisations in which minor local differences are neglected. A knowledge of the influences that have affected soils over many thousands of years, from the period when the parent material became located in a suitable position to give rise to soil, involves not only geology and geomorphology but also palaeoclimates and the recognition of fossil soils which, under changed environmental conditions, are virtually the parent materials for soil formation under existing circumstances. Understanding of these factors may be achieved by research, which enlightens and enlivens the task of achieving proper soil management. If, for instance, it is realised that a heavy soil with difficult working properties may be a relic of a previous period of saline influences, a clear-cut approach to its beneficiation may confidently be attempted.

Wide-ranging, precisely documented and fully competent regional studies of soil formation in many environments is the basis for sound general theories of soil formation. Particular aspects of the topics mentioned above, such as infiltration, water-holding capacity, plant–soil–water relations or soil fertility, are studied by intensive researches on selected soils, both in the field and the laboratory. Occasionally, however, a new concept or new understanding affords the basis for very rapid advances, as with the recognition of the significance of the presence of certain elements as micronutrients in soils. Such elements, which are required only in a few parts per million, profoundly affect plant growth, and, through plants, the stock animals that feed on them. The discovery that soils may be deficient not only in major nutrients such as nitrogen, phosphorus or

[1] Expressed as pH values.

potassium, but also in micronutrients including manganese, molybdenum, zinc or cobalt, has made possible the development as crop land of areas that were previously, at best, pastures of very low carrying capacity. The further discovery that certain chemical compounds containing plant nutrients do not make the nutrient elements available to plants, and that an excess or deficiency of one element may prevent the uptake of another by plants, illustrates the ultimate complexity of most such topics, and the necessity for full understanding if waste and perhaps harmful practices are to be avoided. Again, supposing that a particular area of soil has been so treated that crop or pasture growth has been very greatly stimulated. The additional organic growth itself, and changes in farm practices consequent on this (e.g. greater trampling by animals, or the cutting of grain crops on former pasture land), themselves introduce additional factors which must be understood and directed or redirected into a balanced farming regime. These effects will differ on different soils.

So it is that there is a continuing need for research and investigation on soils. It must also be mentioned that in order to translate knowledge into practice, authorities and individual farmers must be educated in the new knowledge. This is an enormous and difficult task, for which 'extension' services are required on a large scale. It is indeed possible to say that the rate of application of new knowledge lags far behind its acquisition in most countries, especially in relation to soil science and farm practice. Even the ravages of soil erosion may be regarded as normal and unavoidable by farmers who have lived all their lives in an eroded area.

Plant and animal life

The influence of mankind on plant and animal life over the many millennia during which men have lived in arid lands has been incalculable. Destroying here or protecting there, distributing seeds unwittingly or with intent, breeding animals in captivity and so developing new strains, affecting the balance of nature by his actions and works, man had already, long before planned research was undertaken, been a major factor in the scheme of nature, and perhaps more so in the arid and semi-arid regions than elsewhere because of the considerable movements of nomads, the early development of civilised life in some arid parts and the delicacy of the natural balance in such regions. That many past actions of man, even though these were consciously planned, have been adverse is unfortunately true. This applies, for instance, to the introduction of certain animals (including birds) and plants to newly discovered countries such as Australia, which has suffered not only from the rabbit, the population of which at times has multiplied to plague proportions, but also great numbers of weeds and diseases of plants and animals introduced from other lands. Quarantine control having been established, the risks are much less, but inadvertent introductions still occur. However, planned plant and animal introduction, based on careful research, now replaces the haphazard actions of former times.

Many plant and animal introductions were, of course, to the benefit of the country concerned, as with cattle and sheep into Australia, which lacked all the higher mammals, and wheat and barley and other crop plants into many different regions. Again, some introductions have been deliberately planned to control a situation in which a plant or animal species has run rampant and caused great damage to potential farm lands. One of the most successful examples is the introduction of the beetle *Cactoblastis cactorum* into Australia, to destroy vast areas of prickly pear (*Opuntia* spp.) which after its introduction from South America in 1789 made 50–70 thousand square miles of pastoral land in Queensland and New South Wales untenable. Against an animal pest, myxomatosis was introduced into Australia to control the rabbit (also introduced), although this virus, being spread by mosquitoes, is most efficacious in humid rather than arid climates. In both instances of control by infestation, long and strict testing was carried out to make sure that no markedly adverse effects would result. The introduction of *Cactoblastis* was an unqualified success; that of myxomatosis a limited, but nevertheless highly significant, success.

Both plant and animal introduction still hold some promise, although it appears unlikely that many plant introductions on a large scale are likely to succeed. Nevertheless favoured localities with local water supplies or less severe microclimates may be sought and developed where the economic and social conditions will permit of profitable exploitation. Plant and animal breeding from stock already established is of equal potential, sometimes using introduced strains to cross with those present. The principles are fully treated for animals in Chapter XVI, and plant breeding is based on similar genetic principles.

The range of topics in which research of direct economic benefit may be carried out on plants and animals in arid lands is very large, and it must suffice to mention, in addition to the aspects referred to above, matters such as plant–water relations (e.g. the factors controlling the efficiency of water usage by plants; the limiting salinities of soil water for the growth of different plants); plant–animal relations (the ecological balance under pasturage, and the optimum conditions of pasture utilisation; regeneration of depleted pastures; poisonous plants and their control); soil fertility and plant nutrition, including the application of fertilisers.

The main topics of interest in relation to animals have been treated in Chapters IX and XVI, and for human beings in Chapter XVII. The influence of man is, however, paramount in all developed arid zones, and his own artefacts are a part of the geographical entity that each region constitutes.

The works of man

The revolution that man can make in arid regions is perhaps most strongly revealed by the fate of explorers who perished of exhaustion and thirst in attempting 150 years ago to penetrate the arid interior of Australia, where today there are station properties, cars and established tracks or roads, water-bores, aerodromes and ready intercommunication by radio. Although a man will still die if exposed for only some hours to the sun without water, especially if he attempts to walk far, the facilities available, short of breakdowns, are such that traffic and movement are normally as safe as in any other region (perhaps safer than in large cities) and men go about their business in well-set ways, while tourist buses run on regular routes. Urban or rural settlements are established where for any reason men may obtain sustenance, and in lands of older civilisation the very existence of a town, with its facilities for industry, trade, government and private management, education and the like, is a self-perpetuating mechanism.

In learning to live in and with the desert, mankind has adopted vastly different modes of life in different regions, from complete nakedness to heavy clothing,

Fig xx-6. Housing for oilfield workers at Rassudr, Sinai, Egypt. (*Photo E. S. Hills*)

from simple shelters of brushwood, cave dwellings or bark huts to elegant brick, stone or earth (pisé, mud, wattle and daub, adobe) homes. The social and economic limitations on costs that obtain in many cases have induced architects to consider the most favourable designs for houses and larger buildings in hot dry climates, having regard to economy of construction. Such considerations are important where a whole town is to be quickly constructed to serve some particular purpose.

Examples might be quoted from many lands, and the influence of local tradition and architectural evolution in building styles and materials is obvious in comparing the oilfield village of Rassudr in Sinai (Fig xx-6) with the new town of Mount Isa (north-west Queensland) established for Mount Isa Mines (Fig xx-7), or with Broken Hill (New South Wales), which is much older and has many buildings that are simply replicas of domestic or office styles borrowed from the humid zone (Fig xx-8). Much remains to be done in regard to architecture and building appropriate to hot dry lands, having regard to the variety of local materials of construction, and special local conditions such as windiness.

Similar remarks apply also to the construction of roads and airports and indeed all engineering works, for not only must great extremes of temperatures be allowed for, but also special effects such as the liability to 'flash-floods' after long

Fig xx-7. Homes constructed by the Co-operative, Mount Isa, north-west Queensland. (*Photo by courtesy Mt Isa Mines Ltd*)

Fig xx-8. The main street of Broken Hill, NSW, Australia. (*Photo E. S. Hills*)

periods of dryness, and, too, the common fault of stone used as concrete aggregate in arid climates that it may prove to be 'reactive', due to the presence of opaline silica.

Air-conditioning, a *sine qua non* for comfort at some times of the year, may be compared with artificial heating of buildings in cold climates, which is fully accepted as a necessity by all people. Although the initial installation costs are said to be higher, this might be queried, and running costs, given reasonably priced electric power, are certainly relatively low. What is often regarded as a luxury may, in fact, become the norm in building practice in arid zone cities.

With continuing improvements in all forms of transport, in housing, and in services such as electricity supply, education and health, life even in arid regions remote from large centres of urban amenities can be comfortable and rewarding to the individual. There are, however, many social or socio-economic aspects that affect communities very markedly.

If there is no continuing increase in local productivity, the scions of families who live permanently in small communities have few opportunities for employment, and the young tend to move away. Again, there is a tendency in many lands for remote communities to attract unattached persons who virtually seek refuge in them, either for personal or for social reasons. Finally, stable family life and community spirit are difficult to develop, outside the larger towns, and even these are subject to the above-mentioned troubles. Such trends have, however, probably always operated in the pioneer fringe, and there is a compensation, that as well as the troublesome elements, also persons of initiative and skill may find scope there to develop the environment rapidly and profitably. Socio-economic research as a preliminary to planned development is clearly essential, since the economic, social and political aspects of capital investment in arid lands have to be carefully weighed before any large sums are expended. In the long run the position is that there is competition between arid regions and regions of high productivity, for investment and population, and, short of continued government subsidy, land exploitation must be competitive if it is to succeed permanently.

E. S. HILLS

References

[1] VENKITESHWARAN, S. P. and K. P. RAMAKRISHNAN, 'Harnessing the winds of India', *New Scientist*, No. 295, 1962, pp. 75–8.

[2] HOWE, E. D., 'Saline water conversion', *The Problems of the Arid Zone, Proceedings of the Paris Symposium*, Unesco, Paris, 1962, pp. 271–97.

[3] ANON., *Saline Water Conversion Report for 1956*, US Department of the Interior, Washington, 1957.

[4] — ibid. for 1957 (1958).

[5] — ibid. for 1958 (1959).

[6] — *Preliminary Design Study of an Optimum Nuclear Reactor–Saline Water Evaporation Process*, US Department of the Interior (Office of Saline Water Research and Development, Progress Report No. 34), Washington, 1959.

[7] FATT, I., *A Study of the Feasibility of Using an Underground Explosion as a Source of Heat Energy for the Distillation of Sea Water* (University of California, Institute of Engineering Research, Series 75, issue No. 17), Berkeley, California, 1959.

[8] BELL, J. C. *et al.*, *Availability of Geothermal Energy for the Demineralisation of Saline Water*, US Department of the Interior (Office of Saline Water Research and Development, Progress Report No. 27), Washington, 1958.

[9] ANON., ibid., US Department of the Interior (Office of Saline Water Research and Development, Progress Report No. 28), Washington, 1959.

[10] Tass in English, 07.50 GMT, 11/9/62.

[11] HILLS, E. S., 'Geology and geophysics', *Guide Book to Research Data for Arid Zone Development*, Arid Zone Research IX, Unesco, Frankfurt, 1957, pp. 44–50.

[12] HAWKES, H. E. and J. S. WEBB, *Geochemistry in Mineral Exploration*, New York, 1962.

[13] ASGHAR, A. G., 'Use of saline water for irrigation with special reference to saline soils', *Salinity Problems in the Arid Zones*, *Proceedings of the Teheran Symposium*, Unesco, Paris, 1961, pp. 259–65.

[14] KULKARNE, D. G., 'Use of brackish water for irrigation and its effects on soils and crops', ibid., pp. 267–70.

ADDITIONAL REFERENCE

UNESCO, *Wind and Solar Energy*, *Proc. New Delhi Symp.*, Unesco, Paris, 1958 (Arid Zone Research VII).